ANTINEOPLASTIC AGENTS

CHEMISTRY AND PHARMACOLOGY OF DRUGS

A Series of Monographs

Edited by

DANIEL LEDNICER

Volume 1   Central Analgetics
           Edited by Daniel Lednicer

Volume 2   Diuretics
           Edited by Edward J. Cragoe, Jr.

Volume 3   Antineoplastic Agents
           Edited by William A. Remers

In Preparation

Cardiovascular Agents   *Edited by James A. Bristol*
Pulmonary and Antiallergy Drugs   *Edited by John P. Devlin*
Nonsteroidal Antiinflammatory Agents   *Edited by Joseph G. Lombardino*
Antianxiety Agents   *Edited by Joel G. Berger*

ANTINEOPLASTIC AGENTS

Edited by
WILLIAM A. REMERS
College of Pharmacy
University of Arizona, Tucson

A WILEY-INTERSCIENCE PUBLICATION

JOHN WILEY & SONS

New York · Chichester · Brisbane · Toronto · Singapore

Copyright © 1984 by John Wiley & Sons, Inc.

All rights reserved. Published simultaneously in Canada.

Reproduction or translation of any part of this work beyond that permitted by Section 107 or 108 of the 1976 United States Copyright Act without the permission of the copyright owner is unlawful. Requests for permission or further information should be addressed to the Permissions Department, John Wiley & Sons, Inc.

Library of Congress Cataloging in Publication Data:
Main entry under title:

Antineoplastic agents.

   (Chemistry and pharmacology of drugs; v. 3)
   Includes bibliographies and index.
   1. Cancer—Chemotherapy—Addresses, essays, lectures.
2. Antineoplastic agents—Testing—Addresses, essays, lectures.   I. Remers, William A. (William Alan), 1932-   . II. Series. [DNLM: 1. Antineoplastic agents. W1 CH36D v. 3/QV 269 A631]
RC271.C5A677   1984   616.99'4061   83-12411
ISBN 0-471-08080-2

Printed in the United States of America

10 9 8 7 6 5 4 3 2 1

Contributors

RONALD N. BUICK

Ontario Cancer Institute
Toronto, Ontario

WILLIAM T. BRADNER

Bristol-Myers Company
Syracuse, New York

CHARLES A. CLARIDGE

Bristol-Myers Company
Syracuse, New York

WILLIAM A. REMERS

The University of Arizona
Tucson, Arizona

Series Preface

The process of drug development has undergone major changes in the last two decades. To appreciate the magnitude of the change, one needs to think back to the mid-1950s. This was the boom period of pharmaceutical development; better than half the structural classes available to today's clinician had their inception in that era. Yet, in spite of all the demonstrable successes, this was not a period of truly insightful research. Rather, regulations were sufficiently liberal so that novel chemical entities could be—and were—taken to the clinic with only a demonstration of safety and some preliminary animal pharmacology. It is perhaps as a result of this that many of our pharmaceutical mainstays owe their existence to serendipitous clinical findings.

It should, of course, be added that the crude nature of the available pharmacology was a reflection on the state of the art rather than on a desire to skimp on research. A good many of the current concepts in pharmacology postdate the boom era in drug development.

The same applies to medicinal chemistry. With a few notable exceptions, much of the synthesis was aimed at achieving a patentable modification on someone else's drug or consisted in following "interesting chemistry" in the hope of coming up with biological activity. The dialogue between the medicinal chemist and the pharmacologist was in its infancy.

The drug development process in 1982 is an entirely different discipline; though the time and effort involved in taking a drug from the bench to registration has increased enormously, it has, in spite of this, become more intellectually satisfying. The increased knowledge base permits more informed decisions.

The increased stake involved in taking a drug to the clinic means that upper management in drug companies wants greater assurance of success before taking that very expensive step. Consequently, compounds are studied pharmacologically in far greater detail than ever. The gap between animal pharmacology and human therapy is being steadily narrowed by the development of ever more sophisticated tests which may more accurately forecast human responses. Much of this has been made possible by enormous strides in pharmacology. Understanding of drug action is approaching the molecular level.

Medicinal chemistry too has acquired a firmer theoretical underpinning. The general desideratum is rational, or directed, or deliberate, drug development. (*Rational* strongly implies that those who do not follow that design are irrational. There is too large an element of luck, serendipity, and informed intuition involved in drug discovery to use the term irrational for those who choose a more intuitive approach.) This approach has in fact achieved its first success: Cimetidine was developed by studying the interaction of histamine and its congeners with its receptors. Captopril came from a research program motivated by a consideration of the role of the renin angiotensin system in the control of blood pressure.

A hallmark of many laboratories involved in drug development is the existence of the project team. All individuals assigned to research on drugs in a given therapeutic area are expected to interact with a greater or lesser degree of formality and to make their own day-to-day research decisions in close consultation. While the makeup of such teams varies considerably, the medicinal chemist and the pharmacologist are almost obligatory members. It becomes incumbent on each to be able to communicate with the other. The pharmacologist will thus profitably be acquainted with the names and, if possible, structures of compounds relevant to the therapeutic area, be these drugs or endogenous agonists and antagonists. While not expected to actually design analogue series, the pharmacologist may find it appropriate to be able to recognize pharmacophoric groups. The chemist, on the other hand, will certainly want at least nodding acquaintance with the pharmacological basis for drug therapy in an assigned area. An understanding of biological screens, tests, and their limitations will help the chemist better understand the biological significance of test results on compounds being studied.

There are today very few convenient sources to which a scientist can turn for such information. As a rule the pharmacology on any therapeutic area will be scattered in original articles and reviews in the biological literature. An individual seeking the medicinal chemistry background will have to choose between consulting superficial reviews, perusing some sixteen or more

SERIES PREFACE

volumes of highly condensed periodic reports, or going back to the original literature.

Chemistry and Pharmacology of Drugs is a series of books intended to allow scientists involved in drug development to become familiar with specific therapeutic areas by consulting a single volume devoted specifically to that area.

Each of the volumes of the series envisaged treats a fairly discrete disease entity, or sometimes a therapeutic class. Each of the books treats separately the pharmacology, screening, and development methods and medicinal chemistry relevant to the topic. In each book, the first section deals in some detail with both the normal and diseased physiology of the appropriate organ system; it is in this section that the pharmacology and, if pertinent, biochemistry are discussed. The next section deals with the various primary screens that have been used to discover active compounds. More elaborate tests designed to elucidate mechanism of action and the like are discussed as well. The medicinal chemistry section deals with the chemistry used to prepare active compounds; where available, the SAR of active series; and the rationale that led a particular direction to be chosen. Since such a volume is today beyond the scope of any single author, each book will be written by three or, at the most, four authors.

DANIEL LEDNICER
Series Editor

Preface

Progress in the discovery of new cancer chemotherapeutic agents has been unspectacular, considering the amount of money and talent expended in this field. When viewed, however, from the perspective that cancer is an extraordinarily complicated and resilient group of diseases which offers few targets for chemotherapy, the progress is rather admirable. The fundamental difficulty in developing useful antitumor agents is the lack of selectivity between tumor cells and normal cells. Most of the differences on which selective toxicity must be based are quantitative rather than qualitative, a more difficult situation than, for example, the selective killing of bacterial cells whose qualitative differences from mammalian cells in terms of cell walls and ribosomes present highly selective targets for cytotoxic drugs.

Despite the lack of unique targets in cancer cells, it is often possible to reduce their numbers drastically without destroying the contiguous normal cells. A thousandfold reduction in cancer cells can be produced by a number of drugs. Unfortunately, to effect a cure it may be necessary to kill all the cancer cells. Because the drugs kill a constant fraction of the cancer cells, increasing amounts are required to bring the cell count down to low numbers, and there is a high probability that serious toxicity to normal cells will occur in the meantime. In theory, the body's immune system should be able to dispose of the remaining small amounts of aberrant cells. But by their very nature the cancer cells have managed to elude or overcome the immune system. The final eradication of cancer cells is unlikely unless this system is functioning or can be stimulated. Another serious complication in cancer chemotherapy is that many tumors are derived from a number of clones that differ in their drug sensitivity. Initial treatment can produce a remission by destroying the sensitive clones. Nevertheless, the resistant ones can ultimately regenerate the cancer. Combinations of drugs acting by different mechanisms have been developed to forestall the proliferation of resistant clones.

At the present time at least 10 neoplasms can be treated with good expectation of a cure: acute leukemia in children, Burkitt's lymphoma, choriocarcinoma, Ewing's sarcoma, Hodgkin's disease, lymphosarcoma, mycosis fungoides, rhabdomyosarcoma, retinoblastoma in children, and testicular carcinoma. These tumors tend to be the rarer ones. Unfortunately the outlook is not so good for patients with such common tumors as those of the liver, pancreas, colon, and lung. Breast cancer continues to be a serious cause of death, although progress is being made against it. Numerous other tumors give partial responses to chemotherapy, but the beneficial effect often is of short duration. Needless to say, chemotherapy is only one mode of cancer treatment. Surgery is the prime mode for solid tumors and radiation is used for skin cancer and localized deeper tumors. Even in these cases chemotherapy is used as an adjuvant, particularly when metasteses are suspected.

In this volume on antineoplastic drugs we have tried to describe the nature of cancer cells and to relate their special properties to the problem of developing chemotherapeutic agents. Chapter 1 describes cell division and the cell cycle. It considers in detail cell kinetics and stem-cell biology. Opportunities for drug specificity related to differences in these cell properties between cancer and normal cells are discussed. In Chapter 2 the screening systems used for the identification of potential anticancer drugs are described and evaluated for their predictive potential. *In vitro* and *in vivo* systems are considered and problems in correlating antitumor activity in tissue culture or animal screens with human clinical activity are explored. The third chapter is presented from the viewpoint of the medicinal chemist. It considers the large body of research devoted to understanding the chemical interaction of antitumor agents with biological targets such as DNA and enzymes and how investigators have tried to apply this knowledge to the rational design of improved agents. Studies of the preparation of analogs of natural and synthetic lead compounds are described, and recent efforts to place analog selection on a sounder basis by the use of quantitative structure-activity relationships are examined.

Our hope is that this volume will provide a guide and stimulus to chemists and pharmacologists active in or seeking to enter cancer research. Despite the difficulties of curing many types of cancer, progress is being made in all of the essential scientific areas, including tumor biology, screening methodology, drug design, and clinical evaluation. This is a time of rising expectations in the field.

WILLIAM A. REMERS

Tucson, Arizona
December 1983

Contents

1. THE CELLULAR BASIS OF CANCER CHEMOTHERAPY 1
   Ronald N. Buick

2. SCREENING SYSTEMS 41
   William T. Bradner and Charles A. Claridge

3. CHEMISTRY OF ANTITUMOR DRUGS 83
   William A. Remers

INDEX 263

ANTINEOPLASTIC AGENTS

ONE

The Cellular Basis of Cancer Chemotherapy

RONALD N. BUICK
Ontario Cancer Institute
Toronto, Canada

INTRODUCTION AND GENERAL PRINCIPLES	3
CELL DIVISION, CELL CYCLE, AND DRUG SELECTIVITY	5
CELLULAR ORGANIZATION OF NORMAL AND TUMOR TISSUE	9

    Tissue Classification, 9
    Renewal Tissue Behavior Modes, 10
    An Idealized Cell-Renewal System, 11
    The Cellular Organization of Tumor Tissue, 14
    General Implications of Stem-Cell Organization for Tumor Therapy, 16
        Cellular Aspects of Human Tumor Growth, 18

METHODOLOGICAL APPROACHES	19

    Tumor Growth Kinetics, 19
    Measurement of Tumor Size, 19
        Growth Curves, 21
    Cell Kinetics, 23
        Stathmokinetic Technique, 23
        Radioactively Labeled Thymidine, 24
        Labeling Index, 24
        Percent Labeled Mitoses Technique, 25
        Growth Fraction, 26
        Cell Loss, 26
        Summary of Tumor Kinetic Data, 27
        Models of Tumor Growth and Cell Kill, 27
    Stem Cell Biology, 29
        Properties of Clonogenic Tumor Cells, 32

INTERACTION OF CHEMOTHERAPEUTIC AGENTS WITH NORMAL AND TUMOR TISSUE	32

    Cytotoxic Drug Effects on Cells, 32
    Tumor Specificity, 33
    Inherent Drug Sensitivity, 34
    Drug Resistance, 35

CONCLUSIONS	36
REFERENCES	36

INTRODUCTION AND GENERAL PRINCIPLES

Cancer is a disease of cells characterized by the reduction or loss of effectiveness of the normal controlling influences that maintain cellular organization in tissues. Cancer cells have acquired properties that, in simplistic terms, provide them with growth advantages over normal cells; this permits their continuous proliferation not only in their sites of origin but also in other environments. The abnormal behavior of tumor cells leads to damage in the host at a variety of levels; (a) locally by pressure effects, (b) by destruction of involved tissues, both physically and in terms of normal function, and (c) by systemic effects secondary to the localized growths.

As an initiation point in the design of therapy and the understanding of cancer growth, comparisons at different levels have been made of tumor and normal equivalent tissue from which we can summarize the basic features of cancer cells:

1. Uncontrolled cell proliferation.
2. A lack of cellular differentiation features.
3. The ability to invade surrounding tissue.
4. The ability to metastasize (establish new focal growth in distant sites).

An understanding of the mechanisms that underlie these bahavioral characteristics is fundamental to the development of therapy to eradicate cancers. We are concerned in this chapter with those aspects of tumor and normal cell behavior that impact on the use of chemotherapeutic agents.

Cancer therapy with cytotoxic drugs has made enormous progress since the initial application of chemicals in the late 1940s when Farber prescribed methotrexate to treat childhood leukemia. In recent years the emphasis has been on the integration of chemotherapy with other treatment modalities; surgery, radiotherapy, and immunotherapy. It is now clear that chemotherapy's most effective role in solid tumors is as an adjuvant to initial therapy by surgical or radiotherapeutic procedures. This realization has come about through an understanding that failures of primary field therapy are due principally to the existence of occult micrometastases not accessible to surgery or localized radiotherapy. Chemotherapy becomes critical to effective treatment because only systemic therapy can attack micrometastases.

The design of treatment regimens, which include chemotherapeutic drugs, has, in major part, advanced empirically; numerous experimental protocols have been designed, some of which have markedly improved prognosis for a few types of cancer (for a review see Carter et al., 1981); for example, chemotherapy is considered curative in choriocarcinoma in women, Burkitt's lymphoma, Wilm's tumor in children, certain testicular tumors, and childhood leukemia. Unfortunately, in the most common human tumors (colon, lung, breast) chemotherapy has not had a major impact; indications are, however, that significant advances in the treatment of these tumors will derive from increased sophistication in the use of adjuvant chemotherapy.

The most common rationale for the use of chemotherapy is control of the growth of tumor cell populations by cell-kill mechanisms. A major limitation to this approach is the nonselectivity of chemotherapeutic agents. The capacity to use the available cytotoxic agents to kill tumor cells is limited by the effect of these drugs on critical normal tissues; therefore treatment is often limited by cell-kill effects in bone marrow (anemia, thrombocytopenia, neutropenia, and immunodeficiency) and intestinal mucosa and by damage to a variety of other normal tissues (lung, heart, kidney, and brain). A major challenge to the chemotherapist remains the design of tumor-specific therapy. Because cell-kill effects are basically dose-dependent, theoretically a large enough dose of an anticancer agent could eradicate the tumor. This rationale will become reality only when a totally tumor-specific agent can be formulated.

One approach to this problem has been to design regimens that contain combinations of agents that might act synergistically on tumor cells but not on normal tissues (Capizzi et al., 1977). There are a few examples of such effects: the combination of cis-platinum, vinblastine and bleomycin in the treatment of testicular tumors (Einhorn, 1977) and methotrexate, 5-fluorouracil and cyclophosphamide (Canellos et al., 1976) in breast cancer. The combination of chemotherapeutic agents has been attempted rationally by combining agents with different mechanisms of action. A good example is the MOPP regimen (mustard, vincristine, procarbazine, and prednisone) used successfully in the treatment of advanced Hodgkin's disease (Devita et al., 1970).

Because the most obvious characteristic of tumor tissue is its increase in size by an increase in cell number, chemotherapeutic agents have naturally been selected as antiproliferative. Their ability to impede cell proliferation is in general a function of interference with a critical biochemical component of the cell division process. The efficiency of interaction of drugs with tumor tissue is quite obviously dependent to a major degree on noncellular processes, which include pharmacokinetics, drug distribution in tumor and normal tissue, and drug effects on tumor architecture (connective tissue and vascularization). This chapter deals only with cellular effects; it must be acknowledged, however, that these effects may play a minor role in the overall determination of the outcome of chemotherapy.

To discuss adequately the cellular features of tumor and normal tissue that determine drug response it it necessary to compile information from a variety of scientific disciplines. It is beyond the scope of this chapter to provide a critical review of all the scientific areas of this question. I intend therefore to target my discussion to the theme of human tumor cell heterogeneity and its effect on chemotherapeutic outcome. It is my hope that the reader will be able to probe in more detail any particular discipline by use of the bibliography. It is apparent that our understanding of human tumor cell biology has relied heavily on experimentation in animal tumor systems. Wherever possible, however, this discussion emphasizes information on human tumors.

CELL DIVISION, CELL CYCLE, AND DRUG SELECTIVITY

Cancer research has long concentrated on the apparent uncontrolled growth of tumor tissue in relation to the normal equivalent. Study of cell proliferation, as the central process in this growth, has therefore been singled out as holding special importance to the cancer problem. Could differences be detected in the proliferative processes of normal and tumor cells that account for the growth advantages of tumor cells? Although this view does not seem likely now, an understanding of the biochemical events underlying cell division are, of course, fundamental to an understanding of chemotherapy.

The replicate characteristics of mammalian cells are conventionally described in terms of a cell cycle. This concept depends on the events between the birth of a new cell and its subsequent division as a series of ordered, unique biochemical events (Hill & Baserga, 1975; Mueller, 1971).

Morphologically, dividing mammalian cells exist in two basic states; one of actual cell division (mitosis) and a much longer interphase period. It has been possible to identify events in the mitotic process, and the phases during which defined chromosome behavior occurs have been given names: during *prophase* the nuclear membrane disintegrates and the chromosomes condense. The chromosomes align themselves along the central axis of the cell (*metaphase*) and then are segregated to opposite poles by the spindle apparatus (*anaphase*). Finally, cell division occurs by pinching the plasma membrane and reforming new nuclear membranes (*telophase*).

The intermitotic period has been characterized by biochemical events. The doubling of DNA content in preparation for mitosis occurs only in a portion of the interphase. This period has been termed the S phase (synthesis). The period after birth of a new cell and before S phase is termed G_1 (gap 1) and the period after S phase and before mitosis, G_2 or gap 2 phase (Figure 1). In terms of biochemical events, G_1 has proved to be that part of the cell's intermitotic period in which many of the necessary enzymes for DNA synthesis and macromolecules involved in the specialized function of the cell are produced. G_2 phase is characterized by the production of macromolecules necessary for mitosis.

This view of the division of a mammalian cell has derived in major part from the study of cell growth in tissue-culture conditions in which the influence of variables can be minimized. In particular, the complications of concomitant cellular differentiation and proliferation (which exists in all animal proliferating tissues) have been avoided by the use of cell lines adapted to tissue culture in which every cell divides with a closely similar in-

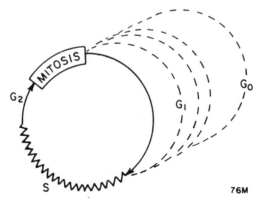

Figure 1 The cell cycle. S, the phase of DNA synthesis; G_2, premitotic phase; G_1, preceding DNA synthesis, an expression of tissue-specific proteins of a highly variable duration; G_0, hypothetical state of reversible exit from cycle.

termitotic time. Thus, in this simple example, if cells are distributed at random in the cell cycle, the extent of heterogeneity (percentage of cells at any phase of the cycle) will be reflected by the fraction of the cell cycle that the particular phase represents. Thus, if S-phase represents 30% of the total cell cycle time, then, in a population distributed at random throughout the cell cycle, 30% of cells will be in S-phase at any point in time.

Certain general conclusions have been drawn in regard to the duration of the phases of the cell cycle. The methodological approaches which resulted in these conclusions and the extrapolation of these concepts to human tumor cell populations are detailed in a subsequent section that deals with the impact of cell kinetic measurements on our knowledge of human tumor behavior. The length of the cell cycle varies considerably from tissue to tissue in proliferating cells. Under optimal culture conditions certain mammalian cell lines can be made to divide every 10–12 h. The majority of cell lines require approximately 1–2 days. Under certain culture conditions, however, cell cycle times can be as long as months. The times required for the S, G_2, and M phases are constant, regardless of cell-cycle time. The differences in cycle times can be accounted for almost entirely by differences in the duration of the G_1 phase which can vary from being nondetectable in rapidly proliferating systems to a number of months. It has been postulated that a high degree of variance in the G_1 phase is in fact due to another phase, namely G_0, wherein the cell enters a resting state and is not actively committed to cell division. The presence of this state is still a matter of considerable debate but, as discussed in a later section, it has considerable implication for the efficiency of chemotherapeutic agents.

Chemotherapeutic agents can be categorized into functional subgroups: alkylating agents, antimetabolites, antibiotics, and antimitotics (Bruce et al., 1966; Madoc-Jones & Mauro, 1974; Ghuyan et al., 1972; Zimmerman et al., 1973). The importance of cell-cycle considerations to the field of chemotherapy relates to the cell-kill specificity of many of these agents, based on intracellular targets present at only defined periods of the cell cycle. These agents are termed cell-cycle phase-specific. Each drug in this class produces a unique biochemical inhibition of a particular reaction crucial to the cell's survival or passage of the cell through a single phase of the cell cycle. Most of these drugs are antimetabolites or antimitotics. A list of the most common phase-specific drugs is given in Table 1 with a description of the particular phase in which their activities are evidenced. The S-phase-dependent antimetabolites act by interference with nucleic acid synthesis; for example, methotrexate binds to dihydrofolate reductase, the enzyme that normally catalyzes the reduction of

Table 1 Common Cell-Cycle (Phase)-Specific Drugs

M-phase dependent	G_2-phase dependent
Vinca alkaloids	Bleomycin
Vincristine	
Vinblastine	S-phase dependent
Colchicine derivatives	Antimetabolites
Trimethylcolchicinic acid (TMC)	Cytarabine (ARA-C)
Podophyllotoxins	Fluorouracil (5-FU)
Etoposide (VP-16)	Mercaptopurine (6-MP)
Teniposide (VM-26)	Methotrexate (MTX)
G_1-phase dependent	Thioguanine (6-TG)
	Hydroxyurea
Asparaginase	Prednisone
Corticosteroids	Diglycoaldehyde

dihydrofolate to tetrahydrofolate. With the inactivation of this enzyme, the cell is deprived of N^5, N^{10}-methylenetetrahydrofolate, a required molecule for the synthesis of thymidylic acid, which, in turn, is a required substrate in the synthesis of DNA. Agents with specificity in the mitotic process (e.g., vincristine) interfere with the mitotic spindle formation, a process essential for cell division.

In contrast, some drugs are described as cell-cycle phase nonspecific. These agents have no toxic properties that are related to a specific part of the cell cycle. Instead, their lethal interaction with cells (often by damage to DNA) is dependent on the cell that attempts DNA replication or DNA repair subsequent to drug exposure. Common agents in this classification, listed in Table 2, include the alkylating agents and antitumor antibiotics.

Table 2 Common Cell-Cycle (Phase)-Nonspecific Dose-Dependent Drugs

Antitumor antibiotics	Alkylating agents
Dactinomycin	Busulfan
Doxorubicin (Adriamycin)	Chlorambucil
Daunorubicin	Cyclophosphamide
Rubidazone	Melphalan (1-PAM)
Nitrosoureas	Mechlorethamine (Nitrogen Mustard)
Semustine (Methyl-CCNU)	Miscellaneous
Carmustine (BCNU)	Dacarbazine (DTIC)
Lomustine (CCNU)	Cisplatin

Certain important therapeutic implications are derived from this classification of drugs. At least from a theoretical standpoint, a tumor in which a large proportion of cells are actively in cell cycle will be more sensitive to cell-cycle phase-specific drugs than a tumor with a high proportion of noncycling cells. On the other hand, cell-cycle phase-nonspecific drugs may be expected to be equally effective in both cases. The phase-specific agents may also be expected to benefit from scheduling; multiple-repeated fractions of a given drug should allow more access to the relevant target cell than one large dose. On the other hand, phase-nonspecific agents should not be schedule-dependent; theoretically, a single large dose should kill the same number of cells as multiple-repeated fractions that total the same amount. These implications have proved to be true in general in animal tumor models, but, as we discuss later on, these theoretical points have not necessarily been of value in terms of therapy of human tumors.

CELLULAR ORGANIZATION OF NORMAL AND TUMOR TISSUE

Tissue Classification

It is apparent that the correct use of the armamentarium of therapeutic drugs (Tables 1 and 2) necessitates a knowledge of the proliferative state of the target tumor cells and the dose-limiting normal tissues incidentally insulted. To describe the vulnerability of both normal and tumor tissue adequately we must concern ourselves with a description of their cellular organization, particularly with respect to growth potential and proliferative state.

Animal tissues can be classified on the basis of cell proliferative biology into three types:

1. *Static populations.* These cells are those in which no cell division takes place in postnatal life. The differentiated neurons of the central nervous system are an example of this type. These cells increase in volume but do not increase in number as the individual matures.
2. *Expanding populations.* These populations are defined as those in which an increase in cell number is matched by an increase in size of the total population; that is, progeny of cell divisions are retained in the cell population indefinitely. The rate of cell proliferation in these tissues decreases with the growth rate of the individual. Few cells divide in normal adult

individuals. Tissue classified in this group includes the parenchymal cells of liver, kidney, and exocrine and endocrine glands.
3. *Cell renewal populations.* Cell renewal is necessary in those tissues in which function is carried out by cells whose life-spans are shorter than the animals in which they reside. In other words, in these populations mitotic activity is not matched by an increase in the size of the cell population. Under normal conditions a steady-state situation exists in which cell production is balanced by cell loss (by normal attrition of differentiated cells in performance of function or by physical loss of cells). Renewal proliferation occurs in the majority of human tissues. The most commonly used examples are gastrointestinal mucosa, epidermis, respiratory epithelium, testicular epithelium, and hemopoietic tissue; all epithelium, however, maintains steady state in this manner, although at vastly different rates proportional to the extent of cell loss (Potten et al., 1979).

Renewal Tissue Behavior Modes

The classification of some tissues as *expanding* may be an oversimplification in that these tissues may undergo renewal proliferation in certain circumstances. If, for example, part of the liver is resected, the remaining tissue proliferates rapidly to regenerate the original tissue mass. It is not known whether the cells of origin of this regenerative proliferative response are true liver stem cells or differentiated hepatocytes that can respond to an appropriate stimulus and divide. Should the former situation exist, it must be assumed that a small subpopulation of undifferentiated cells does exist in the liver but that their presence is obscured morphologically by the vast numerical dominance of differentiated cells. This would also imply that regeneration responses are an example of a normal renewal system behavior and that such tissues are "conditionally renewing."

Renewal tissues also demonstrate behavior patterns in which steady state is not present but nevertheless not in a tumor-forming mode. These situations are classified under the categories of *hyperplasia* and *metaplasia*. In both cases cell proliferation and differentiation of a renewal system results in an increase in the number of cells in the tissue. Hyperplasia is defined as an increase in the number of cells with no concomitant change in their relative frequency at different levels of differentiation. The best documented example of renewal proliferation is the testosterone-induced hyperplasia of the ventral prostate in rats. Administration of testosterone to male rats causes an increase in the number of normal prostatic cells which form additional normal-functioning glands. If testosterone administration is stopped, the tissue reverts to normal size. The proliferative behavior of normal endometrium provides an example of hyperplasia and metaplasia. Hyperplastic proliferation of the endometrium is induced by the presence of estrogens during the first part of the menstrual cycle. Subsequently, in midmenstrual cycle progesterone induces a metaplasia that results in a change in the functional capacities of the endometrium in preparation for implantation of the fertilized ovum. Metaplasia therefore is defined as a proliferative response with a concomitant change in the differentiation state of the tissue. In endometrial metaplasia the changes are reversible when the hormonal induction is removed if no fertilized ovum is implanted.

Thus hyperplastic and metaplastic proliferative responses are maintained only as long as the application of an inductive stimulus. They are a function of the normal stem cells of the tissue in question. The reversibility of the proliferative response is the factor that most clearly distinguishes such proliferation from neoplastic proliferation in which the proliferative response is permanent and inherited by subsequent generations of cells in the absence of the original inductive stimulus. It can therefore be implied that this proliferation is a function of tumor stem cells and that such stem cells differ from normal stem cells in terms of their proliferative response to the inductive stimuli provided by the environment of the tissue. It is the elucidation of the basis for such heritable change that is the fundamental question in cancer research (Mackillop et al., 1983).

An Idealized Cell-Renewal System

A cell-renewal system has certain characteristics that are critical to its function of maintenance of the steady-state size of the tissue (Mackillop et al., 1983). Figure 2 is a simplistic representation of the important functions of an idealized renewal system. The diagram shows a group of cells within a renewal tissue in terms of their common ancestry, but no account is taken of the temporal relationship between parent and progeny. This group of cells, the origin of which can be traced to a single cell, is termed a *clone*. The cells are shown as a cell division pyramid in which the final progeny dominates in a numerical sense. In a normal renewal tissue these cells would be highly specialized for a particular function (i.e., differentiated) and would be produced continuously as need arose by cell loss.

The cell of origin of a clone is termed a *stem cell*. This central cell has two

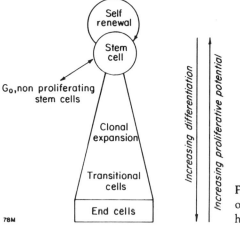

Figure 2 A schematic representation of a clone of cells in a cell-renewal hierarchy.

critical functions: first, it can act to provide cells (by cell division) to a population expansion that will result in a large family of descendants which will perform the function of the particular tissue. Thus a stem cell in a particular renewal tissue is determined; its potential for the provision of cells for differentiation is restricted to that tissue type and no other. The second stem cell function relates to Figure 2 which shows cell loss from the differentiated progeny but no cell input to the stem cell compartment. To take account of the fact that in adult animals no new stem cells are produced from a more primitive cell it is necessary to incorporate in an idealized renewal system the capacity of the stem cell to replace itself during the process of providing cells for clonal expansion. This process is termed self-renewal and is the most defining property of a stem cell. We can imagine the importance of this process by considering the implications of the loss of a stem cell during each clonal expansion combined with a lack of stem cell replacement from a more primitive cell.

Because in normal conditions steady state is maintained with respect to cell number, cell-renewal systems can obviously be tightly controlled. Within a single clone (Figure 2) limitation of cell number is provided in a general sense by the fact that a cell's proliferative potential is related in an inverse fashion to its differentiation state. Proliferation and differentiation which generally seem to be opposing processes result in the tendency toward decreased proliferation as clonal expansion occurs. The molecular mechanisms underlying this tight control are thought to be based on hormonal modulators that act through receptor mechanisms in a tissue specific fashion. It is also important to note that the potential is available for fast or slow response to a requirement for cell production. A modulating activity operative at a stem-cell level would be expected to elicit a larger effect on the size of a population of differentiated progeny than on a cell at an intermediate point on the clonal expansion. However, an effect mediated by a stem cell would also be expected to require a longer time to affect differentiated progeny frequency due to the larger number of cell division events required to express the end result.

These concepts have led to a view of a renewal system as containing cells that exist on a hierarchy with a continuous gradation of properties. This concept is shown in Figure 2 as directional linear changes in properties as clonal expansion occurs. Such changes are best regarded as not occurring in discrete steps but rather as continuous alterations in the probability that a cell will express a given characteristic. Thus the probability of a cell at the stem-cell level of the hierarchy that expresses the property of self-renewal is effectively 1.0. Similarly, the probability of a cell at the base of the hierarchy that expresses differentiated function is 1.0 and the probability that such a cell will express the ability to proliferate is effectively 0. Proliferative potential is defined as a measure of the extent of clonal expansion possible under optimal conditions. Clearly a cell with stem-cell properties has the potential to produce a larger clone than a cell at an intermediate level of differentiation.

The cellular organization of renewal tissues in the normal state has important implications for the determination of cells that are in cell cycle and thus indirectly for the response of these tissues to chemotherapeutic agents. Considerable evidence suggests that the small proportion of stem cells in a normally renewing tissue are less likely to be in cell cycle than those cells in the "amplification" compartment (the transitional compartment, Figure 2). Stem cells therefore seem to act as reserves of growth potential, whereas tissue maintenance in major part derives from cell divisions in less primitive cells with only infrequent input from the stem-cell population. The evidence for this view derives from the study of murine and human hemopoiesis and cell production in the rat small intestine. In hemopoietic tissue direct estimation of the proportion of primitive precursor cells in cell cycle has led to the conclusion that the vast majority of stem cells are normally in a resting phase, whereas most cells in the transitional compartment are rapidly cycling (Till & McCulloch, 1980). Autoradiographic description of cell proliferation in the rat jejunal crypt has provided a similar conclusion for cell renewal in the intestinal mucosa (Cairnie et al., 1965). Stem cells can, however, be triggered quickly into cell cycle in response to specific demands for regeneration; for instance, the work of Lahiri and van Putten (Lahiri & van Putten, 1972) showed that a large fraction of the resting hemopoietic stem cells in mice could be triggered into cycle as soon as 10 min after whole body exposure to 150 rad. Similar regenerative responses have been described in the intestinal mucosa and epidermis. The stem cells of normal tissues may then be regarded as being in a privileged state with regard to susceptibility to cell-cycle specific chemotherapeutic agents.

The Cellular Organization of Tumor Tissue

Human tumors develop only in tissues that normally (or conditionally) proliferate by cell-renewal mechanisms. Further, tumors are tissue-specific (Pierce et al., 1978); that is, a tumor in the intestinal mucosa is identifiable as abnormally proliferating intestinal tissue and not as abnormal lung epithelium. Thus we can speculate that the origin of tumors lies in a cell (or group of cells) that is determined with respect to differentiation and that proliferates under normal conditions. Two general theories of tumor origin account for these obvious characteristics. First, it has long been suggested that a carcinogenic event occurring in a differentiated cell of a particular tissue renders that cell proliferative, although it retains its ability to organize tissue-specific differentiation. The acquisition of proliferative features in a differentiated cell necessitates proposing the process of "de-differentiation." The second class of theories proposes that tumors arise from carcinogenic events that occur in the stem cells of a particular tissue. The proliferative potential and determination in terms of cellular differentiation of this group of cells means that it is not necessary to propose de-differentiation as a mechanism of tumor growth. Rather it is necessary only to propose that the changes associated with the carcinogenic insult cause a loss of control and disassociation of the normal stem-cell functions: cell renewal and cell differentiation (Mackillop et al., 1983).

Evidence for the latter theory is now overwhelming, particularly for the hemopoietic malignancies. The critical circumstance, which has allowed evidence to accumulate, is that the hemopoietic stem cell is not totally determined but rather has the capacity to produce a number of "lines" of differentiation; thus the murine pluripotent stem cell (CFU-S) gives rise to all the lineages of myeloid differentiation as the result of a single clonal expansion. In human myelopoietic malignancies the leukemic clone involves all the myelopoietic lineages. The evidence is derived from analysis of karyotypic abnormalities (e.g., Ph^+ in CML) (Whang et al., 1963) and by study of G-6-PD isoenzymes in heterozygous female patients (Failkow et al., 1977). The description of genetic markers of the malignant clone in all lineages of myelopoiesis is consistent only with carcinogenic transformation that occurs in a single pluripotent stem cell. Similar information is accumulating for epithelial malignancies, although greater technical difficulties are associated with data collection.

A major source of information on the organization of tumor tissue has come from the use of developmental assays: the use of tissue culture or animal transplantation procedures to identify cells with high proliferative capacity by their colony-forming ability. The specific advances in this area are detailed in a subsequent section. In a general sense, however, these procedures have reinforced the view of human tumors as "caricatures" of the cell renewal in their normal tissue equivalent (Pierce et al., 1978). It has been possible to identify cells in human tumor populations with features of the three component populations of the renewal hierarchy: stem, transitional, and end cells. In the tumor types studied so far a minority cell population capable of self-renewal has been identified, whereas most tumor cells are limited in terms of proliferative ability and express markers of tissue specific differentiation. All current data are consistent with a stem-cell origin of human tumors and a stem-cell renewal model for their growth.

We are faced with a view of human tumors that in terms of clonal organization is strikingly similar to the normal tissue equivalent. What, then, is the basis for the lack of steady state of tumor tissue? Normal renewing tissue maintains a steady state by a tightly controlled balance between cell proliferation, cell differentiation, cell loss, and stem-cell renewal. Tumor stem cells and their progeny clearly are not responsive to the same controlling influences; cell population expansion could occur by excess cell proliferation, excess frequency of renewal events within the stem cell population, blockage of cell differentiation, and inadequate removal of cells by normal cell-loss processes. There is evidence for all these mechanisms; the overall affect is likely caused by a combination of such circumstances (Mackillop et al., 1983).

Tumor cells are therefore heterogeneous with respect to proliferative potential and cellular differentiation, as imposed by the relative position on the cell renewal hierarchy of the clone from which they are derived. In addition to this complexity, however, human tumors have frequently undergone clonal progression by the time of clinical intervention (Nowell, 1976). This process is thought to occur because of the genetic instability of tumor cells;

although tumors probably arise from a single stem cell, during growth of the tumor new genetically discrete stem lines are produced, some of which may have sufficient growth advantage to overtake the other stem lines and predominate in the tumor. As we discuss in a later section, this process may be particularly important in the predominance of new stem lines with drug-resistant characteristics during treatment (Barranco et al., 1972). Thus the heterogeneity imposed by the cell-renewal hierarchy may be complicated by several orders of magnitude by the multitude of stem lines operative in the tumor at the time of diagnosis. This problem is described diagrammatically in Figure 3.

General Implications of Stem-Cell Organization for Tumor Therapy

These considerations have immediate impact on our discussion of the cellular basis of therapy. Clearly, if tumor tissue is organized as a renewal hierarchy we can make certain generalizations about the cells to which any therapy should be directed. Figure 4 shows the relationship of a tumor cell's position in a hierarchy to the role of that cell in the pathogenesis of the tumor. Because cure requires eradication of all cells with repopulating capacity, it is a function only of stem-cell death. Likewise, the rate of tumor regrowth will be related to the number of stem cells surviving subcurative therapy. A more usual clinical endpoint in studies of chemotherapy in solid tumors is clinical response (measured by reduction in tumor bulk). This is not necessarily a stem-

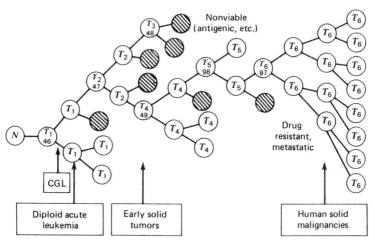

Figure 3 Model of clonal evolution in neoplasia. (After Nowell, 1976.)

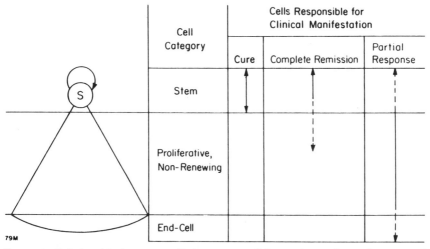

Figure 4 Relationship between cells in a stem-cell hierarchy and manifestation of clinical treatment parameters.

cell function; for instance, cell lysis in the end-stage population could induce short-term reduction in tumor volume without having impaired stem-cell function.

The tumor stem cells are therefore the optimal target for any therapeutic regimen with curative intent. It is important to consider solid tumors and hemopoietic malignancies separately in a discussion of treatment goals because of the inherent differences in their natural history.

In solid tumors the destruction or removal of every stem cell is a feasible objective. The probability of achieving it, however, depends on the number of stem cells present in the patient and the extent to which they are localized in a single site. Radiation therapy and surgery are effective curative modalities in certain cases of geographically limited disease. Systemic therapy is indicated wherever stem cells are widely spread; this spread occurs at variable times in solid tumors. Thus, in devising therapeutic strategies for solid tumors, emphasis is on the local disease and its treatment by surgery or radiation therapy; only if there is evidence of spread is chemotherapy necessary. Recently it has become feasible from studies of the natural history of the disease to identify the patients in whom micrometastases may exist; for them adjuvant chemotherapy is indicated even if dissemination cannot be demonstrated objectively.

Cure is seldom the only goal for treatment of solid tumors: control may be valuable to patients who may experience prolonged survival and acceptable life quality if cancer cells can be reduced in number or growth potential or both. Hormonal manipulation of breast or prostatic cancer cells are examples of this approach. Its success depends on careful patient selection. The use of hormonal receptor assays for this purpose is now well established (McGuire, 1978).

Cure and control have special meanings in hemopoietic malignancies. These diseases not only begin in the treatment-limiting hemopoietic tissue, their very presence suppresses and may extinguish normal hemopoietic stem cells. Thus cure can be obtained in only two ways. First, chemotherapeutic agents may eliminate abnormal clones and allow normal stem cells to repopulate the tissue. This outcome may be the basis for long-term survival of children with acute lymphocytic leukemia (ALL) and young patients with acute myelocytic leukemia (AML). Second, when normal stem cells are absent or damaged beyond recovery they may be replaced by marrow transplantation. When an HLA-DR-matched sibling donor is available transplantation is followed by long-term survival and perhaps cure in approximately 60% of the young patients with acute leukemia in whom the transplant procedure is carried out in first remission.

Control may be achieved in hemopoietic malignancies by influencing the behavior of the dominant tumor clones. The best example is provided by chronic myelocytic leukemia (CML); following treatment, the abnormal clone persists in each patient but the balance between erythropoiesis and granulopoiesis approaches normal. Similar intraclonal changes may be responsible for apparently normal hemopoiesis in AML after chemotherapy. Observations which indicate that differentiation can sometimes be induced in culture of leukemic cells support the possibility that treatment of AML may improve the production of functional cells from leukemic precursors.

Reduction of the number of end cells may also be effective in controlling abnormal hemopoietic clones. The treatment of polycythemia vera by phlebotomy and of CML by leukophoresis are examples of this approach, analogous to the control of solid tumors by reducing total tumor load.

Cellular Aspects of Human Tumor Growth

In the preceding section theoretical aspects of cellular organization in normal and malignant tissues were considered as they affect cancer-treatment strategies. The view was advanced that a minority population of cells in a cancer, those with stem cell properties, are the appropriate targets for curative treatment. Further, the capacity to use the available cytotoxic agents to kill these cells is limited by the effects of these drugs on normal tissues, which are themselves maintained by a minority population of stem cells.

Clearly, detailed knowledge of the organization of tumor and normal tissue is apparently one of the prerequisites for the design of effective cancer chemotherapy. Most information on tumor cellular organization has come from two scientific disciplines; tumor growth kinetics and the developing field of tumor stem-cell biology and kinetics. These areas have received considerable optimistic support in the hope that they might answer the following questions. What is the impact of the respective cellular organizations of tumor and normal tissue on the outcome of chemotherapy? Are there ways in which the investigator can predict, from knowledge of cellular organization, the response of a given tumor? We shall see that a great deal of the knowledge of proliferative cell biology of human tumors has not resulted in a clear answer to either of these questions, although recent methodology in cell kinetics and stem-cell biology show promise of providing a major leap in the potential use of nonempiric chemotherapy.

METHODOLOGICAL APPROACHES

Tumor Growth Kinetics

The science of tumor kinetics is concerned with a description of the growth of tumors in relation to time. With the inclusion of procedures to mark individual cells (tumor cell kinetics), the contribution of proliferating cells to overall population dynamics can be studied. As such, it is a major source of information about the cellular organization of tumor tissue. For an extensive review of this literature the reader is referred to the work of Steel (1977).

Measurement of Tumor Size

The simplest means of constructing a tumor growth curve is by serial measurements of tumor size. For dispersed tumors total tumor burden can be estimated by cell counting with a knowledge of dilution factor. For solid tumors measurements of linear dimensions or volume have been obtained. Linear external dimensions can sometimes be measured by Vernier calipers and estimates of volume made from these measurements or more problematically by direct methods. The extrapolation of linear measurements to volume estimates has considerable theoretical limitations but has been carried out with some success by the use of geometrical formulas or a calibration curve constructed from a set of experimental tumors covering the range of size of interest.

The practicality of making these measurements on human tumors has greatly limited the available information on human-tumor growth rates. Study of the growth of dynamic human tumors is, of necessity, limited to that period of the natural history of the tumor between the time it becomes clinically detectable and the death of the patient. This period normally represents an extremely small portion of the tumor's growth period (Figure 5). In addition, successful study has been limited to the cases in which serial observations of untreated tumors can be made. In particular, measurement of the growth of two categories of tumor has been useful: tumors (primary or secondary) in surgically inaccessible anatomical sites that can be recorded by radiography or ultrasound and, less commonly, skin nodules that can be measured directly in a patient in whom treatment is not considered because of extensive metastatic spread.

An indirect, but extremely useful, determination of tumor size can be made in some instances by the measurement of systemic levels of substances produced by tumor cells. Tumor-specific protein synthesis is associated with myeloma (paraproteins) and choriocarcinoma (human chorionic gonadotrophin, HCG). Carcino-embryonic antigen (CEA) and α-foetaprotein are indicators also of the extent of certain tumors but are seldom sufficiently specific. In human multiple myeloma abnormal immunoglobulins (parapro-

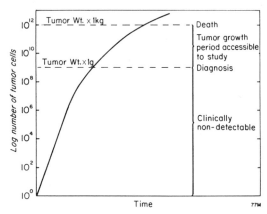

Figure 5 Schematic representation of the relationship between tumor growth and clinical time points.

teins) have a narrow range of electophoretic mobility when derived from an individual patient but differ between patients. Hobbs (1969) demonstrated that when paraprotein levels were measured serially in the serum of untreated myeloma patients they increased in an exponential fashion. Salmon and Smith (1970) showed that paraprotein synthesis from cultured myeloma cells resembled that occurring in situ and used data on culture synthetic rate to derive an estimate of the total body burden of myeloma cells. These estimates have ranged from 5×10^{11} to 3×10^{12} malignant cells.

In comparison to the necessarily rather crude procedures applied to the measurement of size of solid tumors the measurements of serum levels or synthetic rates of tumor-specific products have many advantages. Uniquely, these measurements lead to a realization of the total body tumor burden, not simply that located in one accessible site. It is perhaps more important that they have proved to be useful in the estimation of tumor burden when the level is below the limits of clinical detection. Thus study of tumor growth has been possible in presymptomatic stages and during remission.

Growth Curves

The relation of these measures of tumor size to time provides a growth curve for the tumor in question. Typically, with a measure of size (weight or volume) plotted vertically and of time, horizontally, on semilogarithmic scales, growth curves are convex upward and flatten out with time; that is, in most cases growth rate decreases with time. In others growth curves form straight lines on a semilogarithmic plot; this represents the simplest form of tumor growth in which growth rate as a fraction of tumor size is constant and the logarithm of tumor volume increases linearly with time (i.e., exponential growth). The doubling time can be read off the chart by taking the difference between the times at which the line cuts any two volumes that differ by a factor of 2. Exponential growth occurs in some cell cultures, ascites tumors, and murine leukemias.

For the majority of cases in which growth rate decreases with time a variety of mathematical functions has been proposed to explain such behavior; the most commonly applied, was suggested by Laird (1969). It is a modified exponential law of growth in which the doubling time increases continuously with time. The Gompertz equation is used for this law.

$$Wt = [W_0 e A_0 (1 - e^{-\alpha t})/\alpha]$$

where W_0 is the initial tumor size, A_0, the initial specific growth rate for the period of observation, and α, the rate of exponential decrease of A_0; Wt is the tumor size at time t. This equation describes a sigmoidal curve on a linear scale; 37% of the final volume is the point at which maximum growth rate occurs.

The Gompertz function has proved to be the most applicable predictor of tumor growth in transplantable animal tumor models (Schabel, 1969) and human multiple myeloma (Sullivan & Salmon, 1972). By attempting back-extrapolation to the early nonmeasurable stages of tumor growth, however, it is clear that Gompertzian growth cannot occur over the full life-history of the tumor. Steel (1977), on the basis of tumor-growth measurements and cell titration experiments for B16 melanoma, proposes that the full growth curve can be fitted by applying a curve to data at 10^6 tumor cells or more and by back-extrapolating exponentially from this point to a single cell of origin.

The most frequently reported growth curves of human tumors have been derived from serial radiographic studies of lung tumors or metastases (Steel, 1977). Surprisingly, the majority of these reports showed exponential growth curves of lung metastases up to a volume of 1000 μcm^3 in some cases. In addition, a number of examples of irregular growth have been documented. It is not clear at this time whether generalizations regarding the existence of exponential growth of human tumors can be made from data almost entirely limited to tumors growing in the lung and close to the end of their life-span.

The available data on volume doubling times of human tumors have been summarized and analyzed by Steel (1977), who has accumulated a series of 780 measurements of primary tumors of the lung, breast, colon-rectum, and bone and metastatic tumors in the lung, lymph nodes, and superficial sites. The distribution of volume doubling times of human tumors is lognormal with a large spread around the mean, even for individual types. Yet the differences between volume doubling times of some different tumors are sufficient to be statistically significant. In the series reported by Steel the mean values of volume doubling times range from less than 20 days for lymph node metastases of reticulum cell sarcoma to more than 600 days for primary tumors of the colon-rectum. A typical volume doubling time for lung metastases is 50–100 days. *In general terms*, adenocarcinomas grow more slowly than squamous cell carcinomas or sarcomas. When directly compared, metastatic tumors have a shorter doubling time than the primary tumor of origin. Contrary to popular misconceptions, therefore, human tumors do not grow extremely rapidly. It must be stated, however, that this doubling-time data, by necessity, is collected on clinically measurable tumors, close to the end of their development (Figure 5).

Cell Kinetics

To explain the slow rate of growth of human tumors it is necessary to analyze the fate of the individual cells within the tumor. Some of the possible explanations for slow growth have a bearing on the overall question of therapeutically important cell heterogeneity; for instance, a slow overall growth rate could be variously attributed to a low number of proliferating cells, or slow division time for those cells that are dividing, the presence of dead cells, and a high rate of cell loss. At least in theory, the contribution of these variables would make considerable impact in the response of the tumor to a chemotherapeutic agent with selectivity for cycling cells. To analyze the life history of individual tumor cells the science of cell kinetics takes advantage of a number of markers of cell division.

Stathmokinetic Technique

The simplest method of marking cell division is by the identification of mitotic figures. In general terms, in a rapidly proliferating cell population the proportion of cells in mitosis (mitotic index) is higher than in a slowly proliferating population. The proposal that mitotic index relates to the rate of cell production depends on the assumption that the duration of the mitotic event is identical in different cells. To overcome this source of error quantitative kinetic studies of metaphase arrest have been performed (stathmokinetic). Such procedures use agents that arrest cells in metaphase but have (ideally) no effect on cells outside metaphase. These agents include colchicine, colcemid (demecolcine), and vinblastine, all of which act biochemically by destroying the mitotic spindle, causing failure of chromatid separation. At the time of application of an appropriate concentration of one of these agents to a cell population a number of cells will be in metaphase. The number of cells in metaphase will subsequently increase and the cell birth rate can be computed from the rate of accumulation of cells in metaphase.

Because the rate of metaphase accumulations is, in practice, not linear (Steel, 1977), the stathmokinetic procedure has been superseded largely by techniques that require marking the cell division with radioactively labeled DNA precursors.

Radioactively Labeled Thymidine

The use of labeled thymidine in the study of cell kinetics has provided a unique marker with which to investigate the life-history of individual cells within a heterogeneous population. Because thymidine is incorporated specifically into the DNA of replicating cells, the number of cells that incorporate labeled material (^{3}H or ^{14}C) is an approximation of the number of cells that synthesize DNA at the time of exposure. The β-particles emitted by decay of tritium can be used to expose silver-grain photographic emulsion to form an autoradiograph of a tissue section, cell smear, or cytocentrifuge preparation. The fact that DNA is one of the most metabolically stable cellular constituents means that cells labeled during DNA synthesis become permanently tagged. Thus when tagged cells divide the marker is diluted but still carried by daughter cells.

Labeling Index

The simplest piece of information to be gained from thymidine labeling and subsequent autoradiography of a cell population is the labeling index, or the ratio of labeled cells to total cells. Administration of labeled thymidine can be made to the whole animal/patient, to the tumor itself, or to sections or cell suspensions derived from excised tumor tissue. In terms of human tumor studies, relatively few have been made by in situ administration because of the obvious ethical considerations. In vitro exposure has the added advantage that labeling and autoradiographic exposure times can be kept short by the use of high-specific-activity ^{3}H-thymidine. When comparisons have been made between labeling indices estimated by in situ or in vitro procedures for animal tissues the correlation has been good.

Given certain assumptions, comparisons of labeling indices for different tumors can act as a *rough guide* to the respective rate of cell proliferation and to the responses of the proliferative tumor population to cytotoxic therapy (Thirwell & Mansell, 1976).

Steel (1977) summarizes data from 705 measurements of a thymidine labeling index of human tumors. Mean LI values are listed as varying between 25% for lymphomas to less than 2% for adenocarcinomas. The points are made that for individual tumor types data from different investigators do not agree and that computations of tumor growth rates from LI data (assuming an S-phase of 10–20-h duration) do not explain the observed growth rates. Thus cell loss obviously plays a major role in the determination of tumor growth. The concepts and importance of tumor cell loss are approached later in this section. There is, however, a general consensus that LI is *usually* high in tumors with rapid growth rates.

Percent Labeled Mitoses Technique

The principal method of deriving estimates of cell-cycle phase times is the technique of percent labeled mitoses introduced by Quastler and Sherman (1959) and applied to tumor populations by Mendelsohn (1962). A pulse of tritiated thymidine given in situ is incorporated into the DNA of any cell in the process of DNA replication (S-phase). At intervals after this pulse autoradiography is performed on tissue sections or smears and mitotic figures are scored as labeled or unlabeled. A plot is made of percent labeled mitosis against time after pulse labeling. There is an initial period when all mitotic figures are unlabeled because these mitotic cells were post-S-phase when the thymidine pulse was administered.

As cells labeled in S-phase reach the "window" of mitosis, the labeled mitoses will in theory rise to 100%. The time required for all labeled cells to move through mitosis is indicative of the duration of DNA synthesis (S-phase). The second increase represents the cell division of the progeny of the initial cells. The time between the peaks therefore is a measure of the intermitotic time. In a commonly used procedure the times between the 50% level of the first and second peaks are measured as the intermitotic time. The sharpness of the peaks reflects the synchrony of the original labeled cells. When human tissue is studied variability in phase duration leads to damping of the curve. The extent of damping is related to the spread in intermitotic times within the cell population. As a result, complex mathematical models have been necessary to compute cell-cycle time parameters and computer programs are available for this purpose.

In addition to often necessary complex mathematical computation, PLM curves have detracting features that may lead to lack of accuracy in the data. Because multiple biopsies are required after pulse labeling, it is possible that the experimenter may alter cell-cycle parameters artificially by factors introduced in the sampling procedure. A major assumption is made that there is no significant re-use of labeled breakdown products. The presence of cells with long intermitotic times will be underestimated because they are seen less frequently in the window of mitosis. Despite these reservations, the technique of percent labeled mitoses is extremely powerful and is regarded as the most important for cell cycle analysis of in vivo cell populations. In recent years the advent of flow microfluorimetry apparatus has resulted in the description of cell-cycle parameters in tumors readily converted to single-cell separations.

Human tumor studies with PLM analysis have been limited to sites that allow multiple biopsies. A review of cell-cycle phase information from PLM curves is presented by Tannock (1978) for hemopoietic and solid tumors. Values for intermitotic times for many human tumors are short (T_c, 2–4 days) and S-phase duration varies from 6 to 24 h, approximately. Thus the commonly held misconception of faster cycle times in tumor tissue compared with the normal are clearly not justified.

Growth Fraction

Mendelsohn (1962) introduced the important term, growth fraction (GF), as an indication of the proportion of tumor cells within the proliferative compartment. A PLM curve and a measure of LI permit the calculation of GF defined as equal to the ratio of the measured LI to the value of LI that would be obtained by sampling only proliferating cells. The latter value can be obtained by computation from the PLM-derived measurements of T_s and T_c. Steel has pointed out that the technique is not sufficiently sensitive for use in the classification of cells with long intermitotic times because they will appear as nonproliferative in a computation of GF. Reports of an association between calculated GF and tumor growth rate (Frindel et al., 1968) have been made, but in general there is no consensus in the literature regarding this relationship.

Cell Loss

Cell loss is proposed as the *major factor* that contributes to the discrepancy between the calculated potential growth rate and the actual measured growth rate. In general, all human tumors grow much more slowly in vivo than would be predicted by the kinetic parameters of the individual cells within them. The cell loss factor Φ is defined as

$$1 - \frac{\text{potential doubling time}}{\text{actual doubling time}}$$

Potential doubling time is computed from GF and cycle time (both of which are inaccurate as described); thus the cell-loss factor has limitations in terms of quantitation. It cannot be determined directly.

As calculated, however, cell loss clearly plays a major role in the determination of tumor growth rate. The cell-loss factor can be as much as 90% in human tumors.

Summary of Tumor Kinetic Data

By now it will have become clear that information on the generation of cell kinetics in human tumors has not had a major impact on the effectiveness of chemotherapy in individual patients, probably because the derived data is quantitatively inexact due to the enormous technical difficulties associated with the measurements. On the other hand, it has become obvious that the cell kinetics of the tumor stem-cell population may be the critical variable. The techniques described above have no discriminatory power to assess stem cells alone; they all rely on the ability of single cells to perform DNA synthesis and as such do not discriminate between cells in the transitional cell compartment and those in the stem-cell compartment (Figure 1). However, the literature supports a general relationship between tumor cell kinetics and drug sensitivity (Clarkson, 1974; Valeriote & Edelstein, 1977; Price et al., 1975); for example, Charbit et al. (1971) related drug sensitivity to tumor histology and kinetic behavior and showed that drug sensitivity was closely associated with growth fraction in a spectrum of tumor histologies. These findings have not necessarily translated into improved survival for patients who bear these tumors because the most sensitive tumors (high GF, high growth rate) are also those in which tumor regrowth is fastest.

Models of Tumor Growth and Cell Kill

From the study of tumor kinetics a number of models have been derived which have had an impact on the design of therapy. The most prevalent kinetic consideration applied in cancer chemotherapy is the cell-kill concept. Skipper (1971) and colleagues (Skipper et al., 1964) established a relationship between "cure" rate and tumor cell number in mice bearing the L1210 leukemia. In this simple system the proportion of tumor cells killed by a given drug dose is a constant percentage of the total number of cells present, and tumor growth and regrowth after therapy is exponential, regardless of that number. By scheduling consecutive doses at intervals close enough to prevent significant regrowth cures could be elicited in mice bearing L1210 leukemia. This model implies that chemotherapy can be curative when the number of tumor cells is small. Treatment schedules for human leukemia have been influenced by these concepts. In addition, they have provided the theoretical basis for adjuvant chemotherapy. Because drugs are most likely to be curative when the tumor burden is small, adjuvant therapy is indicated immediately after primary field therapy. The log-kill model is therefore based on studies of L1210 leukemia, a tumor that represents a poor kinetic model for human nonlymphoid tumors. The basic features of this tumor are not seen in human tumors; high-growth fraction, constant first-order kinetics of cell kill, regardless of tumor cell number, and constant exponential rate of growth and regrowth (with a doubling time of 12 h). Norton and Simon (1977) have introduced an alternative concept based on Gompertzian growth kinetics, a situation perhaps more like that in most human tumors. These authors propose that the magnitude of tumor regressions induced by a given chemotherapeutic dose are best explained by the relative growth fraction present in the tumor at the time of treatment. Because the growth fraction varies, as shown in Figure 6 for Gompertzian growth kinetics, tumors will be least sensitive when tumor cell numbers are small or very large. The magnitude of cell-kill effects elicited by a given drug dose will therefore vary with the surviving fraction, unlike the constant cell-kill hypothesis of Skipper et al. (1964). Al-

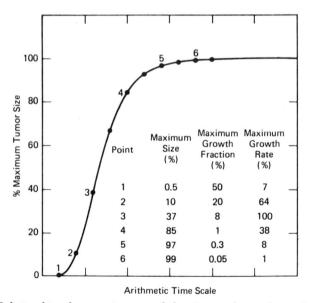

Figure 6 Relationship of tumor size, growth fraction, and growth rate for Gompertzian kinetics. (After Norton and Simon, 1977.)

though the Norton and Simon model has not been confirmed clinically, it would seem to have some attractiveness as a simple first approximation of drug effects in human tumors.

Stem Cell Biology

The second major research area to have an impact on cancer therapy concepts is the field of tumor stem-cell biology. We have discussed at some length the failure of tumor cell kinetics to provide therapeutically important information in individual patients. This has been due in part to logistical problems but, in addition, a growing body of opinion supports the notion that knowledge of the stem-cell compartment specifically may be necessary.

The views expressed earlier of tumors as stem-cell renewal systems have developed over the last two decades. In major part they have relied on artificial methods of estimating cells with stem-cell properties in tumor populations. The basic stem-cell property is the proliferative potential of the cells. As previously described, a cell-renewal hierarchy represents considerable cellular heterogeneity, based on proliferative potential. Because, in an appropriate environment, stem cells can give rise to a large family of descendents (i.e., they are clonogenic), this property can be used to distinguish the cell of origin of a clone from a more differentiated cell with lower proliferative potential.

Stem-cell concepts (both normal and tumor) have evolved in two phases that coincide with two approaches to the provision of adequate environmental influences for the measurement of cell clonogenicity. The first of these approaches was derived from the discovery of the spleen-colony procedure by Till and McCulloch (1961) and the second by adaptation of semisolid culture procedures applied to murine hemopoietic cell populations by Bradley and Metcalf (1966) and Pluznik and Sachs (1965).

The spleen-colony procedure was developed in investigations of the rescue capability of normal bone marrow injected into lethally irradiated mice. Till and McCulloch (1961) noted that the injected cells grew as clones in the spleens of the recipient animals. On counting these clonal growths, they found that their numbers were linearly related to the number of cells injected and that the cloning efficiency was approximately 1/10000. Subsequent studies (Becker et al., 1963; Becker et al., 1965) demonstrated the clonality of these colonies and that they contained cells of different lineages of hemopoietic differentiation. The important conclusion, therefore, was that these colonies represented the endpoint of proliferation from a single cell with multiple differentiation potential; this represented an assay for a class of pluripotent hemopoietic stem cells. Because these infrequent cells could not be visualized on morphological grounds, this assay provided a fundamental advance in the understanding of cell renewal in normal hemopoiesis.

The application of the spleen-colony procedure to tumors was first reported by Bruce and van der Gaag (1963), who demonstrated spleen and kidney colonies in nonirradiated recipient mice after the intravenous injection of cells prepared from the thymus of AKR mice with spontaneous lymphoma. Other experimental tumors that have demonstrated the ability to colonize the spleen are L1210 leukemia (Wodinsky, 1967) and the PC-5 plasmocytoma (Bergsagel & Valeriote, 1968).

Bruce and his colleagues (1966) utilized spleen-colony-forming assays for normal CFU-S and AKR lymphoma cells to investigate the chemotherapy of the lymphoma. The relative sensitivity of CFU-S and lymphoma colony-forming cells to a variety of cytotoxic agents resulted in the design of curative regimens with a high degree of selective toxicity for the tumor. The principle of the spleen-colony procedure has also been applied to nonlymphoid tumors. The predominant organ of colonization in these cases is the lung (Williams & Till, 1966; Hill & Bush, 1969).

Consideration of cell renewal in human tumors is impossible by assays that rely on transplantation. With the advent of immune-deprived mice (congenital or drug-induced) able to accept transplants of human tumor tissue similar assays are now feasible for human cells. These developments are awaited with interest. In the interim studies of human tumor-cell proliferative potential has been forced to rely heavily on artificial-tissue culture procedures.

The principal of assessment of tissue culture clonogenicity is a simple one; heterogeneous tumor cells are placed as a single-cell suspension under tissue culture conditions that (a) provide adequate nutritional support for proliferation and (b) maintain progeny in close association with cells of origin. The clonal units that appear can then be enumerated to obtain an approximation of the frequency of clonogenic cells.

As in animal transplantation clonogenic assays, the major impetus for tissue-culture tumor-clonogenic assays developed in normal murine hemopoiesis. The work of Bradley and Metcalf (1966) and Pluznick and Sachs (1965) established the use of semisolid culture procedures to assess the function of clonogenic hemopoietic cells committed to the granulopoietic lineage of differentiation. The proliferation of these cells is dependent on regulatory molecules (colony-stimulating activity) in culture. Similar colony assays have been developed for human cells. In fact, it is now possible to quantitate colony-forming cells of all the human hemopoietic lineages and a population of cells with multipotency. The view has developed that the colony-forming cells with differentiation limited to one lineage are progeny of the multipotential hemopoietic stem cell. Demonstration of these cells with a commitment to a single line of differentiation outlines the lineage diagram of hemopoiesis (McCulloch, 1979). Clearly, the development of these procedures and the continuing study of controlling influences on hemopoietic stem cells has fundamentally altered the sophistication of our understanding of cell renewal in normal hemopoiesis.

The application of these techniques to human tumors has spread by direct use in the study of hemopoietic neoplasms to the adaptation of procedures for the study of solid tumors. Early studies applied standard culture procedures for granulocytic colony-forming cells to the marrow of patients with acute myeloblastic leukemia. Abnormalities of proliferation and differentiation resulted in the classification of growth characteristics from different patients into categories that proved to be prognostically significant (Moore et al., 1974).

More recently, a number of groups have applied similar technology with the inclusion of more specific growth stimulators for malignant hemopoietic populations. The growth of AML blasts in culture has demonstrated a striking dependence on phytohaemagglutinin (PHA) or medium conditioned by hemopoietic cells in the presence of PHA (Buick et al., 1977; Dicke et al., 1976; Park et al., 1977). Myeloma clonogenic cells have been stimulated by murine-spleen-cell-conditioned medium (Hamburger & Salmon, 1977a) and by T-cell-conditioned medium (Izaguirre et al., 1980). Likewise, lymphoma and ALL cell populations have required the addition of specific additives (Jones et al., 1979; Izaguirre et al., 1981). The quantitative aspects of these procedures are consistent with a stem-cell system for the organization of the malignant hemopoietic cell populations.

By contrast to the defined stimulator requirements for hemopoietic neoplasms colony-forming assays from carcinomas have been developed with no special additives. Hamburger and Salmon (1977b) were the first to report the large-scale use of semisolid culture procedures to quantitate clonogenic cells in epithelial tumors. In the short time since then the list of published reports and different histologies studied has grown impressively. The concept, initiated for normal hemopoietic populations, obviously has broad applicability to the study of human neoplasms.

Properties of Clonogenic Tumor Cells

The hope is prevalent that the study of tumor clonogenic cells, as opposed to total proliferative cells (by cell-kinetic procedures), will allow a greater understanding of cellular organization related to therapy. It is not clear that this has, in fact, come about; a number of technical problems need to be studied in more detail (Selby et al., 1983); for example, are clonogenic cells the stem cells of the tumor? Are those clonogenic cells assessed representative (kinetically or clonally) of the total tumor clonogenic-cell population?

The extent of our knowledge of the biological properties of clonogenic cells can be summarized. These cells represent a tiny minority population in all the tumors studied. Typical frequencies reported are $1:10^5$–$1:10^3$ total tumor cells. They appear to be a kinetically active subpopulation in all tumors studied (Minden et al., 1978; Hamburger et al., 1978), with the possible exception of melanoma. Not all clonogenic cells can be classified as stem cells (Buick & Mackillop, 1981; Thomson & Meyskens, 1982). The property of drug sensitivity and its relation to the outcome of chemotherapy are discussed in the next section.

The biological properties described so far for tumor clonogenic cells have therefore reinforced the view that human tumors are examples of stem-cell renewal systems. Some procedures have allowed classification of patients into prognostic groups (Moore et al., 1974; McCulloch et al., 1982), particularly in the hemopoietic neoplasms, but clearly the development of widespread clinical applicability of these new technologies will require a considerable investment in basic biological research. In particular, the potential is present for major advances in understanding of therapeutic efficacy and failure by studying the cell kinetics of this specialized subpopulation of tumor cells.

INTERACTION OF CHEMOTHERAPEUTIC AGENTS WITH NORMAL AND TUMOR TISSUE

Cytotoxic Drug Effects on Cells

Changes in tumor growth after exposure to a chemotherapeutic agent may come about by two basic mechanisms. Some agents have a direct lethal effect on cells; the alkylating agents serve as examples. The damage to DNA elicited by alkylation precludes further DNA replication and thereby sterilizes the cell effectively. The fact that the damage can be repaired, however, makes the lethal effects dependent on kinetics; for example, holding an alkylation-damaged cell in a state in which it cannot attempt DNA replication could serve to permit repair and overcome the potential lethality.

Antimetabolites block cell-cycle progression by inhibition of a critical biochemical process. Cell death might ensue in this case by unbalanced growth. A corollary of cell-cycle blockade is synchrony of the affected cell population. This property has clear implications for tumor therapy; it has long been attempted to take advantage of drug-induced cell synchrony to maximize lethal effects of a secondary scheduled exposure to a chemotherapeutic agent (Tubiana et al., 1975).

The extrapolation of lethal or cytostatic effects on cell populations to changes in the clinical appearance of the entire tumor tissue is an extremely complicated issue. It is possible theoretically for all cells in a tumor to be sterilized by a given treatment, although the tumor size remains unchanged. Reduction in tumor size, a necessary criterion for the designation of tumor response to therapy, comes about by continuation of the processes of cell loss in the absence or reduction of input of new cells by cell division. Drug effects on cell clearance, vascularization, and stromal tissue thus play a major nonintrinsic role in the determination of in vivo chemotherapeutic response.

Tumor Specificity

Drugs targeted toward proliferating cells obviously have the disadvantage of damage to critical normal cell function. These effects are dose-limiting for most antitumor agents. The fact that antitumor agents are successful at all is due in major part to a favorable therapeutic ratio. Kinetically, we have seen that normal tissue stem cells may spend the majority of their lifespan in a G_0 state, whereas there is a little evidence to indicate that human tumor stem cells are more constantly cycling. Damage to normal tissue therefore is predominantly in the proliferative nonstem-cell compartment and is repairable by the proliferation of surviving stem cells. The role of the chemotherapist is often to apply his or her judgment to the extent and frequency of therapy to allow repopulation of normal tissues while continuing to assault the tumor.

With noncytotoxic drugs specificity can be generated by molecular mechanisms operative in the target tissues. Examples of this specificity are provided by the use of hormonal agents, the action of which is governed by cellular distribution of appropriate receptor molecules. Occasionally cytotoxic agents have a fortunate tissue specificity that renders them useful for particular tumor sites; the use of a dacarbazine (DTIC) in the treatment of human melanoma is an example.

Inherent Drug Sensitivity

Given that cell kill by cytotoxic agents depends on interaction with intracellular biochemical processes or macromolecules and on the efficiency of the repair of such damage, it is not surprising that there should be considerable patient-to-patient variation in the efficacy of chemotherapy. Inherent cellular differences would be based on differences in intracellular concentration or localization of enzymes and intermediates of the various target biochemical processes or repair mechanisms. The relative impact of inherent cellular sensitivity on overall patient response (as opposed to noncellular aspects such as drug distribution) is unknown at this time. However, the challenge of defining an individual patient's tumor in terms of drug sensitivity has inspired many investigators over the last 30 years. As soon as tissue culture methodology allowed laboratory propagation of human tumor cells application was attempted to the goal of predictive chemotherapy. Initial reports of success in this area have almost invariably been followed by lack of reproducibility and failure to extend observations, but renewed interest has been generated by the work of Salmon et al. (1978) and Von Hoff et al. (1981). The crucial development applied to the problem by these investigators is the tissue culture clonogenic assay for the subpopulation of tumor cells with high growth potential. The success they have demonstrated stands out as a major advance in the historical development of chemotherapy and, as such, deserves an extensive description.

The efficacy of measuring culture-clonogenic-tumor-cell survival as a predictive endpoint of drug-cell interactions was demonstrated by McCulloch and coinvestigators in a mouse myeloma model (Park et al., 1971; Ogawa et al., 1976). The survival of mouse myeloma clonogenic cells after exposure to drugs showed that differential sensitivity was present between different myeloma cell lines and that such sensitivity was predictive of response of the tumor growing and treated in vivo. Other studies have reinforced the view that clonogenic assays are the most reliable measures of reproductive death. Given the growing realization that human tumors were examples of cell-renewal systems and that the tumor stem cells were the limiting cell populations in terms of therapy, Salmon et al. (1978) applied the developing technology in the assay of human tumor clonogenic cells to the possibility of measuring drug sensitivity of the tumors in myeloma and ovarian carcinoma patients.

The initial reports of this group have been followed over the last three years by extended series and reports of other tumor types (Meyskens et al., 1981; Alberts et al., 1980) and by corroboration by other groups (Von Hoff et al., 1981). All reports stated an element of concordance between drug responses *in culture* and the patient. The most striking aspect of the data is the high proportion of tumors insensitive both in vitro and in vivo to chemotherapy. This points to a paucity of useful cytotoxic agents for the treatment of advanced disease. Thus the extension of these procedures to the adjuvant chemotherapy setting and to the problem of selection of active agents (drug screening) (Salmon, 1980) are areas of high priority.

Drug Resistance

The acquisition of resistance to chemotherapeutic drugs, a well documented clinical phenomenon (Skeel & Lindquist, 1977; Brockman, 1974), presents a major obstacle to effective therapy. The underlying molecular and cellular mechanisms are not totally understood. Drug resistance appears to involve cellular selection, processes analogous to emerging antibiotic resistance in bacteria. Populations of drug-resistant cells may be inherently produced by clonal evolution (Nowell, 1976) or mutation, perhaps under the influence of a mutagenic cytotoxic agent. The selection process then occurs in the presence of a cytotoxic agent and creates a higher proportion of drug-resistant cells. At the same time it provides them with a growth advantage.

At the molecular level there appear to be a number of major mechanisms by which drug resistance can be generated (Hall, 1977). Access of the active drug to the biochemical machinery of the cell can be inhibited. This inhibition could come about by altered drug disposition or metabolism. At the cellular level alterations of drug transport is a well characterized basis for resistance. In cell-culture systems membrane-altered mutants of CHO cells have extensive cross resistance to unrelated drugs, apparently caused by an excess of a membrane glycoprotein (P-glycoprotein) (Ling, 1982). Although this sophisticated technology has not yet been applied to primary human tumor cells, the implications are indeed great for chemotherapy. Based on the premise that the appearance of drug-resistant mutant cells is a likely cause of clinical drug resistance, Goldie and Coltman (1979) proposed a model that outlined therapeutic strategies designed to avoid therapeutic failure.

Intracellular effects that bring about drug resistance often relate to adaptive enzyme changes. The most often quoted example relates to methotrexate resistance. There is considerable evidence in cell-culture systems that drug resistance in this case can be due to increased intracellular levels of the enzyme dihydrofolate reductase (Schimke et al., 1978). This increased level of protein is secondary to an increased number of the corresponding gene copies. Gene amplification may be a widespread mechanism for the acquisition of drug resistance.

The cellular mechanisms of drug resistance in human tumor cells have not been studied extensively. The clinical experience of drug resistance following treatment with a particular agent has been extended to the in vitro response of clonogenic cells. Salmon (1980) found that tumor clonogenic cells frequently display more resistance to agents after in vivo therapy with those same agents. In addition, significant cross resistance indicates that the multiple drug-resistant phenotypes seen in CHO mutant cells may have a parallel in the human tumor setting.

CONCLUSIONS

Advances in the administration of chemotherapeutic drugs have been made in large part by empiric procedures. It is fair to say that the ever-increasing knowledge of tumor growth biology has not found its way into the design of improved drug regimens.

In this chapter I have attempted to summarize what is understood about the growth of human tumors. There is a growing realization that extensive cellular heterogeneity of function underlies the abnormal growth of tumor tissue. How this heterogeneity relates to tumor response to chemotherapy and in particular how to predict success or failure based on this knowledge remains the major challenge.

REFERENCES

Alberts, D. S., Chen, H. S. G., Soehnlen, B., Salmon, S. E., Surwit, E. A. & Young, L. (1980), *Lancet*, **2**, 340–342.

Barranco, S. C., Ho, D. H. W., Drewinko, B., Romsdahl, M. M. & Humphrey, R. M. (1972), *Cancer Res.*, **32**, 2733–2736.

Becker, A. J., McCulloch, E. A. Siminovitch, L. & Till, J. E. (1965), *Blood*, **26**, 296–308.

Becker, A. J., McCulloch, E. A. & Till, J. E. (1963), *Nature London*, **197**, 452–454.
Bersagel, D. E. & Valeriote, F. A. (1968), *Cancer Res.*, **28**, 2187–2195.
Bradley, T. R. & Metcalf, D. (1966), *Aust. J. Exp. Biol. Med. Sci.*, **44**, 287–299.
Brockman, R. W. (1974), in *Handbook of Experimental Pharmacology* (Sartorelli, A. C. & Johns, D. G., Eds.), Vol. 38, Part 1, Springer-Verlag, Berlin.
Bruce, W. R., Meeker, B. E. & Valeriote, F. A. (1966), *J. Nat. Cancer Inst.*, **27**, 233–245.
Bruce, W. R. & van der Gaag, H. (1963), *Nature London*, **199**, 79–82.
Buick, R. N. & Mackillop, W. J. (1981), *Br. J. Cancer*, **44**, 349–355.
Buick, R. N., Till, J. E., & McCulloch, E. A. (1977), *Lancet*, **1**. 862–863.
Cairnie, A. B., Lamerton, L. F. & Steel, G. G. (1965), *Exp. Cell Res.*, **39**, 528–538.
Canellos, G. P., Pocock, S. J. & Taylor, S. G., III, Sears, M. E., Klaasen, D. J. & Band, P. R. (1976), *Cancer*, **38**, 1882–1886.
Capizzi, R. L., Keiser, L. W. & Sartorelli, A. C. (1977), *Semin. Oncol.*, **4**, 227–253.
Carter, S. K., Bakowski, M. T. & Hellman, K. (1981), *Chemotherapy of Cancer* 2nd ed., Wiley-Interscience, New York.
Charbit, A., Malaise, E. P. & Tubiana, M. (1971), *Eur. J. Cancer*, **7**, 307–315.
Clarkson, B. (1974), in *Handbook of Experimental Pharmacology* (Sartorelli, A. C. & Johns, D. G., Eds.), Vol. 38, Part 1, p. 156, Springer-Verlag, Berlin.
DeVita, V. T., Serpick, A., & Carbone, P. P. (1970), *Ann. Int. Med.*, **73**, 881–895.
Dicke, K. A., Spitzer, G. & Ahearn, M. J. (1976), *Nature London*, **259**, 129–130.
Einhorn, L. H. & Donohue, J. (1977), *Ann. Int. Med.*, **87**, 293–298.
Fialkow, P. J., Jacobson, R. J., & Papayannopoulou, T. (1977), *Am. J. Med.*, **63**, 125–130.
Frindel, E., Blayo, M. C. & Tubiana, M. (1968), *Cancer*, **22**, 611–620.
Ghuyan, B. K., Scheidt, L. G. & Fraser, T. J. (1972), *Cancer Res.*, **32**, 398–407.
Goldie, J. H. & Coltman, A. J. (1979), *Cancer Treat. Rep.*, **63**, 1727–1731.
Hall, T. C. (1977), *Semin. Oncol.*, **4**, 193–202.
Hamburger, A. W. & Salmon, S. E. (1977a), *J. Clin. Invest.*, **60**, 846–854.
Hamburger, A. W. & Salmon, S. E. (1977b), *Science*, **197**, 461–463.
Hamburger, A. W., Salmon, S. E., Kim, M. B., Trent, J. M., Soehnlen, B. J., Alberts, D. S. & Schmidt, H. J. (1978), *Cancer Res.*, **38**, 3438–3443.
Hill, B. T. & Baserga, R. (1975), *Cancer Treat. Rev.*, **2**, 159–175.
Hill, R. P. & Bush, R. S. (1969), *Int. J. Radiat. Biol.*, **15**, 435–444.
Hobbs, J. R. (1969), *Br. J. Haematol.*, **16**, 607–617.
Izaguirre, C. A., Curtis, J. E., Messner, H. A. & McCulloch, E. A. (1981), *Blood*, **57**, 823–829.
Izaguirre, C. A., Minden, M. D., Howatson, A. F. & McCulloch, E. A. (1980), *Br. J. Cancer*, **42**, 430–437.
Jones, S. E., Hamburger, A. W., Kim, M. B. & Salmon, S. E. (1979), *Blood*, **53**, 294–303.
Lahiri, S. K. & van Putten, L. M. (1972), *Cell Tissue Kinet.*, **5**, 365–369.
Laird, A. K. (1969), *Nat. Cancer Inst. Monogr.*, **30**, 15–28.
Ling, V. (1982), in *Drug and Hormone Resistance in Neoplasia*, CRC Cleveland, Ohio.
Mackillop, W. J., Ciampi, A., Till, J. E. & Buick, R. N. (1983), *J. Nat. Cancer Inst.*, **70**, 9–16.

Madoc-Jones, H. & Mauro, F. (1974), in *Handbook of Experimental Pharmacology* (Sartorelli, A. C. & Johns, D. G., Eds.), Vol. 38, Part 1, p. 205, Springer-Verlag, Berlin.

McCulloch, E. A. (1979), *J. Nat. Cancer Inst.*, **63**, 883–891.

McCulloch, E. A., Curtis, J. E., Messner, H. A., Senn, J. S. & Germanson, T. P. (1982), *Blood*, **59**, 601–608.

McGuire, W. L. (1978), *Semin. Oncol.*, **5**, 428–433.

Mendelsohn, M. L. (1962), *J. Nat. Cancer Inst.*, **28**, 1015–1029.

Meyskens, F. L., Jr., Moon, T. E., Dana, B., Gilmarten, E., Casey, W. J., Chen, H. S. G., Franks, D. H. & Salmon, S. E. (1981), *Br. J. Cancer*, **44**, 787–797.

Minden, M. D., Till, J. E. & McCulloch, E. A. (1978), *Blood*, **52**, 592–600.

Moore, M. A. S., Spitzer, G., Williams, N., Metcalf, D. & Buckley, J. (1974), *Blood*, **44**, 1–18.

Mueller, G. C. (1971), *The Cell Cycle and Cancer* (R. Baserga, Ed.), p. 269, Dekker, New York.

Norton, L. & Simon, R. (1977), *Cancer Treat. Rep.*, **61**, 1307–1317.

Nowell, P. C. (1976), *Science*, **194**, 23–28.

Ogawa, M., Bergsagel, D. E. & McCulloch, E. A. (1976), *Blood*, **41**, 7–15.

Park, C. H., Bergsagel, D. E. & McCulloch, E. A. (1971), *J. Nat. Cancer Inst.*, **46**, 411–420.

Park, C. H., Savin, M. A., Hoogstraten, B., Amare, M. & Hathaway, P. (1977), *Cancer Res.*, **37**, 4595–4601.

Pierce, G. B., Shikes, R. & Fink, L. M. (1978), in *Foundations of Developmental Biology Series*, Prentice-Hall, Englewood Cliffs, New Jersey.

Pluznik, D. H. & Sachs, L. (1965), *J. Cell. Comp. Physiol.*, **66**, 319–324.

Potten, C. J., Schofield, R., & Lajtha, L. G. (1979), *Biochim. Biophys. Acta.*, **560**, 281–299.

Price, L. A., Hill, B. T., Calvert, A. H., Shaw, H. J. & Hughes, K. B. (1975), *Br. Med. J.*, **3**, 10–11.

Quastler, H. & Sherman, F. G. (1959), *Exp. Cell. Res.*, **17**, 420–438.

Roper, P. R. & Drewinko, B. (1976), *Cancer Res.*, **36**, 2182–2188.

Salmon, S. E. (1980), *Progress in Clinical and Biological Research*, **48**. Liss, New York.

Salmon, S. E., Hamburger, A. W., Soehnlen, B. S., Durie, B. G. M., Alberts, D. S. & Moon, T. E. (1978), *N. Engl. J. Med.*, **298**, 1321–1327.

Salmon, S. E. & Smith, B. A. (1970), *J. Clin. Invest.*, **49**, 1114–1121.

Schabel, F. M., Jr. (1969), *Cancer Res.*, **29**, 2384–2389.

Schimke, R. T., Kaufman, R. J., Alt, F. W. & Kellems, R. F. (1978), *Science*, **202**, 1051–1055.

Selby, P., Buick, R. N., & Tannock, I. F. (1983), *N. Engl. J. Med.*, **308**, 129–134.

Skeel, R. T. & Lindquist, C. A. (1977), in *Cancer, a Comprehensive Treatise* (Becker, F. F. Ed.), Vol. 5, Plenum, New York.

Skipper, H. E. (1971), *Cancer Res.*, **31**, 1173–1180.

Skipper, H. E., Schabel, F. M. & Wilcox, W. S. (1964), *Cancer Chemother. Rep.*, **35**, 1–111.

Steel, G. G. (1977), *Growth Kinetics of Tumors*, Clarendon, Oxford.

Sullivan, P. W. & Salmon, S. E. (1972), *J. Clin. Invest.*, **51**, 1697–1708.
Tannock, I. F. (1978), *Cancer Treat. Rep.*, **62**, 1117–1133.
Thomson, S. P. & Meyskens, F. L., Jr. (1982), *Cancer Res.*, **42**, 4606–4613.
Thirswell, M. P. & Mansell, P. W. A. (1976), *Proc. Am. Assoc. Cancer Res.*, **17**, 307.
Till, J. E. & McCulloch, E. A. (1961), *Radiat. Res.*, **18**, 96–105.
Till, J. E. & McCulloch, E. A. (1980), *Biochim. Biophysica Acta.*, **605**, 431–459.
Tubiana, M., Frindel, E. & Vassort, F. (1975), *Rec. Results Cancer Res.*, **52**, 187–205.
Valeriote, F. A. & Edelstein, M. B. (1977), *Semin. Oncol.*, **4**, 217–226.
Van Hoff, D. D., Casper, J., Bradley, E., Sandbach, J., Jones, D. & Makuch, R. (1981), *Am. J. Med.*, **70**, 1027–1032.
Whang, J., Frei, E., Tjio, J. H., Carbone, P. P. & Brecher, G. (1963), *Blood*, **22**, 664–673.
Williams, J. F. & Till, J. E. (1966), *J. Nat. Cancer Inst.*, **37**, 177–184.
Wodinsky, I., Swiniarsky, J. & Kensler, C. J. (1967), *Cancer Chemother. Rep.*, **51**, 415–421.
Zimmerman, A. M., Padilla, G. M. & Cameron, I. L. (1973), *Drugs and the Cell Cycle.*, Academic, New York.

TWO

Screening Systems

WILLIAM T. BRADNER
CHARLES A. CLARIDGE
Bristol-Myers Company
Syracuse, New York

INTRODUCTION	43
IN VITRO	43
Microbial, 44	
Microbial Inhibition, 44	
Viral Inhibition, 47	
Prophage Induction, 48	
Tissue Culture, 49	
Cytotoxicity, 49	
Human Tumor Clones, 51	
Phase-Specific Screening, 51	
Malignant Transformation, 52	
Biochemical, 53	
Enzymes, 53	
Macromolecular Synthesis, 57	
In Vitro Screening Strategies, 57	
IN VIVO	59
Transplanted Tumors, 60	
Nonspecific Tumors, 60	
Syngeneic Tumors, 61	
Organ Specific Tumors, 64	
Autochthonous Tumors, 65	
Spontaneous and Viral Tumors, 66	
Carcinogen-Induced Tumors, 67	
Special Systems, 68	
Resistant Tumors, 68	
Host Response Modification, 68	
Side Effects, 69	
Screening Strategies, 69	
Primary Screening, 70	
Secondary Screening, 73	
CONCLUSION	75
REFERENCES	76

INTRODUCTION

For many years it has been possible to test pharmacologic or antiinfective agents in animals and obtain a fairly accurate prediction of the potential efficacy of treatment of a related disease in man. In neoplastic diseases the closest equivalent to this testing has been the use of spontaneous or carcinogen-induced tumors of specific organ systems in animals. These autochthonous tumors have proved to be too slow-growing and cumbersome to serve as practical drug-screening systems, thus leading to a long history of test systems, in vitro and in vivo, which have been tried as models for drug screening against neoplastic disease. In this chapter we have *not* given a comprehensive review of the history or the literature that pertains to screening systems. It is our purpose instead to outline the state of the art as it is today, with emphasis on methods, rationale, and strategies, and to cite its applications and limitations. Our attention is directed more toward the scientists who are unfamiliar with anticancer testing and would like to know how to go about it rather than toward those who want a compilation and index of knowledge already familiar to them. It is also directed more toward the early discovery phase of screening, that is, prescreening and primary screening, rather than secondary evaluation and advanced studies.

As public awareness of the rapid advances now being made in cancer chemotherapy increases, more and more pharmaceutical concerns and university chemists are asking what compounds are perhaps already on their shelves or could be obtained that might have tumor-inhibitory potential and how best to test them. It is hoped that this chapter will provide some answers.

IN VITRO

The selection of appropriate in vitro systems for the screening of compounds with antitumor activity produced by synthetic means, natural product isolation, or fermentation processes is an important phase in the development of new antineoplastic chemotherapeutic agents. The generally recognized excessive cost of in vivo systems alone has led to the design and development of a large number of in vitro assays that may be used with some degree of predictability for the selection of compounds that will have a specific mode of action. When compounds or fermentation extracts have been demonstrated initially to possess in vitro activity of one degree or another confirmatory in vivo activity can be sought. If, however, there is no obvious correlation between the two tests, a reevaluation of the system must be made. Thus a large number of potentially useful in vitro prescreening systems described in the literature have been valuable in certain instances in the preselection of antitumor agents before secondary evaluation. Some have been found wanting, some are limited in their usefulness to the selection of a minor group of compounds with a unique mode of action, and some have found widespread use in the field in the development of novel antineoplastic agents.

Hanka et al. (1978c) cited three prerequisites for an ideal in vitro prescreen: The in vitro prescreen should select potentially in vivo active samples and be fairly inexpensive and easy to operate in large volume. Bradner (1978a) states that the most important attribute of a fermentation prescreen is sensitivity; specificity is of secondary importance.

A considerable number of reviews in recent years have been devoted to the development of prescreen tests for the selection of antitumor agents (Bradner, 1978a; Coetzee & Ove, 1979; Hanka et al., 1978a,b,c; Douros, 1978; and Bunge et al. 1978). Because the site of action of many antitumor agents is cellular DNA (Rosenkranz, 1973), a good number of the tests exploit this fact. Many are discussed in more detail in the following pages.

Microbial

Microbial Inhibition

A variety of microbial species has been tested for use in inhibition-type assays as a potential measure of in vivo antitumor activity. Foley et al. (1958) reported studies in which 16 bioassay microorganisms, including strains of protozoa, were used on a selected number of compounds known to be in vivo tumor active and inactive. They concluded that although a large number of false-positive results were obtained microbiological assay systems for the detection of potential antitumor agents were useful. Hutner et al. (1958) urged the use of particle-ingesting protozoa organisms for prescreens, and Bradner (1958) and Bradner & Clark (1958) suggested the use of anaerobic organisms as indicators of antitumor activity. Szybalski (1958) noted that a high percentage of antineoplastic agents acted as mutagens in a test that used a streptomycin-dependent mutant of *Escherichia coli*. These agents produced a back mutation to prototrophy. Thus this author established a high correlation between the mutagenic and antineoplastic properties of more than 400 substances.

Some specific microorganisms that are used in various laboratories in which fermentative research on antitumor agents is being conducted include *E. coli* and *Bacillus subtilis* in an antimetabolite assay (Hanka et al., 1978b). Fermentation liquors are tested against these organisms grown in complex and minimal media. Samples that inhibit the sensitive microorganism more when it is cultivated in the synthetic medium are considered potential leads.

Similarly, mutants of these two species have served as assay tools. An *E. coli* DNA polymerase A-deficient strain (Slater et al., 1971) and a DNA recombinant deficient mutant of *B. subtilis* (Kada et al., 1972) have been reported to show greater sensitivity to a number of antitumor agents; for example, mitomycin, bleomycin, and sibiromycin.

Bunge et al. (1979) screened fermentation broths that showed cytotoxicity to L1210 cells in vitro, against four species of microorganisms. The combinations of activity observed served as a rough classification of the type of antitumor agent. The four species they used were *E. coli, B. subtilis, Sarcina lutea,* and *Torulopsis albida*. Hanka et al. (1978a) reported on the use of a multiendpoint in vitro system for the selection of active fermentation broths. They used *B. subtilis* cultivated in a complex and synthetic medium, two mutants of *E. coli* blocked in the pathway of carbohydrate metabolism, and *S. lutea* and *Streptomyces flavotricini*, two strains they consider sensitive to certain activities not detected by the other organisms.

Some even more specific activities can be measured by *Penicillium avellaneum*, which has been shown by Hanka and Barnett (1974) to measure antitumor compounds related to maytansine, an antileukemic ansa macrolide isolated by Kupchan et al. (1972). *Candida albicans*, generally considered resistant to most antibiotics except certain antifungal agents such as amphotericin B, trichodermin, and nystatin, has been used recently as a screening agent for the selection of the polyene antibiotic rapamycin (Singh et al., 1979). Rapamycin also has in vivo activity (Douros, personal communication). The sterol-nonrequiring mycoplasma strain *Acholeplasma laidlawii* has been shown by Kitame et al. (1974) to be useful for the isolation of new antibotics, some of which have antitumor activity. Agents such as mitomycin C, actinomycin D, anthracyclines, and sibiromycin are active against this organism.

Non-antitumor agents such as the tetracyclines, chloramphenicol, and erythromycin also are active.

A number of *Xanthomonas* sp., particularly *X. citri* and *X. campestris*, have been suggested as potentially useful organisms for an antitumor prescreen (Douros, personal communication). *X. citri* has unusual sensitivity to a number of mycotoxins, some of which have reported antitumor activity. *X. campestris* has an unusual cell-wall composition (Volk, 1968) and its growth is inhibited by such nucleoside antibiotics as tunicamycin. Tunicamycin prevents the formation of a cell-wall disaccharide lipid intermediate (Bettinger & Young, 1975). Another system of recent interest that may be useful as a potential antitumor prescreen is the crown-gall producing *Agrobacterium tumefaciens* which infects dicotyledonous plants (Drummond, 1979; Nester & Montoya, 1979; Zambryski et al., 1980). Infectious strains of this organism transfer information from a Ti plasmid to override the plant growth-control mechanism and produce a class of compounds, called opines, on which the organism grows. Interference of this DNA transfer may select for agents that could have antitumor activity.

Although the United States National Cancer Institute (NCI) would prefer to discover antineoplastic agents that are not mutagenic (Douros, 1978), there is no question that a considerable number of drugs useful in therapy have mutagenic potential. Devoret (1979) estimates that about half the drugs administered in an effort to arrest cancer are DNA-damaging agents; and so a significant number of people are being regularly exposed, for therapeutic purposes, to these agents. Numerous mutagen tests have been studied. The best known test was devised by Ames who used select strains of *Salmonella* (Ames et al., 1975). Bradner (1978a) has listed several positive antitumor agents as well as some that are negative in this test (Table 1). The correlation between the test and published data on the prophage induction assay are remarkable, although there are a few notable exceptions such as sterigmatocystin and the aureolic acids. Other suggested reverse mutation tests use selected mutants of *E. coli* (Mohn et al., 1974; Iyer & Szybalski, 1958; Green et al., 1976) and *Neurospora crassa* (Ong, 1978).

A number of species of yeast have been studied as possible organisms for prescreen of antineoplastic compounds. Gause et al. (1976) used a permeability mutant of *Saccharomyces cerevisiae* which he showed was more sensitive to a number of antitumor agents than its parent. Speedie et al. (1980) combined these two strains in a radiometric assay to measure differential growth. The benzodiazepine antitumor antibiotics, anthramycin, tomaymycin, and sibiro-

Table 1 Testing of Antitumor Agents by *Salmonella* Mutagenicity Testing According to Ames et al. (1975)

Positive Agents	Negative Agents
Adriamycin	Aclacinomycin
Carminomycin	Actinomycin D
Daunomycin	Anguidine
Hedamycin	Bleomycin
Mitomycin C	Marcellomycin
Neocarzinostatin	Tallysomycin
Olivomycin	Vincristin
Sterigmatocystin	

From Bradner, 1978a, by permission of S. Karger AG, Basel.

mycin, have shown activity against a *S. cerevisiae* mutant defective in DNA repair but not against a repair-proficient strain (Hannan & Hurley, 1978).

The killer proteins elicited by certain strains of yeast (Kandel & Stern, 1979), the fungus *Ustilago maydis* (Koltin & Day, 1975), and *Paramecium aurelia* (Preer, 1977) are controlled by double stranded RNA-containing viruslike particles (Wickner & Liebowitz, 1976). Perhaps an assay seeking inhibition of reversal of this phenomenon may identify compounds with antineoplastic activity. It has already been shown that treatment of a killer strain of *S. cerevisiae* with cycloheximide but not ethidium bromide converts it into a sensitive nonkiller (Fink & Styles, 1972).

Viral Inhibition

A number of antitumor agents have been isolated by the use of an in vitro screen established for antiviral activity. Herrmann et al. (1960) developed a rapid assay in which chick embryo cells were infected with West Nile, vaccinia, herpes simplex, or Newcastle disease viruses. Fermentation broths could be tested for the inhibition of plaque formation. Tunicamycin was isolated by the use of this screen (Takatsuki et al., 1971) and subsequently shown to have antitumor activity and to inhibit SV-40-transformed mouse kidney cells (Takatsuki et al., 1977). Siminoff (1961) developed a similar plaque suppression assay with vaccinia-infected chick kidney monolayers.

The inhibition of plaque formation in monolayer cultures, when infected by rhino, measles, herpes simplex, and vaccinia viruses, was also used by

Streightoff et al. (1969) to isolate pyrazomycin. The antitumor agent alanosine was isolated by examining culture broths for the inhibition of the cytopathic effect of a human epithelial cell line when challenged by polio, neurovaccinia, sheep pox, or cow pox viruses. Alanosine is active in vivo against a transplanted fibrosarcoma in hamsters (Murthy et al., 1966).

The inhibition of bacterial viruses, such as the SP-10 phage of *B. subtilis* or the T-3 phage of *E. coli*, was used to screen for and isolate the antitumor agent tomaymycin from the fermentation of *Streptomyces achromogenes* var. *tomaymyceticus* (Arima et al., 1972). Hori (1972) also used the inhibition of the reproduction of an *E. coli* phage to screen for active agents.

Nakamura et al. (1967) measured the mutation of an *E. coli* phage T_2 from h^+ to h (hazy to clear plaques) for screening actinomycete fermentation broths. Known agents that cause this mutation include alazopeptin, the mitomycins, and roseolic acid.

The infectious bronchitis virus (RNA) cilia test has been used to screen for antiviral agents but to date has not been productive for screening antitumor activity (Douros, 1978).

Prophage Induction

In 1953, before the structure of DNA had even been established, Lwoff of the Pasteur Institute in Paris speculated that the use of inducible lysogenic bacteria might be a good test for carcinogenic and perhaps anticarcinogenic activity (Lwoff, 1953). His postulation proved to be true and since that time a number of variations of a prophage induction test have been reported with applications directed toward the identification of carcinogens or antitumor agents (Lein et al., 1962; Heinemann & Howard 1964; Price et al., 1964; Gadó et al., 1965; Fleck, 1968; Moreau et al., 1976; Elespuru & Yarmolinsky, 1979). With the exception of the test described by Elespuru and Yarmolinsky, the induction of the prophage releases virulent particles that are measured as plaques by a sensitive tester strain. An example of the response of a number of antitumor fermentation products to the assay described by Price et al. (1964) is given in Table 2. Those that fail at tolerated levels are listed in Table 3.

The test by Elespuru and Yarmolinsky involves a "genetically engineered" strain of *E. coli* in which a β-galactosidase synthesized from a lac Z gene is fused to an operon under lambda repressor control. One merely measures for β-galactosidase activity colorimetrically after induction; the whole assay takes only 3 h. This test would appear to be one of the most promising of the

Table 2 Antitumor Agents with Prophage Induction Activity

Compound	Minimum Inducing Concentration (μg/mL)	Compound	Minimum Inducing Concentration (μg/mL)
Mitomycin C	0.004	Baumycin	0.2
Neocarzinostatin	0.006	Sibiromycin	0.2
Phleomycin	0.006	Pluramycin	0.5
Streptonigrin	0.008	Adriamycin	0.8
Bleomycin	0.0125	Carminomycin	0.8
Hedamycin	0.0125	Daunomycin	0.8
Porfiromycin	0.032	Streptozotin	1.0
Azaserine	0.032	Iyomycin	4.0
Figaroic acid	0.05	Valinomycin	12.5
Macromomycin	0.05		

From Bradner, 1978a, by permission of S. Karger AG, Basel.

prophage induction tests reported to date for rapid screening of chemical compounds or fermentation broths for antitumor activity.

Tissue Culture

Cytotoxicity

Tissue culture has been widely used in the United States as a prescreen and a fractionation assay. Thayer et al. (1971) compared the activity of more than 3800 compounds against three microorganisms with their growth inhibition in the KB cell culture cytotoxicity assay. Their data revealed that the KB cell

Table 3 Antitumor Agents Without Prophage Induction Activity at Tolerated Levels

Actinomycin D	Aclacinomycin
Anthramycin	Pyrromycin type anthracyclines
Aureolic acids	Pyrromycin
Borrelidin	Musettamycin
Kundrymycin	Marcellomycin
Anguidine	Cinerubin A
Muconomycin A	Typanomycin
Sterigmatocystin	Violamycin

From Bradner, 1978a, by permission of S. Karger AG, Basel.

culture system appeared to be the in vitro system of choice. Bunge et al. (1979) have reported on the screening of more than 10,000 soil isolates for cytotoxic activity in vitro against L1210 and KB9 tumor cells. Because the KB9 test failed to detect worthwhile antitumor leads other than those detected by L1210 cells alone, it was deleted for routine screening. Garretson et al. (1981) have described an agar-diffusion cytotoxicity test in which activity against the P388 cell line is measured by an inhibition of dye reduction. Among approximately 3000 actinomycete cultures tested about 4.1% were positive.

In our laboratory (Bradner, 1978a) we have shown why tissue culture assays (HeLa cells) are used for characterizing products and as an assay system but not as a prescreen. The results are summarized in Table 4. In a 10-year interval tissue culture was used first in parallel with in vivo testing, second on a series of untested broths as a true prescreen, and third as a characterization of in vivo actives. Although certain highly potent antitumor antibiotics could be detected by tissue culture or animal tumors in vivo, three highly insoluble compounds which had marginal in vivo activity when isolated in pure form were revealed by tissue culture alone. Moreover, two fungal products, anguidine and sterigmatocystin, which have potent cytotoxic effects in pure form, were undetected in tissue culture in tests of the same primary fermentation broth that was positive in tests on animal tumors in vivo. Tests of active fermentation broths at a later time in development and theoretical considerations based on potency suggest that the phleomycin-bleomycin and neocarzinostatin classes of antibiotics would be missed by tissue culture.

Hanka et al. (1978b) have outlined their experience with various tissue-culture prescreen tests in which they used KB cells initially in an assay described by Smith et al. (1959), but, although a number of antitumor antibiotics were discovered, only one (tubercidin) showed any effect in human cancer.

Table 4 Tissue Culture (HeLa) Sensitivity to Fermentation Products

Detected by tissue culture and in vivo systems	Detected by tissue culture alone
Anthracyclines	Peliomycin
Actinomycins	Ossamycin
Aureolic acids	Demetric acid
Quinoxalines	
	Probable in vivo alone
Detected in vivo alone	
	Phleomycin
Anguidine	Bleomycin
Sterigmatocystin	Neocarzinostatin

From Bradner, 1978a, by permission of S. Karger AG, Basel.

When an in vitro prescreen based on the ability of test samples to inhibit growth of L1210 mouse leukemia cells in suspension culture was attempted, a relatively large number of in vivo active leads resulted. Nineteen percent of the broths tested were active in this L1210 assay; among them, 28% became confirmed in vivo actives.

Human Tumor Clones

An increasing interest has been shown in the possibility of developing in vitro screening tests to make the best use of existing drugs by the use of human tumor biopsy specimens (Dendy, 1976). It is suggested that the tumor cells from each patient will respond only to certain drugs that must be selected by sensitivity tests carried out on an individualized basis. Effectiveness of the selected agents against the human tumor-tissue culture cells is measured by an inhibition of the incorporation of ^{3}H thymidine or ^{125}iododeoxyuridine.

Although this test does not appear to apply to the direct in vitro screening of fermentation broths for the detection of new antitumor agents, it is useful as a secondary evaluative test for new chemotherapeutic agents; however, many difficulties are still associated with the use of human tumor biopsy specimens for experimental work. Certain culture procedures may select only one cell type from the biopsy population, the proportion of cells in cycle may change, and the duration of the cell phases, particularly S-phase, may be different in culture (Dendy, 1976).

Cowan and Von Hoff (1980) used the in vitro human-tumor stem-cell assay system of Hamburger and Salmon (1977) to screen new, less toxic, synthetic retinoids for antineoplastic activity. Finally, Salmon et al. (1978) demonstrated a highly significant degree of correlation between the response of certain tumors in patients to drug therapy and the in vitro sensitivity of neoplastic stem cells derived from these tumors to the same drugs.

Phase-Specific Screening

Bruce et al. (1966) have classified chemotherapeutic agents into three groups; namely, phase-specific, cycle-specific, and nonspecific agents. Fonken (1972) defined a phase-specific agent as one that kills a cell preferentially in one phase of its traversal of the cell cycle, a cycle-specific agent as one that kills only those cells that are cycling, and a nonspecific agent as one that kills both cycling and resting cells. Examples of the three classes are cytosine arabinoside, actinomycin, and nitrogen mustard, respectively.

For phase-specific screening synchronized DON cells with a cycle that consists of about 2 h in G_1 phase, 8 h in S phase, 2 h in G_2 phase, and 30 min in M phase have been used. The obvious length that would be required to monitor cells by the full cycle could be shortened if cells in different phases of the cell cycle were suspended in DMSO or glycerol and frozen in liquid N_2 (Bhuyan & Fraser, 1974). Some results of this type of screening are shown in Table 5. As Hanka et al. (1978c) pointed out, however, this technique did not prove to be practical for routine operation.

Synchronous Yoshida lymphosarcoma, HeLa, and Chinese hamster cells have also been used to detect the phase-specific activity of compounds that produce chromosome damage and cell lethality (Scott, 1976).

Malignant Transformation

When normal fibroblasts are transformed by oncogenic viruses or by carcinogenic chemicals their morphology and growth characteristics are profoundly altered. These changes include decreased adhesiveness to plastic and a more rounded shape, increased agglutinability by lectins, lower intracellular ATP levels, and growth to a higher density. There is evidence that some or all of these changes are induced transiently in normal fibroblasts by treatment with low concentrations of proteolytic enzymes such as trypsin, chymotrypsin, and pronase. Thus the search for inhibitors of these enzymes for antitumor activity. But not all transformed cells behave this way, and assays for direct inhibition of malignant transformation may detect distinct, yet valuable, new antineoplastic agents. SV-40-transformed mouse kidney cells were inhibited more than normal cells by the antibiotic tunicamycin (Takatsuki et al., 1977).

McIlhinney and Hogan (1974) tested compounds on the effect of normal and polyoma-transformed baby hamster kidney cells in culture. Pepstatin and antipain had no differential effect on the growth of normal or transformed cells, whereas leupeptin and trasylol inhibited the growth of normal cells more than the transformed. Leupeptin, however, inhibited tumorigenesis in mouse skin induced by a single noncarcinogenic dose of 7,12-dimethylbenzanthracene followed by repeated applications of croton oil. Tumors already induced were not affected by leupeptin (Hozumi et al., 1972). The use of a tissue-culture assay that compared the activity of agents against normal and transformed cells would obviously be a valuable in vitro prescreen test.

Langen (1980) has reviewed recent work on the protein kinases responsible for the transformation of virally infected cells to the malignant state. These kinases have the novel property of catalyzing the phosphorylation of tyrosine

Table 5 Phase-Specific Actives and Inactives[a] in the 2-h DON Test (Fonken, 1972)

Compound	Actives — Most Sensitive Phase	Inactives
Nitrogen mustard	G_1, G_1/S	Methotrexate
Actinomycin D	G_1/S	Busulfan
Chlorambucil	M, G_1	6-Thioguanine
ThioTEPA	M, G_1	6-Mercaptopurine[b]
Melphalan	M, G_1	Triethylenemelamine
5-Fluorouracil	All	Methyl-GAG
Mitomycin C	M, G_1[c]	NSC-45388
Hydroxyurea	S[c]	6-Azauridine triacetate
Mitotane (o,p'-DDD)	S	Procarbazine
Vinblastine	Late G_1 S, M[c]	Dibromomannitol
Cytosine arabinoside	S[d]	L-Asparaginase
Vincristine	S[c]	
Daunomycin	S[c]	
Streptozotocin	All	
BCNU	G_1/S	

[a]Inactive: <25% cell-kill when ascynchronous cells were exposed to drug at 100 μg/mL for 2 h.
[b]Questionable.
[c]References: Mitomycin C, Djordevic et al. (1968); Mauro et al. (1970).
[d]Madoc-Jones and Mauro (1968); Kim et al. (1968).
Phase-specific: killed cells in one phase only, even at high drug levels.

residues in their substrate proteins rather than the usual serine or threonine residues. Inhibition of these kinases from viruses with this transforming ability, such as Rous sarcoma, adenovirus, SV-40, polyoma, Abeleson murine leukemia, and Fujinama viruses, may lead to a compound with clinical potential.

Biochemical

Enzymes

Most antitumor antibiotics have been discovered by screening microbial culture filtrates for activity against experimental animal tumors. There is no direct relationship, however, between a particular type of human tumor and these experimental animal tumors and all antitumor antibiotics thus found exhibit cytotoxic action (Umezawa, 1976). Umezawa reasoned that if a specific enzyme played a role in cell division of cancer cells it would be possible to screen microbial culture filtrates for inhibition against it. Egyud and Szent-Gyorgi believed that cancer cells had lost the ability to maintain a proper balance of methylglyoxal and continued to grow at an uncontrolled rate (1966a,b). Umezawa drew from this hypothesis and isolated two inhibitors of glyoxylase from *Actinomycete* filtrates that showed inhibition against Yoshida rat sarcoma cells and Ehrlich carcinoma. One of the agents was easily hydrolyzed and the other had no sustained activity; the ascites increased seven days after the last injection (Umezawa, 1976). Nevertheless, these results suggested the possibility of finding antitumor agents by screening enzyme inhibitors. Weber (1980) pointed out that certain key enzymes, particularly those in the biosynthetic pathways of purine and pyrimidine metabolism, are likely sensitive targets for screening clinically useful anticancer drugs.

In recent years the mechanisms of action of a number of known antitumor agents has been more clearly defined and, for some, their activity against specific enzymes has been noted. Thus a search for other inhibitors of a variety of enzymes may reveal novel structures of antitumor agents with useful chemotherapeutic properties. Some suggested enzyme systems that may be useful follow.

Aminopeptidases. These enzymes are located on the cell surface, and Umezawa and his group reasoned that any interference with them would have a profound influence on the cell (Aoyagi et al., 1976, 1978a). Amastatin, an inhibitor of aminopeptidase A (Aoyagi et al., 1978b), has been isolated and shown to modify the immune response-participating cells in mice. Bestatin, which inhibits aminopeptidase B and leucine aminopeptidase on the cell surface, enhances delayed-type hypersensitivity and also increases the effect of adriamycin on Ehrlich carcinoma and L1210 leukemia in mice (Aoyagi, 1978). Bestatin has also recently been shown to retard the growth of slow-growing solid tumors of Gardner lymphosarcoma and IMC carcinoma, to improve the antitumor action of bleomycin, and to retard the induction of skin cancer by 20-methylcholanthrene (Ishizuka et al., 1980).

Esterase. Esterastin has been isolated from the fermentation broth of *Streptomyces lavendulae* as an inhibitor of this cell-surface enzyme (Umezawa et al., 1978). Although no antitumor activity has yet been demonstrated, it does suppress delayed-type hypersensitivity and antibody formation.

Cyclic-AMP Phosphodiesterase (cAMP). Lack of contact inhibition is a characteristic of cancer cells and, in general, when contact inhibition occurs cAMP increases. In the screening of inhibitors of cAMP phosphodiesterase prepared from rabbit brain three compounds were found as metabolites of streptomyces: reticulol (reported by Mitscher et al., 1964), genistein, and orobol (Umezawa, 1976).

β-Galactosidase. This enzyme is widely distributed in microorganisms, plants, and animals. Bosmann (1972) showed that its activity in fibroblasts is markedly increased by infection with oncogenic viruses, which suggests an involvement in oncogenesis. Aoyagi et al. (1975) screened culture filtrates of various microorganisms for inhibitors of β-galactosidase and found several active against virus-mediated tumors.

S-Adenosylmethionine Decarboxylase. It is felt that inhibition of methionine will kill certain tumor cells which require this amino acid; thus inhibition of S-adenosylmethionine decarboxylase may be a means of searching for desirable compounds (Douros, 1978).

Reverse Transcriptase Inhibition. Reverse transcriptase (Baltimore, 1970; Temin & Mizutani, 1970) is necessary to make RNA virus replicate and cause viral oncogenesis. Compounds that inhibit RNA-dependent DNA polymerase may therefore have antitumor potential. Numata et al., (1975) described an assay procedure based on the inhibition by a test material of the incorporation of ^{3}H-dTMP into DNA synthesized by the reverse transcriptase of the oncogenic RNA virus of murine leukemia. Screening of fermentation broths led to the isolation of revistin from the fermentation broth of *Streptomyces filipinesis*.

DNA Gyrase. This enzyme produces supercoiling of DNA, and although inhibitors such as coumermycin A_1, nalidixic acid, and oxolinic acid have not yet been shown to have antitumor activity others may (Denhardt, 1979). Because this enzyme is involved with nucleic acid metabolism and because many antitumor agents produce their effects by their interaction with DNA, inhibitors at this site may be of value.

Ornithine Decarboxylase. Inhibition of this enzyme may represent a different class of compounds with possible antitumor activity. Chapman, et al. (1980) showed that cultured C1300 neuroblastoma cells and C6 glioma cells are sensitive to inhibitors of ornithine decarboxylase.

Phospholipase. This represents another cell-surface enzyme as noted with aminopeptidase and inhibitors may offer a compound with differing specificity.

Tubulin Binding. The known target for antitumor agents of the "spindle poisons" group is tubulin, an ubiquitous protein of the eukaryotic cells (Snyder & McIntosh, 1976). Tubulin undergoes polymerization to microtubules which constitute the mitotic spindle. Because of their dramatic effect on mitosis, drugs that bind tubulin were cited a number of years ago as potentially useful antineoplastic agents (Savel, 1966; Sartorelli & Creasey, 1969). It has been shown that the alkaloids of the vinblastine group fall into this category of drugs. Their high affinity for tubulin prevents its polymerization (Zavala et al., 1978). Kelleher (1977) showed that with a mouse-brain tubulin binding assay a correlation exists between the degree of tubulin binding of podophyllotoxin analogues and the activity of these compounds in vivo against mouse sarcoma. Zavala et al. (1980) reported on the activity of synthetic analogues of the antitumor lignan steganacin in a similar assay.

Dihydrofolate Reductase. The inhibition of rat-liver dihydrofolate reductase has been used to measure potential antineoplastic activity (Abdul-Ahad et al., 1980), particularly with quinazoline analogues.

Adenosine Deaminase. Coformycin, a product of the same streptomyces that produces formycin, is a specific inhibitor of this enzyme (Sawa et al., 1967).

Microbial Hydroxylation of Biphenyl. Certain carcinogenic agents are known to enhance biphenyl-2-hydroxylation but not biphenyl-4-hydroxylation (McPherson, et al., 1974). Because so many antineoplastic agents are also carcinogenic, isolation of agents that stimulate the 2-hydroxylation of biphenyl by *Helicostylum piriforme* may lead to compounds from fermentation broths with the desired activity (Davis & Smith, 1978).

Circular DNA. The interactions of antitumor agents with covalently closed circular DNA obtained from the bacteriophage PM-2 have been studied by Mong et al. (1979). Antibiotics such as actinomycin D and adriamycin induce conformational changes in the PM-2 DNA that can readily be measured by a decrease in the fluorescence of ethidium bromide DNA complexes. It may be possible to extend this activity to a search for compounds with an effect on RNA complexes obtained from the viruses of filamentous fungi (Lemke, 1977).

Tyrosine Hydroxylase. Bovine adrenal tyrosine hydroxylase has been shown by Oka et al. (1980) to be inhibited by bleomycin. A screen for inhibitors of this enzyme may reveal other interesting compounds that could have antitumor activity in vivo.

Macromolecular Synthesis

Protein Synthesis Inhibition. The antitumor agent emetine is known to block eukaryotic but not prokaryotic protein synthesis (Rosenkranz, 1973). This observation may be exploited to seek compounds more active than emetine with a similar mode of action.

Nucleic Acid Polymerases. RNA synthesis in isolated nucleoli from Novikoff hepatoma ascites cells is inhibited by several anthracycline antitumor agents (DuVernay & Crooke, 1980). The DNA-dependent RNA polymerase is inhibited by sibiromycin, therefore the inhibition of this enzyme system would be a useful means of seeking new analogues (Gause & Dudnik, 1980).

Inhibition Mitochondrial Functions. The cytotoxic antibiotics duclauxine and bikaverin (lycopersin) are inhibitors of respiration and uncouplers of oxidative phosphorylation of tumor cells (P388) and isolated rat-liver mitochondria, respectively (Kovăc et al., 1978).

Inhibition DNA Methylation. Certain carcinogenic agents inhibit DNA methylation (Salas et al., 1979). Because it is known that many antitumor agents are also carcinogenic, a search for other inhibitors of DNA methylation from fermentation broths may lead to interesting compounds. DNA methylation can be quantitated with a DNA methyl transferase from rat-brain tissue. The transfer is measured from adenosyl-L-methyl^{3}H-methionine to the DNA from chicken erythrocytes or *Micrococcus lysodeikticus* (Salas et al., 1979).

In Vitro Screening Strategies

With the exception of human tumor clones, most in vitro screen tests in cancer research bear little direct relation to neoplastic disease in humans.

Each test, at best, represents a single biochemical process that is most sensitive to interference and, it is hoped, is representative of similar effects on tumor tissue. In vitro tests at all levels nevertheless have the advantage over in vivo screens of speed, high sensitivity, economy, and need for small amounts of sample for test. In vitro tests are applied in three major ways. Examples drawn from our experience are outlined in Table 6.

Prescreens. These prescreens may be defined as tests run for the purpose of detecting biological activity before performing a primary screen which is the first definitive test of the type of activity sought. Prescreens are particularly useful in antitumor antibiotic screening and have made it possible, in our laboratory, to test more than 10,000 new cultures per year compared with 500 per year with direct primary broth screening in vivo.

The prescreens listed in Table 6 in general terms are usually run in groups of three or more and a variety of metabolic types is covered. The first three are straight microbial inhibition tests of organisms with high sensitivity in each category. The phage tests may involve lysogenic induction, prevention of transduction, direct phage inhibition, or chemical disruption. The particular event measured would relate to some specific aspect of nucleic acid metabolism. Protein inhibition is representative of still more specialized tests that might be applied if, for example, inhibitors of tubularin formation were sought. Prescreens are usually rotated or replaced on a regular basis. For purposes of evaluation of the performance of a prescreen in detecting antitumor fermentation products we consider that a minimum of 1000 cultures should be tested, or, better still, about 5000. Prescreens that fail to detect antitumor substances or duplicate an existing screen are replaced.

Table 6 Applications of In Vitro Tests and Types Used

Prescreens	*Assays*
1. Highly sensitive gram positive microorganisms	1. Antimicrobial
2. Sensitive DNA repair deficient mutants	2. Phage induction
	3. Enzymatic
3. Perturbed cell wall mutants	4. Interaction with nucleic acid
4. DNA phage	*Evaluation*
5. RNA phage	1. Tissue culture cell cycle
6. Protein inhibition	2. Molecular pharmacology
	3. Human tumor clones

Assays. In general, in vitro assays used in guiding fermentation improvement and chemical fractionation of fermentation products are confined to those that are simple and efficient to operate. Correlation with in vivo activity can often be in terms of overall biopotency; however, frequent checking against in vivo antitumor activity is essential until specific chemical markers are established.

Evaluation. The examples given are the most frequently applied in vitro tests for advanced study of a new agent. Determination of cell cycle and phase specificity can be especially useful in suggesting application of the agent in combination therapy. Studies of molecular pharmacology (i.e., the interaction of an agent in protein and nucleic acid metabolism) can give insight into its mechanism of action. Finally, screening in human tumor clones may prove to be the most critical evaluation of all for predicting clinical utility—at least for cytotoxic agents.

IN VIVO

The review by Goldin et al., 1979, and references contained therein, details the history, drug responses, and application of most of the animal tumor systems studied by the NCI. Another important resource is the publication of protocols for NCI tumor testing (Geran et al., 1972). Although now dated because of the introduction of many new systems, it nevertheless continues as a basic publication of testing methodologies. References from these two articles are generally not relisted in our bibliography but are supplemented by a variety of reports. With the numbers of transplanted animal tumor systems running in the hundreds, it is totally impractical to review the merits of each because the majority has not been the subject of sufficient drug testing to develop a reasonable response profile. Instead we describe a limited selection of tumors that seems to us to provide a variety of test characteristics suitable for today's needs for cancer chemotherapy screening, recognizing that this is an evolving process as systems with potentially greater clinical relevance are devised. The tumors we describe are presented in increasing order of complexity. Thus the simpler and more rapidly growing allogeneic and syngeneic systems are most suited to primary screening and the more complex systems for advanced drug evaluation. Finally, testing strategies, as applied in our laboratory and several others, are described to provide an overview of how various combinations of tumors meet differing test needs.

Transplanted Tumors

Nonspecific Tumors

These are tumors that usually developed spontaneously in animals and were transplanted serially in the same species but not in in-bred strains. Dedifferentiation of histologic type with some loss of histocompatibility antigens often occurred and this, together with rapid growth, could overcome the immune response of animals that might reject the tumor as an allogeneic tissue graft. Although now largely replaced by transplanted syngeneic tumors that tend to retain histologic characteristics, the nonspecific tumors still serve some useful purposes.

Sarcoma 180 Crocker (S-180). This pleomorphic cell tumor resulted from the transplantation of a carcinoma that arose spontaneously in the axilla of a male mouse in 1914. It became widely used as a primary screen for anticancer drugs in the late 1940s into the 1960s, particularly at Sloan-Kettering Institute, and for a long time as part of the NCI 3 tumor screening system (1955–1965). It has been largely supplanted by syngeneic systems.

S-180 is rapidly growing and 95–100% lethal in most mouse strains. Hosts closest to $H2^d$ histocompatibility configuration show the least regression (Bradner & Pindell, 1965), whereas certain other strains (notably the $H2^b$ C57BL/6, Bradner & Pindell, 1964) manifest a high incidence of regression. Thus S-180 is a sensitive tool for the study of immunological effects and helped in discovering the action of Lentinan (Chihara et al., 1969), which is used clinically in Japan. Like many of the nonspecific tumors, S-180 is sensitive to caloric restriction, a property that reduces its utility as a screen for cytotoxic agents. The ascitic form of S-180 is somewhat more sensitive than the solid form to drug treatment (especially ip tumor-ip treatment) but is less responsive to immunologic alterations.

Ehrlich Tumor. This may be the oldest continuously transplanted tumor. It probably arose as a mammary carcinoma, although the exact origin cannot be traced (Ehrlich & Apolant, 1905). It is nonspecific for host strain and has been widely used throughout the world for cancer drug screening, particularly in Japan, where it was popular for testing natural products. Its high sensitivity tends to yield a high incidence of "false positive" (clinically useless) leads. Like S-180, it has been largely replaced by syngeneic tumors but remains of some interest because of the extensive body of test data that has accumulated over the years, especially for Ehrlich ascites tumor (EA).

Walker 256 Carcinosarcoma (W-256). This tumor of mixed cell type arose as an adenocarcinoma in a pregnant albino rat in 1927. As a subcutaneous (sc) implant it is exquisitely sensitive to alkylating agents and has been used in this country and Great Britain for screening these agents. For a limited period of time it was used with L1210 in intramuscular (im) site as the official NCI screen but has now given way to syngeneic mouse tumors.

Other Tumors. Two other widely used tumors are Sarcoma 37, which resembles S-180 in many respects and served in studies by Shear et al. (1958) of tumor necrosis factors, and the Yoshida tumor, an undifferentiated neoplasm of rat used for chemotherapy studies in Japan.

In addition to the immunological properties of the nonspecific tumors in solid form, S-180, W-256, and EA ascitic forms are extremely sensitive to the phleomycin-bleomycin class of antibiotics and could serve as rapid in vivo screens if one were searching for these agents in particular among fermentation products.

Syngeneic Tumors

These tumors arose and are maintained in inbred strains of animals. Because inbred strains are genetically like identical twins, there is no natural tissue rejection of a grafted tumor. Spontaneous regression is virtually unknown in the strain of origin; however, hybrid mice, which are often the subjects of large-scale chemotherapy testing, usually manifest a low level of hybrid resistance (see Clark & Harmon, 1980). Syngeneic tumors tend to maintain histologic and biochemical characteristics on repeat transplant and are much more difficult to affect by immunological means than nonspecific tumors. Thus on a worldwide basis syngeneic tumors, particularly the leukemias P388 and L1210, are now used almost exclusively for cancer chemotherapy screening.

P388 Leukemia. This tumor was induced in 1955 in a DBA/2 mouse by carcinogen painting. It is maintained in passage in the DBA/2 and therapeutic tests are performed in the CDF_1 hybrid (BALB/c × DBA/2). Animals implanted ip with 10^6 cells survive 9–14 days and survival time of ip-implanted mice treated with drug by the same route forms the basis of the overwhelming majority of all screening studies. P388 is similar to L1210 in its responses to various types of drug but in many cases is much more sensitive and has become a mainstay for the testing of natural products, especially in crude form. It is sometimes referred to as an in vivo prescreen but we prefer to reserve the term prescreen for in vitro systems and call the first test against any tumor in animals the primary screen. P388 is unresponsive to hormonal alterations but can be affected by treatment with *Corynebacterium parvum*, which suggests some susceptibility to immunologic manipulations (Houchens et al., 1976).

L1210 Leukemia. This tumor was also induced by carcinogen painting of the skin of a DBA/2 mouse in 1948. The ascites form is one of the most widely used tumor test systems in the world. The NCI protocols call for tests to be performed in CDF_1 hybrid mice with a cell dose of 10^5 cells per mouse. This inoculum gives control survival routinely in the 8–11-day range. Use of a 10^6 cell dose produces slightly higher sensitivity and a 6–8-day median survival time (MST). This shortens the experimental time interval significantly in which to determine antitumor activity and is particularly useful for assays of fermentation products in early development stages. L1210 is a rapid and efficient primary screen for most alkylating agents, antimetabolites, and certain natural products (e.g., camptothecin). It is popular as a primary screen of *cis*-platinum analogs. In immunotherapeutic studies L1210 is quite resistant and requires special manipulation. Unlike P388, L1210 is not affected by host treatment with *C. parvum*. The fact that cytotoxic therapy is less effective in immune suppressed hosts (Mathé, 1977a) and enhanced in BCG-stimulated hosts (Mathé, 1977b) infers some host defense interaction. L1210 cells altered by exposure to neuraminidase (Sethi, 1973) or concanavalin A (Killion, 1977) demonstrate increased immunogenicity.

B16 Melanoma. This melanotic melanoma arose spontaneously on the skin of a C57BL/6 mouse in 1954. It is carried in this strain by sc trocar implant and testing is typically performed in the BDF_1 (C57BL/6 × DBA/2) hybrid, or $B_6D_2F_1$. The solid tumor is homogenized to produce a tumor-brei inoculum and implanted sc to produce a solid tumor that can be measured or ip which causes a more diffuse tumor and more rapid death. Alternatively, solid tumors may be produced by trocar implant. The MST for animals with ip tumors is about 14–22 days and for those with sc tumors about eight days longer.

B16 was originally incorporated into the NCI screening program as an example of a slow-growing solid tumor. It has become popular in recent years because of its more consistent responses compared with many other syngeneic solid tumors. It responds well to antibiotics and is used in a number of laboratories in studies of anthracyclines. B16 responds moderately to bleomycin and related analogs if the tumor is implanted sc but only slightly if the tumor is given ip. Although its drug sensitivities are similar to those of the leukemias (P338 and L1210), in absolute terms B16 will often show a different rank order in the activity of analogs when compared with tests on one of the leukemias.

B16 does respond to immunotherapy alone (Proctor, 1976) and combined with other modalities (Pendergrast, 1976), provided that the timing of treatments is carefully staged.

Lewis Lung Carcinoma (LL). This tumor arose spontaneously in the lung of a C57BL/6 mouse in 1951. It is maintained by passage of tumor fragments sc and, like B16, is used for chemotherapy screening with im or sc fragment or ip tumor-brei implant sites. The C57BL/6 is host for propagation and BDF_1, for testing. Control MST is 18–28 days by im implant.

LL was incorporated into the NCI screen as a solid tumor at the same time as B16, but it has fallen from favor somewhat as a screening system because of higher variability than a number of other tumors. As an example, we have reported data from this laboratory on repeated testing of *cis*-diamminedichloro platinum II (*cis*-DDP) in conjunction with the screening of new platinum analogs (Bradner et al., 1980a). In multiple experiments on the same test regimen at optimum doses LL responded with increases in lifespan (ILS) of 44–243% and long-term survivor (LTS) incidences of 0/10–8/10. Similar parameters for B16 were ILS, 57–137%; LTS, 0/10–1/10; and for L1210 leukemia, ILS, 57–185; LTS, 0/10. LL is resistant to vinca alkaloids and most antimetabolites and antibiotics, although it is inhibited in sc form by bleomycins. A noted sensitivity is its differentiation of methyl-CCNU from CCNU when therapy was delayed (Mayo et al., 1972).

LL is responsive alone or combined with other modalities to agents that stimulate host defenses like *C. parvum* (Sadler & Castro, 1976), BCG (Mathé, 1977b; Dubois, 1976; Kurata, 1977), and levamisole (Renoux & Renoux, 1972) if conditions of treatment are carefully defined. As noted in many of these references, LL is often used as a model of metastatic disease.

Other. Numerous other syngeneic transplanted tumors have been studied with varying degrees of success. The Ridgeway osteogenic sarcoma (ROS), developed by Dr. Sugiura at Sloan-Kettering Institute, is highly responsive to most antitumor antibiotics (except bleomycin) and to cyclophosphamide (CPA). It is less sensitive to antimetabolites and nitrosoureas. In our laboratory the myeloid leukemia C-1498 was studied as a potential chemotherapy model (Bradner & Pindell, 1966). It proved to be too resistant to drug therapy to serve as a practical screen. The Madison 109 lung carcinoma (M109), evaluated recently in this laboratory (Rose, 1981), has some unique responses, especially to immune modifiers such as pyran copolymer, and is currently being used in an immunology program.

Organ Specific Tumors

Transplanted Rodent Tumors. Although long-transplanted syngeneic tumors tend to retain histologic characteristics and some functional properties (B16 forms melanin, LL metastasizes to the lungs), they apparently lack or do not maintain sufficient biochemical conformity to be good predictors of the effect of drugs on analogous human disease. In recent years the NCI has made a concerted effort to address this problem by supporting the isolation and propagation of animal tumors of breast, lung, and colon origin to test the theory that these neoplasms could be the equivalents of the clinical diseases that are representative of the highest incidence of solid tumors in humans. At the same time a series of xenografts of human tumors was studied (see below). Although it is too early to draw definitive conclusions, our impression so far has been that the mouse tumors may contribute to a tumor spectrum profile of responses to a drug but not as a predictor for the human disease equivalents; for example, the most sensitive tumor to 5-FU was one designated Colon 38. This tumor was responsive to the mycotoxin, anguidine (Corbett et al., 1977). In clinical trials, however, anguidine treatment failed to give a partial response (50% or more regression) in any of 50 patients treated who had colorectal cancers (Belt et al., 1979; DeSimone et al., 1979; Diggs et al., 1978; Murphy et al., 1978). These were Phase I trials and it is hoped that some responses can be anticipated if broad Phase II trials are conducted. Nevertheless, the available data at this point suggest a lack of correlation between Colon 38 and its human disease equivalent. One explanation for this correlation failure is that human colorectal cancer is not one disease but many because these tumors are considered to be caused by carcinogens. Typically, each carcinogen-induced tumor may differ antigenically from others, and it follows that this could extend to biochemistry and drug responses as well. The whole problem is the subject of investigations with xenografts (see the next section). In the meantime research will continue on the isolation of syngeneic rodent tumors that retain all functional properties on transplantation and, hopefully, responses to drugs as well. One such tumor under study in our laboratory is the MXT mouse mammary tumor, a urethane-induced ductal cell carcinoma of BDF_1 mice that retains estrogen dependence for growth (Watson et al., 1977). It is responsive to the antiestrogen drug tamoxifen and to those cytotoxic drugs tested so far that are known to have activity in human breast cancer (Rose, unpublished observation). The real test will be the clinical results in human breast carcinoma with a new drug predicted to be active by MXT.

Human Tumors. The possibility of growing human tumors in animals and using them as a drug-screening tool was the subject of considerable study more than two decades ago. The New York Academy of Sciences symposium on cancer chemotherapy screening (Stock, 1958) has considerable information on a variety of rat and mouse systems tested. In general, these tests used "conditioning" of the animals—treatment with x-irradiation and glucocorticoids to suppress xenograft rejection. More recently this conditioning has been improved by thymectomy of young animals, followed by x-irradiation. Alternatively, genetically athymic ("nude") mice which may require some added immune suppression, usually in the form of antilymphocyte or antithymocyte serum, are used. Because of the high susceptibility of the hosts to infection and possibly greater sensitivity to the toxicity of various drugs, we do not recommend human tumor systems in animals for primary antitumor screening. Instead, their most likely utility will be in secondary testing, as is already being done by the NCI in the operation of its tumor panel which includes xenografts of human colon, breast, and lung tumors. Studies by Houghton and Houghton (1978) have shown that six human colorectal tumors carried as xenografts in mice had six different patterns of drug response for single-agent therapy. Similar information reported by Nowak et al. (1978) suggests that, at least for this class of tumors, a rather extensive battery of neoplasms may be needed to have available all possible types of human disease.

Autochthonous Tumors

Tumors that arise as a result of an oncogenic stimulus are termed autochthonous in the original host. Most of these tumors have such long latent periods of development and are so slow-growing that they are impractical as drug-screening tools. Therefore our discussion is limited to a few examples in each category.

Spontaneous and Viral Tumors

The term spontaneous, as applied to tumorigenesis, needs some clarification. A true spontaneous tumor would be one that arose in an animal without the interposition of any outside agent. In fact, there are few examples of tumors that are actually genetically programmed (e.g., retinoblastoma in humans). Thus the term spontaneous is extended to tumors arising in hosts that carry an etiological agent but have not been subject to direct inoculation of the agent as a laboratory procedure. The commonest examples are certain viral tumors which are transmitted from parent to offspring and are usually manifested in the later life stages of the host. Among the many systems that have been investigated two appear to have been studied most extensively in chemotherapeutic tests: AKR leukemia and $CD8F_1$ mammary carcinoma.

AKR Leukemia. The original AK strain was established by Furth et al. (1933) as a line of mice with a high incidence of spontaneous leukemia. It is caused by a vertically transmitted murine leukemia virus (see Gross, 1970). Extensive chemotherapeutic testing has been carried on at the Southern Research Institute with spontaneous tumors and early transplants. In a pattern similar to human disease remissions could be induced by treatment with vincristine plus prednisone and maintained by various sequences of drug therapies (Schabel et al., 1974). With this particular model viral reinduction of the disease can occur even in the face of apparent complete eradication of tumor cells by chemotherapy. To combat this condition combined modality therapy, which included treatment with mouse interferon that demonstrably reduced viral titer, has delayed the appearance of primary lymphoma significantly (Bekesi et al., 1976).

$CD8F_1$ Mammary Carcinoma. This is a viral-associated spontaneous tumor that arises in 60–70% of the $CD8F_1$ hybrid mice at an average age of 11 months. It has been studied in depth at the Catholic Medical Center in Jamaica, New York, in experiments that included combined surgery, chemotherapy, and immunotherapy (Fugmann et al., 1970) and treatment of metastatic disease (Fugmann et al., 1977). This tumor has also been studied as a first-generation transplant, a system that provides far more tumor material

for chemotherapy studies than is available by using animals with spontaneous tumor directly. Martin and his associates (1975) showed that this system responded to all eight drugs tested and known to have activity in human breast cancer. This system has now been incorporated into the NCI Tumor Panel and will be the subject of study over several years to determine its predictive accuracy as a drug screen for human disease.

The concept of using first-generation transplants of spontaneous tumors as a more efficient test system has been under investigation for many years. Scholler et al. (1955) tested both first- and second-generation transplants of spontaneous mammary tumors from C3H mice for therapeutic response to a series of drugs and found that they did not differ from common patterns of response of long-transplanted tumors. It is our assessment at this time that the $CD8F_1$ and AKR tumors may follow the same course in early transplant and are not likely to offer significant advantage over standard transplanted tumors. Scholler (1958) also reviewed the practical considerations (logistics and costs) of operating a drug-screening program on spontaneous tumors. The costs quoted still look formidable after 24 years!

Carcinogen-Induced Tumors

In her analysis Scholler also demonstrated that the problems inherent in screening with spontaneous tumors can also be extended to those that are carcinogen-induced because of similar long latent periods for the appearance of the tumors. There may be a place for such tumors, however, in secondary testing in which an association exists between the carcinogenic action and a specific target organ. Two models are discussed as examples. The first is the mouse bladder cancer model under investigation by Soloway and Murphy (1979). Transitional cell carcinomas are formed in C_3H/He mice as a result of dietary incorporation of the carcinogen N-[4-(5-nitro-2-furyl-)-2-thiazolyl] formamide (FANFT). Tumor incidence is close to 100% by 11 months and chemotherapy studies can be conducted. A high correlation of response exists between this experimental system and the disease as it occurs in humans. Soloway's group also uses a transplantable form, which has maintained its histologic characteristics, as a prescreen for tests against the autochthonous tumor.

The second example is the 7,12-dimethylbenz-(a)anthracene-(DMBA)-induced mammary cancer in the rat initiated by a single feeding of the carcinogen (Huggins et al., 1961). The tumors formed differ in estrogen receptor (ER) concentration and respond similarly to estrogen antagonists or ablation, in accordance with ER status. These tumors also have progesterone and prolactin receptors. In early stages of growth they are inhibited by ergolines (Cassady & Floss, 1977) and respond well to androgens (Zava & McGuire, 1977) and nonandrogenic antiestrogens such as tamoxifen (Jordan, 1976). Typically, 50-day-old female SD rats are given 20 mg of DMBA by intubation and mammary tumors appear between two and four months. Treatment can be started at the time of DMBA administration to assess prevention of tumor appearance or be given to animals with established tumors to evaluate regressive effects (Peters & Lewis, 1976).

Special Systems

Resistant Tumors

It is well known that many human tumors treated clinically become rapidly resistant to certain drugs and regrow or metastasize from resistant clones. Thus for many years laboratories throughout the world have developed lines of transplanted tumors resistant to treatment, usually with a single chemotherapeutic agent, and tested them for cross resistance to related agents and collateral (increased) sensitivity to other agents. The NCI contract screeners maintain tumor lines, mostly L1210 and P388 leukemias, that are resistant to most of the widely used drugs. In our own laboratory we have studied tumors made resistant to neocarzinostatin (Bradner & Rossomano, 1968) and mitomycin C (Reich & Bradner, 1979) and have obtained a line of L1210 highly resistant to *cis*-platinum therapy (Burchenal et al., 1978). This line has identified *cis*-platinum analogs that are just as effective against the resistant tumor as against the parent. By using the two resistant lines (L1210/mitomycin C and L1210/*cis*-platinum) and submitting new compounds to the NCI for testing on tumors they carry, which are resistant to agents such as adriamycin, vinca alkaloids, and antimetabolites, we can obtain a rather broad indication of the potential effectiveness of each agent in combination and sequential therapy.

Host Response Modification

This title is now used instead of immunotherapy because it can encompass agents like interferon that are not fully defined in immunologic terms. Nevertheless, the basic principle of seeking agents that alter the response of the host to a tumor and the need to identify appropriate screening tools continue. The relevance of any screens to clinical disease is even more uncertain for host response modifiers than for cytotoxic drugs (see Hewitt's critical review, 1978). Most host response modifiers, particularly immune stimulants, yield multiphasic responses; that is, over a range of doses and/or timing of administration such agents can enhance or inhibit tumor growth. Nevertheless, both active and passive immunotherapy of tumors by a variety of mechanisms can be demonstrated in laboratory animals (see Ferguson & Schmidtke, 1979). In the past we have used a highly sensitive tumor such as solid S-180, which may respond by multiple mechanisms (Bradner & Pindell, 1965), as a primary screen and various syngeneic tumors such as B16 melanoma, Lewis Lung carcinoma, and Madison 109 as more resistant secondary tests. More recently we have been using syngeneic tumors like Lewis Lung and Madison 109 as direct primary screens by implanting a small inoculum ($\sim 5 \times 10^5$) of trypsin-separated tumor cells into the footpads of appropriate strains of mice (Reed, F. C., Siminoff, P. & Issell, B. F., unpublished work). These systems have a high sensitivity and consistency and are suitable for use directly as primary screens.

Side Effects

Although the purpose of this chapter is to describe systems of detecting antitumor effects, the more recent emphasis on preparing analogs of known agents with reduced toxicity has made early evaluation of the side effects a critical part of the total antitumor efficacy profile. It is for this reason that we have developed a number of screens for side effects in small animals that can be operated with the same simplicity and efficiency as animal tumor screens. These are being applied primarily in analog programs and have been described in detail elsewhere (Bradner et al., 1980b; Bradner & Schurig, 1981). Our discussion is limited to showing where these tests fit in at the primary and secondary screening levels (see next section).

Screening Strategies

The strategies used in the selection and application of in vivo tumor systems for cancer chemotherapy screening are closely defined by their objectives. In maintaining in vivo tumor systems, the scope of these tumor programs tends to be limited by the resources available. Probably the largest program currently operated is that of the NCI, which utilizes P388 as a primary screen and a panel of eight tumors for secondary evaluation. Over recent years special paths of testing have evolved at the NCI for screening analogs. Other laboratories, including our own, develop and continue to refine analog prescreening strategies that promote always faster and more definitive test results. Even though the widely used transplanted tumors have no exact clinical correlates, the general impression is that the more experimental tumors that respond to a drug, the greater the chances that the drug will have clinical utility. A drug that is highly active in one tumor system but has little activity in any others apparently has much less chance of clinical success. Thus it would seem that entirely new chemotypes, especially those with a novel mechanism of action, are the most amenable to broad-scale tumor spectrum studies to predict and rank the probabilities of clinical activity. Analogs of existing agents, however, can be screened far more efficiently. It should be possible in the future to select three or even two tumor systems that could predict a rank order of clinical utility. The biggest drawback at the present time is the lack of sufficient clinical information. The only drug classes that have a large body of clinical data are the alkylating agents and the anthracyclines. Unfortunately, the wide variety of experimental tumor systems used as screens and the lack of comparative clinical trials make the correlation of alkylating agents difficult, if not impossible. Anthracyclines, however, present a better opportunity because substantial comparative data exist for L1210 and B16 with adriamycin as a prototype drug and clinical data are now accumulating for such drugs as carminomycin, rubidazone, aclacinomycin, AD-32, and quelamycin to add to that already available on daunorubicin. Careful correlative analysis should establish whether L1210 and B16 can indeed predict clinical activity of anthracyclines and whether one or more tumors from the panel might augment the index of correlations.

Primary Screening

Table 7 is based on retrospective analysis of the clinically active drugs tested in several tumor systems. This is similar to that published by Johnson and Goldin (1975, Table 2) but is much more condensed and has been modified according to our own experience. The sensitivity of the leukemias is immediately apparent, and it is for this reason, and their more rapid growth and ease of handling, that they are used almost universally as primary antitumor screens. Table 8 shows our currently applied primary in vivo screening and includes the initial tests of side effects. Because the side effects screens are performed in normal (nontumor-bearing) animals and are based on the LD_{50} and fractions,

Table 7 Response of Mouse Tumors to Various Classes of Agent

Types of Agent	P-388	L-1210	B16	LL	M109	MXT
Alkylating agents (other than nitrosoureas)	++	++	+	±	+	++
Nitrosoureas	++	++	+	++	+	++
Antimetabolites	+	++	±	±	+	++
Alkaloids	+	+	+	−	++	±
Antibiotics	++	+	++	−	+	++
Miscellaneous agents (e.g., DTIC, procarbazine)	±	+	±	±	±	+
Hormonal agents	−	−	−	−	−	+
Immunological	(+)	(±)	(±)	(±)	(+)	NT

Key
++ Responds strongly to most materials.
+ Responds weakly/moderately to most materials.
± Responds to a few materials.
− Generally unresponsive.
( ) Requires special conditions.
NT Not tested.

it is therefore necessary to establish the acute LD_{50} in mice of each substance at some point in its testing. Whether this is done before or after antitumor screening is dependent on two factors: (1) the amount of compound supplied and (2) the expectations for the observance of antitumor effects; for example, with platinums we usually receive ample supplies of the compound and expect that nearly all will be highly active. Thus an LD_{50} is performed first and the antitumor test uses the information to set doses for primary screening. With mitomycins the amount of compound available is often small and many fail to meet activity criteria; therefore performance of the LD_{50} first is impractical (see Bradner, 1979).

In selecting the tumor for primary screening of analogs, a sensitivity that gives a survival response range of T/C 150–200% (T/C = treated survival time/control survival time × 100) is best. It has been our experience with platinums and anthracyclines that P388 is too sensitive to be discriminating among analogs of these chemotypes [Casazza (1979) has reported similar impressions in screening anthracyclines] and that L1210 is the superior system. It is true that P388 could be staged for reduced sensitivity by delaying treatment but such staging usually results in a more highly variable experiment. With mitomycin analogs we have found that P388 is superior as a primary screen to

Table 8 Primary in Vivo Screening

Compounds	Schedule	Side Effects	References
Platinums	L-1210 1X and QD 1→9[a]	BUN, WBC[b]	Prestayko et al., 1979 Bradner et al., 1980a Schurig et al., 1980b
Mitomycins	P-388 1X	WBC	Bradner, 1979
Anthracyclines	L-1210 1X and QD 1→9	WBC, CPK-MB	Bradner and Rose, 1980 Schurig et al., 1980a
Bleomycins	P-388-J[c] QD 1→9	HDP BUN	Bradner, 1978b Bradner and Schurig, 1981
Cyclic depsipeptides	P-388 1X and QD 1→9	WBC	
Tricothecanes	P-388 QD 1→9	WBC	Claridge et al., 1978 Doyle and Bradner, 1980
Other natural products	P-388 Q4D and QD 1→9	WBC	Unpublished
Alkylating agents	L-1210 Q3D	WBC	Unpublished
Antimetabolites	L-1210 QD 1→9	WBC	Unpublished
Hormonal agents	MXT, sc Q2d, 10X	WBC	Unpublished
Host response modifiers	M109,[e] LL[e]	f	Unpublished

[a]Schedules 1X = single dose, day 1.
    QD 1→9 = daily for nine days.
    Q2D, 10X = every second day for 10 injections.
    Q3D, Q4D = every third or fourth day for three injections.
[b]BUN = Blood urea nitrogen.
  WBC = White blood cell, total count.
  CPK-MB = Creatine phosphokinase isozyme MB.
  HDP = Total lung hydroxyproline.
[c]P-388-J = a line of P-388 from Japan-sensitive to bleomycin-type agents.
[d]Unpublished = unpublished in-house screening.
[e]Footpad implant, treatment days 1, 5, and 8.
[f]Effects on various models of humoral and cellular immunological response are tested.

L1210 because the response of L1210 appears to be at the low end of the activity range (T/C ~ 150%) and may be lacking in discriminatory function for this reason (Bradner, 1979; Imai et al., 1980).

Suitable dose schedules are used for each class of compound based on experience. For unknown natural products the single injection is usually avoided because so many would be missed on this schedule (e.g., bleomycins, aclacinomycin, trichothecanes). In those cases in which the primary antitumor screen is run without previous LD_{50} determination, the test provides a reasonable estimate of the potency and toxicity of the compound. This can be an important determinant because most analogs become economically unfeasible as they pass the point of being 20-fold less potent than the parent compound.

In assessing the outcome of the primary antitumor screen, we require that analogs have activity within the historical range of the prototype compounds to be considered of further interest. The argument can be made in cases in which only marginal activity is observed that perhaps a new analog has shifted in spectrum and that we might miss an important new compound by not testing it in additional systems. This is a calculated risk that must be taken to avoid being inundated with too many compounds. Furthermore, we reduce our risk by accepting for further testing (1) analogs at the low end of historical ranges and (2) analogs that have special properties, such as high potency or broad therapeutic range, even though the optimum survival increase is below our criterion for that class. Completely novel chemicals and natural products that show even slight activity in a primary screen are always tested further.

Secondary Screening

We define secondary screening as the next level of testing performed after a new material passes criteria established in primary screening. The wide variety of tests that can be performed at this stage includes those in a broad spectrum or panel of experimental tumors, studies of dose regimens and different routes of administration, staging of tumor growth, varying implant site, and performing combined modality therapy with new materials in a variety of different settings. Although there may be some merit in having in-depth information on every active compound, the vast number of available compounds and the limitations of test laboratories and funding have made such an undertaking impractical. We have therefore designed our secondary screening as a rather limited expansion of testing that includes what we believe to be the minimum number of experiments to characterize a compound sufficiently to judge its suitability as a clinical candidate. This is outlined in Table 9. The order of tests

Table 9 Secondary In Vivo Screening

Compounds	Tumors	Side Effects	Routes	Regimens
Platinums	B16, LL, M109	GI stasis,[a] emesis	IV, sc oral	Q3D
Mitomycins	B16, L-1210	Differential[b] BUN, SGPT	IV, sc oral	QD 1→9 Q3D
Anthracyclines	B16, LL, M109	EKG, SGPT	IV, sc oral	Q4D
Bleomycins	B16, LL, M109	Pulmonary mechanics[c]	IV, sc oral	Q4D
Cyclic depsipeptides	B16, L-1210	Differential SGPT	IV, sc oral	Q4D
Tricothecanes	B16, L-1210 Colon 38[d]	Differential SGPT	IV, sc oral	Q4D
Other natural products	B16, L-1210	TBD[e]	IV, sc	1X
Alkylating agents	B16, LL	Differential BUN, SGPT	IV, sc oral	QD 1→9
Antimetabolites	B16, M109 L-1210	TBD	IV, sc oral	1X
Hormonal agents	DMBA[d]	TBD	IV, sc sc	
Host response modifiers	P-388, M109 and/or LL sc	TBD	IV, sc oral	Varied

[a]Florczyk et al., 1980.
[b]wbc, differential count.
[c]Schurig et al., 1979.
[d]Tests by outside laboratory.
[e]To be developed.

does not follow a rigid pattern but usually proceeds with added tumor- and side-effects testing that would be expected to be most informative for the compounds under study. Dose regimen studies are usually confined to gathering information not already generated in the primary screen. Considerable emphasis has been put on dose-schedule-dependency studies in the past because this information contains an early clue to the pharmacokinetic properties of a drug and has often had direct clinical relevance (e.g., methotrexate, Goldin et al., 1956, and arabinosyl cytosine, Ellison et al., 1968, based on Kline et al., 1966; see also Venditti, 1971). Dose schedule information, however, has highly practical applications for the screening laboratory. As an example, the discovery that the active principle sought in a crude fermentation is sufficiently active, given as a single injection, can reduce sample-size requirements and the labor involved in administering injections in the running of in vivo fractionation assays. It can also suggest appropriate schedules for other tumor systems and add information on toxicity.

Studies of routes of injection are probably less critical because the convenient ip route for screening and the iv route, for preclinical toxicology are so widely used. The determination of iv, sc, and oral antitumor effectiveness of a drug we usually reserve for those more advanced in development unless other information suggests that studies should be performed at an earlier stage.

Experiments occasionally performed but not as a routine part of our secondary screening are the use of intercerebrally inoculated tumors which, if responsive to a new drug, suggest that the blood-brain barrier is being crossed, and combination therapy studies that might signal an unusual synergism when a new drug is given with standard agents. As an example of a report of the affect of tumor inoculation route, the recent publication by Neil et al. (1979) is of interest. These authors found that the anthracycline 7-con-0-methylnogarol was comparable to adriamycin in antitumor effectiveness when the tumor P388 was implanted ip but was far more active than adriamycin when the tumor was inoculated iv. Because 7-con-0-methylnogarol is being readied for clinical trial by the NCI, there may in the future be clinical feedback that will establish the relevance of this observation.

CONCLUSION

In this chapter we have described a number of in vitro and in vivo screens and have outlined how they may be applied for the detection of antineoplastic properties. Several examples have been given in which the response of a particular experimental tumor has failed to predict for response of human disease. In other cases tests in animal systems have closely correlated with clinical response. This is especially true for pharmacologic properties of a given drug such as routes of absorption and dose schedule dependency. In vitro systems have particular utility as prescreens for detecting bioactivity, as assays in natural product isolation, and as evaluation tools for defining the biochemical properties of drugs. Human tumor clones appear to be the in vitro system of most promise for the future in predicting the clinical use of cytotoxic agents. In vivo systems, although still leaving much to be desired in correlating with specific human neoplasms, nevertheless generate essential information in antitumor drug development, particularly in terms of tumor respone in relation to host toxicity, host mediated actions, and pharmacologic properties. Validation of the test systems currently applied will come from feedback of clinical data on drug responses, especially to close analogs of existing agents.

REFERENCES

Abdul-Ahad, P. G., Blair, T. & Webb, G. A. (1980), *Int. J. Quantum Chem.*, **17**, 821–831.
Ames, B. M., McCann, J. & Yamasaki, E. (1975), *Mutat. Res.*, **31**, 347–364.
Aoyagi, T. (1978), in *Bioactive Peptides Produced by Microorganisms* (Umezawa, H., Taketa, T., & Shiba, T., Eds.), pp. 129–151, Wiley, New York.
Aoyagi, T., Kumagai, M., Hazato, T., Hamada, M., Takeuchi, T. & Umezawa, H. (1975), *J. Antibiot.*, **28**, 555–557.
Aoyagi, T., Suda, H., Nagai, M., Ogawa, K., Suzuki, J., Takeuchi, T. & Umezawa, H. (1976), *Biochim. Biophys. Acta*, **452**, 131–143.
Aoyagi, T., Suda, H., Nagai, M., Tobe, H., Suzuki, J., Takeuchi, T. & Umezawa, H. (1978a), *Biochem. Biophys. Res. Commun.*, **80**, 435–442.
Aoyagi, T., Tobe, H., Kojima, F., Hamada, M., Takeuchi, T., & Umezawa, H. (1978b), *J. Antibiot.*, **31**, 636–638.
Arima, K., Kohsaka, M. Tamura, G., Imanaka, H. & Sakai, H. (1972), *J. Antibiot.*, **25**, 437–444.
Baltimore, D. (1970), *Nature London*, **226**, 1209–1211.
Bekesi, J. G., Roboz, J. P. Zimmerman, E. & Holland, J. F. (1976), *Cancer Res.*, **36**, 631–639.
Belt, R. J., Haas, C. D., Usha, J., Goodwin, W., Moore, D. & Hoogstraten, B. (1979), *Cancer Treat. Rpt.*, **63**, 1993–1995.
Bettinger, G. E. & Young, F. E. (1975), *Biochem. Biophys. Res. Comm.*, **67**, 16–21.
Bhuyan, G. K., & Fraser, T. J. (1974), *Prog. Chemother.*, **3**, 111–115.
Bosmann, H. B. (1972), *Biochim. Biophys. Acta*, **264**, 339–343.
Bradner, W. T. (1958), *Ann. N.Y. Acad. Sci.*, **76**, 469–474.
Bradner, W. T. (1978a), *Antibiot. Chemother.*, **23**, 4–11.
Bradner, W. T. (1978b), in *Bleomycin: Current Status and New Developments* (Carter, S. K., Crooke, S. T. & Umezawa, H., Eds.) Academic, New York, pp. 333–342.
Bradner, W. T. (1979), in Mitomycin C: *Current Status and New Developments* (S. K. Carter & S. T. Crooke, Eds.), Academic, New York.
Bradner, W. T. & Clarke, D. A. (1958), *Cancer Res.*, **18**, 299–304.
Bradner, W. T. & Pindell, M. H. (1964), *Proc. Am. Assoc. Cancer Res.*, **5**, 7.
Bradner, W. T. & Pindell, M. H. (1965), *Cancer Res.*, **25**, 859–864.
Bradner, W. T. & Pindell, M. H. (1966), *Cancer Res.* (supp), **26** (No. 4, Part 2), 375–390.
Bradner, W. T., & Rose, W. C. (1980), Antitumor testing in animals. In *Anthracyclines: Current Status and New Developments* (Crooke, S. T. & Reich, S. D., Eds.), Academic, New York.

Bradner, W. T., Rose, W. C. & Huftalen, J. B. (1980a), in *Cisplatin: Current Status and New Developments* (Prestayko, A. W., Crooke, S. T. & Carter, S. K., Eds.), Academic, New York, pp. 171–182.
Bradner, W. T. & Rossomano, C. A. (1968), *Proc. Am. Assoc. Cancer Res.*, **9**, 8.
Bradner, W. T. & Schurig, J. E. (1981), *Cancer Treatment Rev.*, **8**, 93–102.
Bradner, W. T., Schurig, J. E., Huftalen, J. B. & Doyle, G. J. (1980b), *Cancer Chemother. Pharmcol.*, **4**, 95–101.
Bruce, W. R., Meeker, B. E., & Valeriote, F. A. (1966), *J. Nat. Cancer Inst.*, **37**, 233–245.
Bunge, R. H., McCready, D. E., Balta, L. A., Graham, B. D., French, J. C. & Dion, H. W. (1978), *Recent Results Cancer Research*, pp. 77–84, Springer Verlag, Berlin.
Bunge, R. H., McCready, D. E., Balta, L. E., Graham, B. D., French, J. C. & Dion, H. W. (1979), *Dev. Ind. Microbiol.*, **20**, 393–407.
Burchenal, J. H., Kalaher, K., Dew, K., Lokys, L. & Gale, G. (1978), *Biochimie*, **60**, 960–965.
Casazza, A. M. (1979), *Canc. Treat. Rep.*, **63**(5), 835–844.
Cassady, J. M. & Floss, H. G. (1977), *Lloydia*, **40**, 90–106.
Chapman, S. K., Glant, S. K., & Breiner, J. F. (1980), *Fed. Proc.*, **39**, 438.
Chihara, G., Maeda, Y. Y., Hamuro, J., Sasaki, T., & Fukuoka, F. (1969), *Nature London*, **222**, 687–688.
Claridge, C. A., Bradner, W. T. & Schmitz, H. (1978), *J. Antibiot.*, **31**, 485–486.
Clark, E. A., & Harmon, R. C. (1980), in *Advances in Cancer Research*, Vol. 31 (G. Klein and S. Weinhouse, Eds.), Academic, New York.
Coetzee, M. L. & Ove, P. (1979), *Process Biochem*, October 26–37.
Corbett, T. H., Griswold, D. P., Roberts, B. J., Peckham, J. C. & Schabel, F. M., Jr. (1977), *Cancer*, **40**, 2660–2680.
Cowan, J. D., & VonHoff, D. D., (1980), *Abstracts 20th Interscience Conference on Antimicrobial Agents and Chemotherapy*, Abstract No. 29.
Davis, P. P. & Smith, R. V. (1978), *Lloydia*, **41**, 650.
Dendy, P. P. (1976), *Chemotherapy*, Vol. 7 (Hellman, K., Connors, T. A. Eds.), pp. 341–350, Plenum, New York.
Denhardt, D. T. (1979), *Nature London*, **280**, 196–198.
DeSimone, P. A., Greco, F. A. & Lessner, H. F. (1979), *Cancer Treat. Rep.*, **63**, 2015–2017.
Devoret, R. (1979), *Sci. Am.*, **241**, 40–49.
Diggs, C. H., Scoltock, M. H. & Wiermik, P. H. (1978), *Cancer Clin. Trials*, 297–299.
Djordjevic, B. & Kim, J. H. (1968), *J. Cell. Biol.*, **38**, 477–482.
Douros, J. D. (1978), *Recent Results Cancer Res.* pp. 33–48, Springer-Verlag, Berlin.
Doyle, T. W. & Bradner, W. T. (1980), in *Anticancer Agents Based on Natural Product Models* (Cassady J. M. & Douros, J. D., Eds.), Academic, New York.
Drummond, M. (1979), *Nature London*, **281**, 343–347.
Dubois, J. B., & Serrou, B. (1976), *Cancer Res.*, **36**, 1731–1734.
DuVernay, V. H. & Crooke, S. T. (1980), *J. Antibiot.*, **33**, 1048–1053.
Együd, L. G., & Szent-Györgi, A. (1966a), *Proc. Nat. Acad. Sci. U.S.A.*, **55**, 388–393.

Együd, L. G., & Szent-Györgi, A. (1966b), *Proc. Nat. Acad. Sci. U.S.A.*, **56**, 203–207.
Ehrlich, P. Apolant, H. (1905), *Berl. Klin. Wochenschr.*, **42**, 871–874.
Elespuru, R. K., & Yarmolinsky, M. B. (1979), *Environ. Mutagenesis*, **1**, 65–78.
Ellison, R. R., Holland, J. F., Weil, M., Jacquillet, C., Bociou, M., Bernard, J., Sawitsky, A., Rosner, F., Gussoff, B., Silver, R. T., Karanas, A., Cuttner, J., Spurr, C. L., Hayes, D. M., Blom, J., Leone, L. A., Haurani, F., Kyle, R., Hutchison, J. L., Forcier, R. J. & Moon, J. H. (1968), *Blood*, **32**, 507–523.
Ferguson, R. M. & Schmidtke, J. R. (1979), *Surg. Clin. North Am.*, **59**, 349–369.
Fink, G. R., & Styles, C. A. (1972), *Proc. Nat. Acad. Sci. U.S.A.*, **69**, 2846–2849.
Fleck, W. (1968), *Z. Allg. Mikrobiol.*, **8**, 139–144.
Florcyzk, A. P., Schurig, J. E., Rivers, K. A., Bradner, W. T. & Crooke, S. T. (1980), *Pharmacologist*, **22**(3), 240.
Foley, G. E., McCarthy, R. E., Binns, V. M., Snell, E. E., Guirard, B. M., Kidder, G. W., Dewey, V. C., & Thayer, P. S. (1958), *Ann. N.Y. Acad. Sci.*, **76**, 413–438.
Fonken, G. S. (1972), *Cancer Chemother. Rep. Part 3*, **3**, 7–12.
Fugmann, R. A., Anderson, J. C., Stolfi, R. L. & Martin, D. S. (1977), *Cancer Res.*, **37**, 496–500.
Fugmann, R. A., Martin, D. S., Hayworth, P. E. & Stolfi, R. L. (1970), *Cancer Res.*, **30**, 1932–1936.
Furth, J., Siebold, H. R. & Rathbone, R. R. (1933), *Am. J. Cancer*, **19**, 521–604.
Gadó, I., Savtchenko, G. & Horvath, I. (1965), *Acta. Microbiol. Acad. Sci. Hung*, **13**, 363–365.
Garretson, A. L., Elespuru, R. K., Lufriu, I., Warnick, D., Wei, T. & White, R. J. (1981), *Dev. Ind. Microbiol.*, **22**, 211–218.
Gause, G. F. & Dudnik, Y. V. (1980), *Antibiot. Chemother.*, **28**, 102–108.
Gause, G. F., Laiko, A. V. & Selesneva, T. I. (1976), *Cancer Treat. Rep.*, **60**, 637–638.
Geran, R. I., Greenberg, N. H., Macdonald, M. M., Schumacher, A. M. & Abbott, B. J. (1972), *Cancer Chemother. Rep. Part 3*, **3**(2), 1–103.
Goldin A., Schepartz, S. A., Venditti, J. M. & DeVita, V. T., Jr. (1979), in *Methods in Cancer Research* (DeVita, V. T., Jr. & Busch, H. Eds.), pp. 164–245, Academic, New York.
Goldin, A., Venditti, J. M., Humphreys, S. R. & Mantel, N. (1956), *J. Nat. Cancer Inst.*, **17**, 203–212.
Green, M. H. L., Muriel, W. J. & Bridges, B. A. (1976), *Mutat. Res.*, **38**, 33–41.
Gross, L. (1970), *Oncogenic Viruses*, 2nd ed., Pergamon, New York.
Hamburger, A. W. & Salmon, S. E. (1977), *Science*, **197**, 461–463.
Hanka, L. J. & Barnett, M. S. (1974), *Antimicrob. Agents Chemother.*, **6**, 651–652.
Hanka, L. J., Bhuyan, B. K., Martin, D. G., Neil, G. L. & Douros, J. D. (1978a), *Antibiot. Chemother.*, **23**, 26–32.
Hanka, L. J., Kuentzel, S. L., Martin, D. G., Wiley, P. F. & Neil, G. L. (1978b), *Recent Results Cancer Research*, pp. 69–76, Springer-Verlag, Berlin.
Hanka, L. J., Martin, D. G. & Neil, G. L. (1978c), *Lloydia*, **41**, 85–87.
Hannan, M. A. & Hurley, L. H. (1978), *J. Antibiot.*, **31**, 911–913.
Heinemann, B. & Howard, A. (1964), *Appl. Microbiol.*, **12**, 234–239.

Herrmann, E. C., Jr., Gabliks, J., Engle, C. & Perlman, P. L. (1960), *Proc. Soc. Exp. Biol. Med.*, **130**, 625–628.
Hewitt, H. B. (1978), *Adv. Cancer Res.*, **27**, 149–200.
Hori, M., Takemoto, K., Honma, I., Takeuchi, T., Kondo, S., Hamada, M., Okazaki, T., Okami, Y. & Umezawa, H. (1972), *J. Antibiot.*, **25**, 393–399.
Houchens, D. P., Johnson, R. K., Ovejera, A., Gaston, M. R. & Goldin, A. (1976), *Cancer Treat. Rep.*, **60**(7), 823–828.
Houghton, P. J. & Houghton, J. A. (1978), *Br. J. Cancer*, **37**, 833–840.
Hozumi, M., Ogawa, M., Sugimura, T., Takeuchi, T. & Umezawa, H. (1972), *Cancer Res.*, **32**, 1725–1728.
Huggins, C., Grand, L. C. & Brillantes, F. P. (1961), *Nature London*, **179**, 204–207.
Hutner, S. H., Nathan, H. A., Aaronson, S., Baker, H. & Scher, S. (1958), *Ann. NY Acad. Sci.*, **76**, 457–468.
Imai, R., Ashizawa, T., Urakawa, C., Morimoto, M. & Nakamura, N. (1980), *Gann*, **71**, 560–562.
Ishizuka, M., Masuda, T., Kanbayashi, M., Fukasawa, S., Takeuchi, T., Aoyagi, T. & Umezawa, H. (1980), *J. Antibiot.*, **33**, 642–652.
Iyer, V. N. & Szybalski, W. (1958), *Appl. Microbiol.*, **6**, 23–29.
Johnson, R. K., & Goldin, A. (1975), *Cancer Treat. Rep.*, **2**, 1–31.
Jordan, V. C. (1976), *Eur. J. Cancer*, **12**, 419–424.
Kada, T., Tutikawa, K. & Sadaiey, Y. (1972), *Mutat. Res.*, **16**, 165–174.
Kandel, J. S. & Stern, T. A. (1979), *Antimicrob. Agents Chemotherap.*, **16**, 568–571.
Kelleher, J. K. (1977), *Mol. Pharmcol.*, **13**, 232–241.
Killion, J. J. (1977), *Cancer Immunol. Immunother.*, **3**, 87–91.
Kim, J. H., Gelbard, A. S., Djordjevic, B., Kim, S. H. & Perez, A. G. (1968), *Cancer Res.*, **28**, 2437–2442.
Kitame, F., Utsushikawa, K., Kohama, T., Saito, T., Kikuchi, M. & Ishida, N. (1974), *J. Antibiot.*, **27**, 884–888.
Kline, I., Venditti, J. M., Dyrer, D. J. & Goldin, A. (1966), *Cancer Res.*, **26**, 853–859.
Koltin, Y. & Day, P. R. (1975), *Appl. Microbiol.*, **30**, 694–696.
Kováč, L., Böhmerová, E. & Fuska, J. (1978), *J. Antibiot.*, **31**, 616–620.
Kupchan, S. M., Komoda, Y., Court, W. A., Thomas, C. J., Smith, R. M., Karim, A., Gilmore, C. J., Haltiwanger, R. C. & Bryan, R. F. (1972), *J. Am. Chem. Soc.*, **95**, 1354.
Kurata, T. & Mickche, M. (1977), *Oncology*, **34**, 212–215.
Langen, T. (1980), *Nature London*, **286**, 329–330.
Lein, J., Heinemann, B. & Gourevitch, A. (1962), *Nature London*, **196**, 783–784.
Lemke, P. A. (1977), *Microbiology-1977*, pp. 568–570.
Lwoff, A. (1953), *Bacteriol. Rev.*, **17**, 269–337.
McIlhinney, A. & Hogan, B. L. M. (1974), *Biochem. Biophys. Res.*, **60**, 348–354.
McPherson, F., Bridges, J. W. & Parke, D. V. (1974), *Nature London*, **252**, 488–489.
Madoc-Jones, H. & Mauro, F. (1968), *J. Cell Physiol.*, **72**, 185–195.
Martin, D. S., Fugmann, R. A., Stolfi, R. L. & Hayworth, P. E. (1975), *Cancer Chemother. Rep. Part 2*, **5**(1), 89–109.

Mathé, G., Halle-Pannenko, O. & Bourut, C. (1977a), *Cancer Immunol. Immunother.*, **2**, 139-141.
Mathé, G., Halle-Pannenko, O. & Bourut, C. (1977b), *Eur. J. Cancer*, **13**, 1095-1098.
Mauro, F. & Madoc-Jones, H. (1970), *Cancer Res.*, **30**, 1397-1408.
Mayo, J. G., Laster, W. R., Jr., Andrews, C. M. & Schabel, F. M., Jr. (1972), *Cancer Chemother. Rep.*, **56**, 183-195.
Mitscher, L. A., Andres, W. W. & McCrae, W. (1964), *Experientia*, **20**, 258-259.
Mohn, G., Ellenberger, J. & McGregor, D. (1974), *Mutat. Res.*, **25**, 187-195.
Mong, S., Strong, J. E., Bush, J. A. & Crooke, S. T. (1979), *Antimicrob. Agents Chemother.*, **16**, 398-405.
Moreau, P., Bailone, A. & Devoret, R. (1976), *Proc. Nat. Acad. Sci. U.S.A.*, **73**, 3700-3704.
Murphy, W. K., Burgess, M. A., Valdivieso, M., Livingston, R. B., Bodey, G. P. & Freireich, E. J. (1978), *Cancer Treat. Rep.*, **62**, 1497-1502.
Murthy, Y. K. S., Thiemann, J. K., Coronelli, C. & Sensi, P. (1966), *Nature London*, **211**, 1197-1199.
Nakamura, S., Omura, S., Hamada, M., Nishimura, T., Yamaki, H., Tanaka, N., Okami, Y. & Umezawa, H. (1967), *J. Antibiot.*, **20**, 217-222.
Neil, G. L., Kuentzel, S. L. & McGovern, J. P. (1979), *Cancer Treat. Rep.*, **63**(11-12), 1971-1978.
Nester, E. W. & Montoya, A. (1979), *Am. Soc. Microbiol. News*, **45**, 283-287.
Nowak, K., Peckham, M. J. & Steel, G. G. (1978), *Br. J. Cancer*, **37**, 576-584.
Numata, M., Nitta, K., Utahara, R., Maeda, K. & Umezawa, H. (1975), *J. Antibiot.*, **28**, 757-763.
Oka, K., Kato, T., Takita, T., Takeuchi, T., Umezawa, H. & Nagatsu, T. (1980), *J. Antibiot.*, **33**, 1043-1047.
Ong, T-M. (1978), *Mutat. Res.*, **53**, 297-308.
Pendergrast, W. J., Jr., Drake, W. P. & Mardiney, M. R., Jr. (1976), *J. Nat. Cancer Inst.*, **57**, 539-544.
Peters, T. G. & Lewis, J. D. (1976), *Surg. Forum*, **27**, 97-98.
Preer, J. R., Jr. (1977), *Microbiology 1977*, pp. 576-578.
Prestayko, A. W., Bradner, W. T., Huftalen, J. B., Rose, W. C., Schurig, J. E., Clear, M. J., Hydes, P. C. & Crooke, S. T. (1979), *Cancer Treat. Rep.*, **63**, 1503-1508.
Price, K. E., Buck, R. E. & Lein, J. (1964), *Appl. Microbiol.*, **12**, 428-435.
Proctor, J. W., Auclair, B. G. & Lewis, M. G. (1976), *Eur. J. Cancer*, **12**, 203-210.
Reich, S. D. & Bradner, W. T. (1979), *Proc. Am. Assoc. Cancer Res.*, **20**, 867.
Renoux, G. & Renoux, M. (1972), *Nature London*, **240**, 217-218.
Rose, W. C. (1981), *Cancer Treat. Rep.*, **65**, 299-312.
Rosenkranz, H. S. (1973), *Ann. Rev. Microbiol.*, **27**, 383-401.
Sadler, R. E. & Castro, J. E. (1976), *Br. J. Surg.*, **63**, 292-296.
Salas, C. E., Pfohl-Leszkowicz, A., Lang, M. C. & Dirheimer, G. (1979), *Nature London*, **278**, 71-72.
Salmon, S. E., Hamburger, A. W., Soehnlen, B., Durie, B. G. M., Alberts, D. S. & Moon, T. E. (1978), *N. Engl. J. Med.*, **298**, 1321-1327.

Sartorelli, A. C. & Creasey, W. A. (1969), *Ann. Rev. Pharmcol. Toxicol.*, 9, 51–72.
Savel, H. (1966), *Prog. Exp. Tumor Res.*, 8, 189–224.
Sawa, T., Fukugawa, Y., Homma, I., Takeuchi, T. & Umezawa, H. (1967), *J. Antibiot.*, 20, 227.
Schabel, F. M., Jr., Skipper, H. E., Trader, M. W., Laster, W. T., Jr. & Cheeks, J. B. (1974), *Cancer Chemother. Rep. Part 2*, 4(1), 53–72.
Scholler, J. (1958), *Ann. N.Y. Acad. Sci.*, 76(3), 855–860.
Scholler, J., Philips, F. S. & Bittner, J. J. (1955), *Cancer Res. Suppl.*, 13 (Part 2), 32–37.
Schurig, J. E., Bradner, W. T., Huftalen, J. B. & Doyle, G. J. (1980a), in *Anthracyclines: Current Status and New Developments* (Crooke, S. T. & Reich, S. D. Eds.), pp. 141–149, Academic, New York.
Schurig, J. E., Bradner, W. T., Huftalen, J. B., Doyle, G. J. & Gylys, J. A., (1980b), *Cisplatin: Current Status and New Developments.* (Prestayko, A. W., Crooke, S. T. & Carter, S. K., Eds.), pp. 227–236, Academic, New York.
Schurig, J. E., Florczyk, A. P., Hirth, R. S., Bradner, W. T. & Buyniski, J. P. (1979), *Fed. Proc.*, 38, 522.
Scott, D. (1976), *Chemotherapy* Vol. 7 (Hellman, K., Connors, T. A. Eds.), pp. 95–103, Plenum, New York.
Selawry, O. S. (1965), *J. Am. Med. Assoc.*, 194(1), 187–193.
Sethi, K. K. & Brandis, H. (1973), *Br. J. Cancer*, 27, 106–113.
Shear, M. J., Achinstein, B. & Pradhan, S. N. (1958), *J. Nat. Cancer Inst.*, 21, 585–595.
Siminoff, P. (1961), *Appl. Microbiol.*, 9, 66–72.
Singh, K., Sun, S. & Vézina, C. (1979), *J. Antibiot.*, 32, 630–645.
Slater, E. E., Anderson, M.D. & Rosenkranz, H. S. (1971), *Cancer Res.*, 31, 970–973.
Smith, C. G., Lummis, W. L. & Grady, J. E. (1959), *Cancer Res.*, 19, 843–846.
Snyder, J. A. & McIntosh, R. J. (1976), *Ann. Rev. Biochem.*, 45, 699–720.
Soloway, M. S. & Murphy, W. M. (1979), *Semin. Oncol.*, 6(20), 166–183.
Speedie, M. K., Fique, D. V. & Blomster, R. N. (1980), *Antimicrob. Agents Chemother.*, 18, 171–175.
Stock, C. C. (Ed.). (1958), *Ann. N.Y. Acad. Sci.*, 76(3), 409–970.
Streightoff, F., Nelson, J. D., Cline, J. C., Gerzon, K., Williams, R. H. & DeLong, D. C. (1969), *Abstracts 9th Interscience Conference.* Antimicrob. Agents & Chemotherapy Abstract No. 18.
Szybalski, W. (1958), *Ann. N.Y. Acad. Sci.*, 76, 475–489.
Takatsuki, A., Arima, K. & Tamura, G. (1971), *J. Antibiot.*, 24, 215–223.
Takatsuki, A., Munekata, M., Nishimura, M., Kohno, K., Onodera, K. & Tamura, G. (1977), *Agric. Biol. Chem.*, 41, 1831–1834.
Temin, H. M. & Mizutani, S. (1970), *Nature London*, 226, 1211–1213.
Thayer, P. S., Gordon, H. L. & MacDonald, M. (1971), *Cancer Chemother. Rep.*, 2, 27–55.
Umezawa, H. (1976), *Chemotherapy*, Vol. 7 (Hellman, K. & Connors, T. A. Eds.), pp. 294–298, Plenum, New York.
Umezawa, H. Aoyagi, T., Hazato, T., Uotani, K., Hamada, M., & Takeuchi, T. (1978), *J. Antibiot.*, 31, 639–641.
Venditti, J. M. (1971), *Cancer Chemother Rep.*, Part 3, 2(1), 35–59.

Volk, W. A. (1978), *J. Bacteriol.*, **95**, 782–786.
Watson, C., Medina, D. & Clark, J. H. (1977), *Cancer Res.*, **37**, 3344–3358.
Weber, G. (1980), *Antibiot. Chemother.*, **28**, 51–61.
Wickner, R. B. & Liebowitz, M. J. (1976), *J. Mol. Biol.*, **105**, 427–443.
Zambryski, P., Holsters, M., Kruger, K., DePicker, A., Schell, J., Montagu, M. V. & Goodman, H. M. (1980), *Science*, **209**, 1385–1391.
Zava, D. T. & McGuire, W. L. (1977), *Cancer Res.*, **37**, 1608–1610.
Zavala, F., Guénard, D. & Potier, P. (1978), *Experientia*, **34**, 1497–1499.
Zavala, F., Guénard, D., Robin, J. P. & Brown, E. (1980), *J. Med. Chem.*, **23**, 546–549.

THREE

Chemistry of Antitumor Drugs

WILLIAM A. REMERS
The University of Arizona
Tucson, Arizona

INTRODUCTION	85
ALKYLATING AGENTS	86
The Chemistry of Alkylation, 86	
Interaction of Alkylating Agents with Biopolymers, 93	
Alkylating Agent Design and Structure-Activity Relationships, 98	
Synthesis of Alkylating Agents, 123	
ANTIMETABOLITES	127
The Biochemistry of Antimetabolite Action, 127	
The Design and Discovery of Antimetabolites, 140	
Synthesis of Antimetabolites, 162	
INTERCALATION OF ANTITUMOR DRUGS INTO DNA	165
The Intercalation Process, 165	
Antitumor Drugs that Intercalate into DNA, 168	
OTHER COMPOUNDS THAT INTERFERE WITH DNA STRUCTURE AND FUNCTION	199
INHIBITORS OF PROTEIN SYNTHESIS	205
MITOTIC INHIBITORS	210
MISCELLANEOUS NATURAL AND SYNTHETIC ANTITUMOR AGENTS	219
HORMONES	223
Effects of Hormones on Tumors, 223	
Synthesis of Hormones, 228	
IMMUNOSTIMULANTS	233
RADIOSENSITIZING AND RADIOPROTECTING COMPOUNDS	236
REFERENCES	237

INTRODUCTION

The purpose of this chapter is to present a broad picture of the approaches and methods used in the design, synthesis, and structure-activity correlation of compounds used against neoplastic disease. At the same time a substantial depth of coverage has been attempted to make the material valuable to experienced investigators. Major emphasis is placed on ideas used in the design of antitumor agents. This activity always has received clever ideas, but in the early days they often were based on invalid or naive assumptions. More successful approaches have emerged as the basic biological sciences developed; for example, the elucidation of nucleoside biosynthesis led to the rational design of antimetabolites such as 5-fluorouracil. Recent advances in our understanding of the structure and function of DNA, the mitotic process, and tumor immunology promise to provide more clearly defined targets for the design of other types of antitumor agent. Nevertheless, the complexity of biological systems and the subtle differences between normal and neoplastic cells still make rational design a formidable challenge.

This chapter is organized on the basis of modes of antitumor action rather than on other arbitrary bases such as chemical structure or natural versus synthetic compounds. The integrity of chemical classes, however, is largely preserved in this system; for example, the purines, pyrimidines, and pteridines occur in the section on antimetabolites, whereas the polycyclic aromatic compounds are grouped as intercalating agents. Considerable attention is given to chemistry related directly to the modes of antitumor action, including alkylation of DNA, intercalation, and enzyme inhibition. The chemical transformation of natural products into active analogs is emphasized. It has not been possible to cover in depth the substantial literature on the chemical synthesis of important structural types such as nucleosides and steroids, but references are given to authoritative reviews of this literature. Similarly, the total synthesis of complex natural products such as anthracyclines is treated only by reference to the literature. Practical syntheses of those compounds used clinically in the treatment of cancer are described in detail. Structure-activity relationships, including tables of screening data, are given for series of compounds tested against rodent tumors.

ALKYLATING AGENTS

The Chemistry of Alkylation

Alkylation of biological macromolecules involves reactions in which nucleophilic atoms (N, S, or O) on the macromolecules displace leaving groups such as halogen, sulfonate, or ammonium from the alkylating agent. The reaction rate depends on the nucleophilicity of the atom that is alkylated. Assuming equal concentrations and accessibility, the relative rates of reaction at physiological pH (7.4) would be in the order of ionized thiol > amine > ionized phosphate > carboxylate (Montgomery et al., 1970). The reaction order depends on the chemical structure of the alkylating agent. Methanesulfonates, ethylenimines, and epoxides give second-order reactions that are first order each in the alkylating agent and nucleophile (Sn2) (Connors, 1974); for example, the alkylation of a nucleophile (nu-H) by ethylenimine can be expressed as in Equation 1.

$$\text{Nu-H} + \triangleright\text{N-R} \longrightarrow \text{Nu-CH}_2\text{CH}_2\text{NHR} \qquad (1)$$

2-Haloalkylamines (nitrogen mustards) and 2-haloalkyl sulfides (sulfur mustards) form strained three-membered ring "onium" intermediates in water or other solvents polar enough to solvate the halide ion. These intermediates result from a neighboring group reaction in which the nucleophilic nitrogen or sulfur atom displaces the halide (Price, 1974). In 2-diethylaminoethyl chloride cyclization to the aziridinium ion (Equation 2) is complete within 1 min at 0°C and pH 10 in water. The formation of this species can be observed by NMR spectroscopy (Sowa & Price, 1968). Reaction of an aziridinium ion with a nucleophile would be second order; however, the overall rate of reaction between a 2-haloalkylamine and nucleophile would depend on the rate of formation of the aziridinium ion. Because the example just cited has a fast rate of aziridinium ion formation, the reaction with the nucleophile would be rate determining and the overall rate would be second order (Equation 2). In contrast, an aryl nitrogen mustard such as melphalan would form the aziridinium ion slowly because the nitrogen has decreased nucleophilicity. This rate-determining step would make the overall reaction first order (Sn1) (Price, 1974).

$$(C_2H_5)_2NCH_2CH_2Cl \xrightarrow[fast]{-Cl^-} (C_2H_5)_2\overset{+}{N}{\triangleleft} \xrightarrow[slower]{Nu-H} (C_2H_5)_2NCH_2CH_2Nu \qquad (2)$$

Sulfur mustards also react with nucleophiles in first-order processes (Equation 3) because the cyclic sulfonium ions are relatively unstable, high-energy species whose formation is the rate-determining step. Despite the high reactivity of cyclic sulfonium ions, they display selectivity in their reactions with various nucleophiles (Price, 1974).

$$RSCH_2CH_2Cl \xrightarrow[slow]{-Cl^-} \overset{+}{RS}{\triangleleft} \xrightarrow[fast]{Nu-H} RSCH_2CH_2-Nu \qquad (3)$$

Alkylating onium ions are also produced by the decomposition of N-alkyl-N-nitrosoureas in water. These compounds have half-lives ranging from 15 to 200 min under physiological conditions. In addition to the carbonium ions that cause alkylation, these decompositions form isocyanates that can carbamoylate nucleophiles; for example, the decomposition of methylnitrosourea initially provides methyldiazohydroxide and isocyanic acid. Further decomposition of methyldiazohydroxide leads through methyldiazonium ion to methyl carbonium ion (Equation 4; Montgomery et al., 1967). This mode of decomposition differs from the alkaline decomposition of methylnitrosoureas wherein diazomethane is produced. Evidence that diazomethane is not the alkylating agent under physiological conditions was provided by the use of 1-trideuteriomethyl-3-nitro-1-nitrosoguanidine. This agent alkylated nucleophiles with an intact trideuteriomethyl group (Wheeler, 1974).

$$\begin{array}{c} \overset{O}{\underset{\parallel}{}} \\ CH_3N-C-NH \\ | \quad \ \ | \\ N \qquad H \\ \ \diagdown O \diagup \end{array} \longrightarrow \begin{array}{c} CH_3N=NOH \\ + \\ O=C=NH \end{array} \longrightarrow \begin{array}{c} CH_3\overset{+}{N}{\equiv}N \\ \downarrow \\ CH_3{}^+ \end{array} \qquad (4)$$

The decomposition of BCNU and certain related compounds is complicated by the possibility that cyclic intermediates can be formed from the displacement of halide by nucleophiles within the molecule. Three competing pathways have been proposed to account for the products of decomposition in water (Weinkam & Lin, 1979; Lown et al., 1980). They are given in modified form in Scheme 1. Path I, which involves the cleavage of BCNU into chloroethyldiazohydroxide and chloroethylisocyanate, is thought to be the principal one related to biological activity because of the formation of chloroethyldiazohydroxide. The chloroethylcarbonium ion, written with charge delocalization, is responsible for alkylating a nitrogen base of DNA, probably the N^4-position of cytosine, and then crosslinking with another base (Lown & McLaughlin, 1979; Ludlum et al., 1975). Alternative pathways in Scheme 1 illustrate the formation of various byproducts that in-

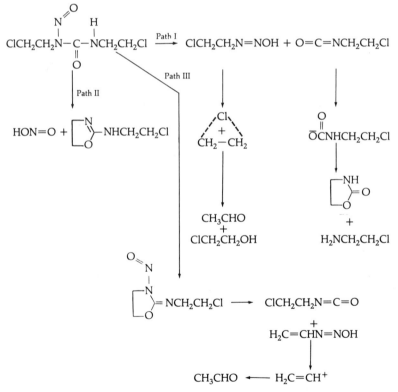

Scheme 1 Decomposition pathways for BCNU.

clude 2-chloroethylamine, 2-oxazolidone, 2-(2-chloroethylamino) oxazoline, and acetaldehyde.

Aryltriazines also appear to alkylate by way of onium ions formed on hydrolysis. Thus in a series of aryl-substituted dimethyltriazines a correlation was made between the half-life of hydrolysis to the aryldiazoniium ions and therapeutic effects on mice with Lewis lung carcinoma (Sava et al., 1979). It is uncertain whether the aryldiazonium ions react as such or lose nitrogen to yield the corresponding carbonium ions. The hypothetical decomposition process is illustrated in Scheme 2 for dacarbazine, a clinical agent. An alternative mode of decomposition that involves metabolic demethylation followed by hydrolysis to methyl carbonium ion has also been suggested (Skibba et al., 1970) and it is included in this scheme.

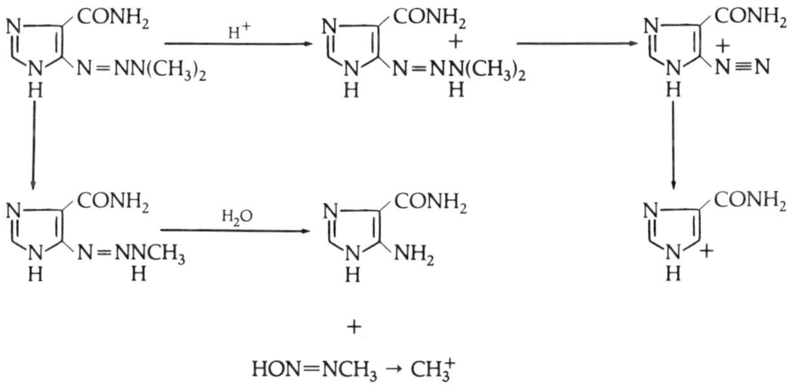

Scheme 2 Hydrolysis of dacarbazine.

The alkylating agents described so far have not required activation other than hydrolysis. Other kinds of molecules, however, do not become effective for alkylation until they have been modified enzymatically. One important group includes those molecules that undergo bioreductive alkylation (Lin et al., 1972). A typical example of this process occurs when a benzoquinone bearing adjacent halomethyl or acetoxymethyl groups is reduced to the corresponding hydroquinone (Equation 5). In the hydroquinone form the elements of hydrogen halide or acetic acid can leave to give a highly reactive quinonemethide. Although the demonstration of these quinonemethides has not yet been made in vivo, they have been trapped by amines when generated in vitro (Lin et al., 1974).

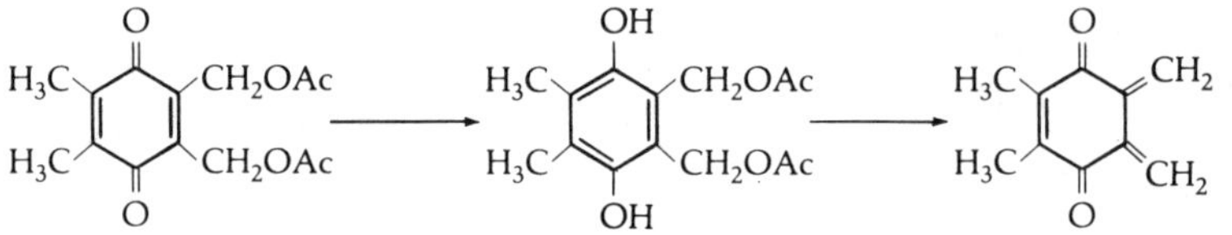

$$(5)$$

Activation of mitomycin C represents a more complex example of the bioreductive process (Scheme 3). The hydroquinone initially formed loses the elements of methanol to give the corresponding indolohydroquinone. In the latter species conjugation of the carbamoyloxy function with the indole nitrogen enhances its rate of elimination and the resulting stabilized carbonium ion alkylates with its 10-position (Szybalski & Iyer, 1967). The aziridine ring also is activated for alkylation with its 1-position. Thus mitomycin C becomes a bifunctional alkylating agent (Szybalski, 1964). Alkylation by the aziridine ring is believed to be the first process because of an observed pH dependence for the interaction of mitomycin C with nucleic acids (Lown et al., 1976).

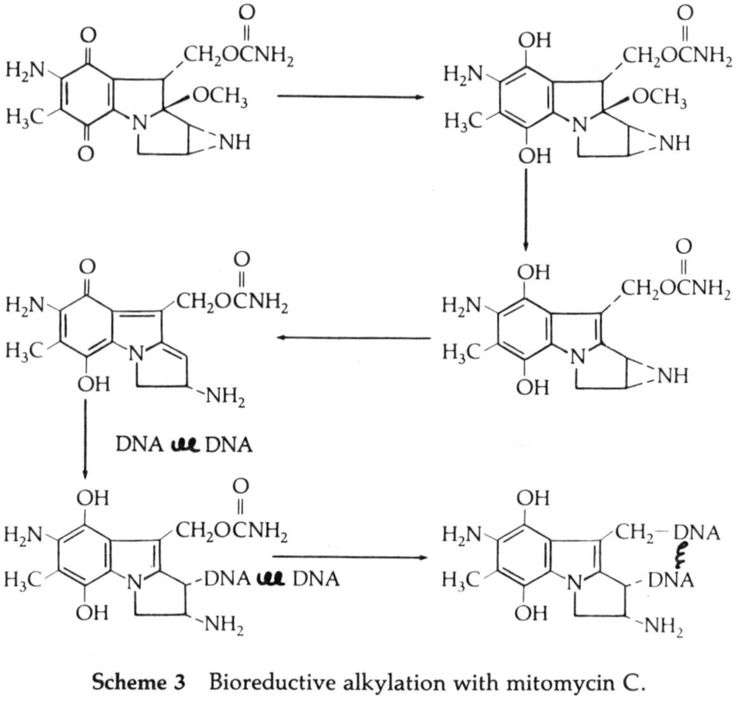

Scheme 3 Bioreductive alkylation with mitomycin C.

CHEMISTRY OF ANTITUMOR DRUGS

The concept of bioreductive alkylation has been applied to the mode of action of many different antitumor agents and a thought-provoking review of this topic has been written (Moore, 1977).

Bioactivation of cyclophosphamide is a complex and remarkable process (Scheme 4). It is first coverted by hepatic cytochrome P-450 into a hydroperoxide which is subsequently reduced to the corresponding alcohol (Takamizawa et al., 1975). This alcohol is actually an aldehydeamine that is in equilibrium with the open-chain amino aldehyde form. Nonenzymatic decomposition of this form generates phosphoramide mustard and acrolein. Phosphoramide mustard is in equilibrium with its monoanion, which can cyclize to the corresponding aziridinium ion with elimination of chloride ion. It is this aziridinium ion that is the main alkylating species derived from cyclophosphamide. Its rate of formation from phosphoramide mustard is optimal at pH 7, according to NMR studies with ^{31}P (Engle et al., 1979). Acrolein also is cytotoxic but it has only $1/40$ the activity of phosphoramide mustard.

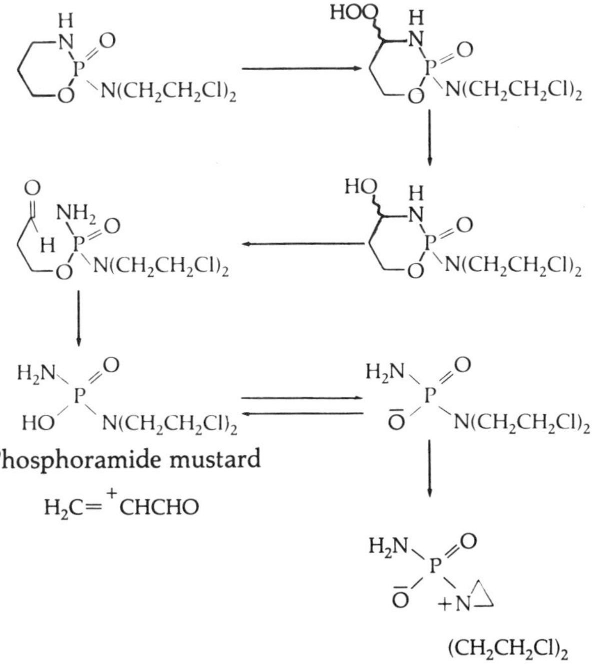

Scheme 4 Bioactivation of cyclophosphamide.

Acrolein alkylates nucleophiles by conjugate addition of the nucleophile to the β-carbon atom (Equation 6). A number of natural products alkylate by way of their α,β-unsaturated carbonyl systems. In particular, cyclopentenone and α-methylenelactone functions are potent alkylators. An example of a cytotoxic agent that contains both functions is helenalin (Figure 1). A study of the stepwise reduction of helenaline revealed that the α-methylenelactone function is more important than the cyclopentenone for antitumor activity (Lee et al., 1981).

$$\text{Nu}-\text{H} + \text{H}_2\text{C}=\text{CHCHO} \rightarrow \text{NuCH}_2\text{CH}_2\text{CHO} \qquad (6)$$

Alkylation also might occur by free radical processes. This is thought to be the case with methylhydrazine derivatives such as procarbazine (Scheme 5). The oxidation of procarbazine to azoprocarbazine is rapid in man; its half-life is 7–10 min (Reed, 1975). It is catalyzed by metalloproteins. Azoprocarbazine undergoes isomerization to the methylhydrazone of p-formyl-N-isopropylbenzamide, which is hydrolyzed into methylhydrazine and the aldehyde (Chabner et al., 1969). The major metabolite in urine is N-isopropylterephthalamic acid, an oxidation product of this aldehyde. The fate of methylhydrazine is complex, as indicated by the isolation of the methyl group as carbon dioxide, formaldehyde, and methane. Methane, however, proves the existence of methyl radicals. One mode of decomposition of methylhydrazine appears to be oxidation to methyldiazine, which undergoes homolytic cleavage to methyl radical, hydrogen radical, and nitrogen (Tsuji & Kosower, 1971).

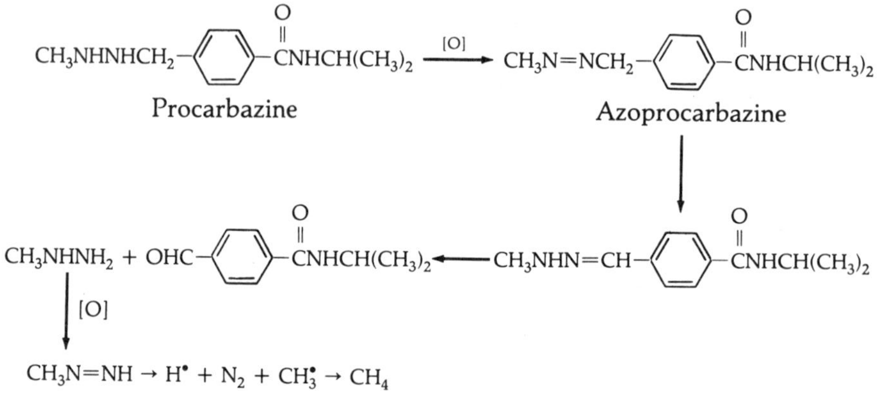

Scheme 5 Decomposition of procarbazine.

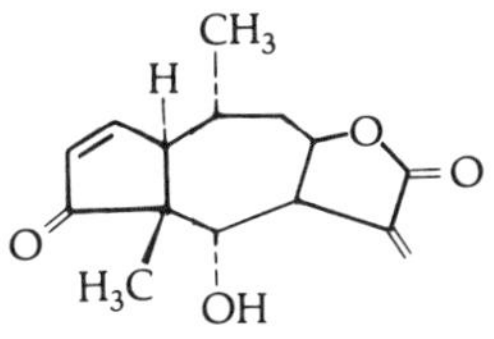

CHEMISTRY OF ANTITUMOR DRUGS

Figure 1 Helenalin.

Interaction of Alkylating Agents with Biopolymers

The interaction of alkylating agents with cell constituents has been extensively studied in a variety of systems which include viruses, bacteria, and human tumor cells in culture. It was observed in the earliest studies that their cytotoxic and mutagenic effects resembled those produced by ionizing radiation, and the term radiomimetic was applied to them (Gilman & Phillips, 1946). The action of alkylating agents is relatively cycle nonspecific (Figure 2). Even cells in the nondividing stage (G_0) can react with alkylating agents, but the polynucleotides of cells are most susceptible to alkylation when their structures are changed or unpaired in the process of replication. Thus alkylation of polynucleotides will be more effective in the late G_1- and S-phases than in the other phases. Although alkylation of cell biopolymers can occur at any stage in the cell cycle, the resulting toxicity usually is expressed when cells enter the synthesis phase (S). This toxicity prevents them from progressing through the premitotic phase (G_2). The result is that a backup of cells occurs behind the blockade. Some of these

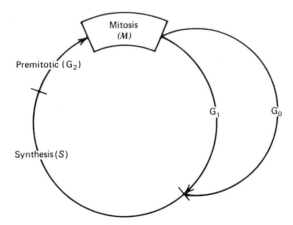

Figure 2 The cell mitotic cycle.

cells develop double complements of DNA and become giant cells. In any event, cell division is prevented (Levis et al., 1965; Ludlum, 1965).

Despite the abundant information obtained on the mode of action of alkylating agents, there still is no universal agreement on which of their actions on cancer cells is most important. A good working model for the alkylation of viruses and bacteria has been established, but there are problems in extrapolating it to more complicated mammalian cells; for example, the crosslinking of double helical DNA is a lethal event for bacteria, but it has not yet been demonstrated in intact mammalian cells. Another problem occurs in extending patterns of DNA alkylation from studies of mutagenesis, a nonlethal event, to cell-killing alkylations (Ludlum, 1974).

Although many cell constituents, including DNA, RNA, proteins, and cell membranes, are alkylated, the present working hypothesis for antitumor activity is that most alkylating agents produce their lethal effects by reacting with cellular DNA. This hypothesis correlates observations over a wide variety of systems and can be tested experimentally. All data do not fit it, however, and alternative hypotheses, such as the alkylation of RNA or proteins, cannot be ruled out (Ludlum, 1974).

A variety of functionalities on DNA can interact with alkylating agents. The phosphate groups are readily alkylated but the significance of the resulting phosphotriesters is uncertain (Ludlum, 1967). It has been suggested that they afford transalkylation of the bases. Studies of the rate of DNA alkylation support direct alkylation at sites such as N-7 of guanine. The most active position on any base toward agents such as nitrogen mustard (mechlorethamine) and methyl methanesulfonate is N-7 of guanine (Ross, 1962). Alkylation at this position produces significant changes in the chemical properties of the guanine residue (Scheme 6). The positive charge generated in the imidazole ring renders the 8-hydrogen labile to deuterium exchange. A test for N-7 alkylation is based on NMR study of this process (Tomasz, 1970). Delocalization of the positive charge favors the enol tautomer in the pyrimidine ring, which has been suggested as the cause of miscoding with thymine instead of cytosine in replication. It now appears that N-7 alkylation is much less mutagenic than O-6 alkylation of guanine. More serious consequences of N-7 alkylation result from the increased lability of the ribonucleotide fragment to hydrolysis. Thus attack of water on C-1' of the ribose unit produces depurination (Scheme 6). The resulting "apurinic" structure is relatively unstable and can undergo chain scission. Attack of water at C-8 of guanine can cleave the imidazole ring and the remaining substituted pyrimidine might also undergo depurination (Montgomery et al., 1970).

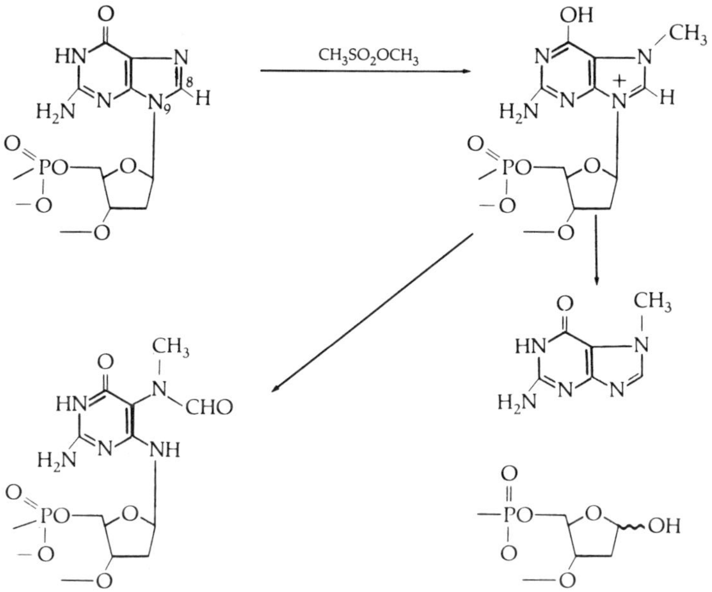

Scheme 6 Hydrolysis of N-7 methyl guanine in DNA.

Bifunctional alkylating agents such as mechlorethamine can react with N-7 of two different guanine residues in DNA. Di(guanin-7-yl) derivatives have been identified among the hydrolysis products of this DNA (Brooks & Lawley, 1961). The process of bifunctional alkylation is known as crosslinking (Kohn, 1966) and can be intrastrand or interstrand with double helical DNA. Interstrand crosslinking can be verified by a test based on the thermal denaturation and renaturation of the DNA. When double-stranded DNA is heated in water it unwinds and the strands separate. Renaturation on cooling is a slow and difficult process. If the strands are crosslinked intermolecularly, they cannot separate and renaturation on cooling is rapid (Kersten, 1974). Mechlorethamine produces intermolecular crosslinks. In contrast, the bismethanesulfonate of 1,4-butanediol, busulfan, gives intramolecular crosslinking. Thus 1′,4′-di(guanin-7-yl) butane can be obtained from the hydrolysate after reaction with double helical DNA, but the crosslinked product is not resistant to denaturation (Kohn et al., 1966).

The N-7 position of guanine is not the only site alkylated on bases in DNA. Certain alkylating agents, including ethyl nitrosourea, alkylate O-6 of guanine. The products appear to be relatively mutagenic. This is especially true for the O-6-ethyl derivative, which pairs abnormally with thymine

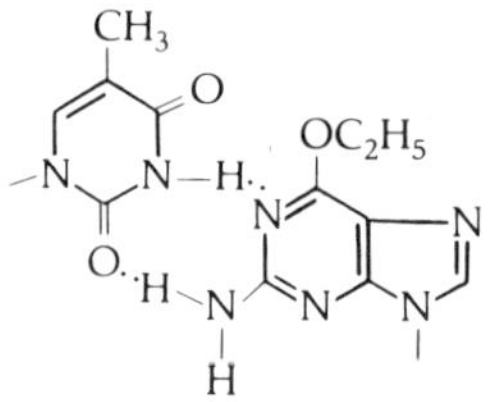

Figure 3 Pairing of O-6-ethyl guanine with thymine.

(Figure 3; Lawley & Martin, 1975). An atom alkylated appreciably by most agents is N-3 of adenine. Atoms alkylated to only a small extent are N-3 of guanine, N-1 and N-7 of adenine, N-3 and O-6 of thymine, and N-3 of cytosine (Ludlum, 1967). It is difficult to assess the biological importance of these minor alkylations. Table 1 shows the distribution of DNA alkylation sites when He La cells in culture are treated with three different reagents.

As already noted, the crosslinking of double helical DNA by BCNU and related nitrosoureas is thought to involve the initial chloroethylation of N^4 of cytidine (Kohn, 1977). The resulting species is close enough to a guanine in the complimentary strand that it should be able to alkylate the guanine O^6-position without causing a significant purturbance in the DNA structure (Equation 7; Lown & McLaughlin, 1979).

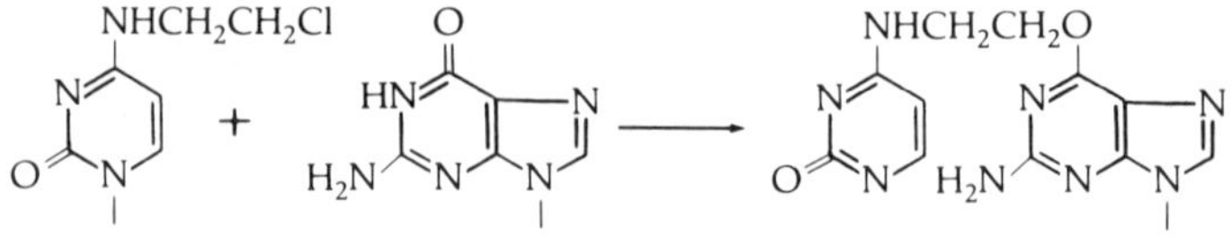

$$\quad (7)$$

The effect of histones in the nucleosome core particle on methylation of purines in nucleosomal DNA from rooster erythrocytes has been investigated. The investigators found that almost no differences existed between the reaction of dimethyl sulfate with N-7 of guanine or N-3 of adenine in nucleosomal DNA and that in naked (histone-free) DNA. These results suggest

Table 1 Alkylation Sites in DNA of He La Cells

Reagent	Guanine (%)			Adenine (%)			Cytosine (%)
	N-3	O-6	N-7	N-1	N-3	N-7	N-3
$(C_2H_5)_2SO_4$	0.2	1.6	71	0.3	4	0.5	0.5
$C_2H_5OSO_2CH_3$	0.4	0.3	81	0.1	2.2	0.7	0.1
$C_2H_5NONHCONH_2$	1.5	7.5	17	0.1	4.6	1.3	0.3

that the bases of DNA in nucleosomes are remarkably accessible (McGhee & Felsenfeld, 1979).

Alkylation of the bases in RNA follows a pattern similar to that found for DNA; N-7 of guanine is the predominant site of alkylation and the other ring-nitrogens of guanine, adenine, and cytosine, plus O-6 of guanine, are alkylated. In contrast to the DNA pattern (Table 1) the extent of alkylation of N-1 of adenine and N-3 of cytosine exceeds that of N-3 of adenine (Singer, 1974).

Enzymes and other proteins react with alkylating agents, and α-methylenelactones bind with enzymes such as phosphofructokinase and glycogen synthetase by a rapid Michael-type addition of nucleophilic groups present in the enzymes. In vitro α-methylenelactones readily alkylate the thiol groups of cysteine and reduced glutathione. Busulfan also reacts with the sulfhydryl groups of glutathione and various proteins by removing the sulfur atom from these molecules in a process known as "sulfur stripping." In this process, as exemplified by cysteine, bisalkylation of the sulfur atom produces a cyclic sulfonium compound (Scheme 7; Parham & Wilbur, 1961). These compounds are stable in vitro at physiological pH, but in vivo they decompose to tetrahydrothiophene and α-aminoacrylic acid, possibly in an enzymic reaction. The latter compound is hydrolyzed to pyruvate, whereas the former is metabolized mainly to 3-hydroxytetrahydrothiophene-1,1-dioxide (Roberts & Warwick, 1961).

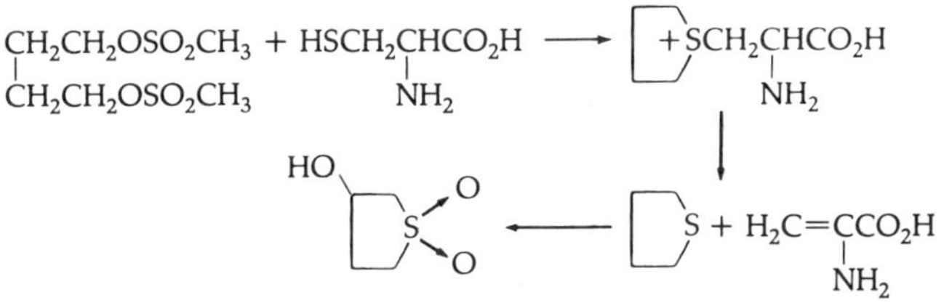

$$\begin{array}{c} CH_2CH_2OSO_2CH_3 \\ | \\ CH_2CH_2OSO_2CH_3 \end{array} + HSCH_2\underset{\underset{NH_2}{|}}{C}HCO_2H \longrightarrow \left[\right]\overset{+}{S}CH_2\underset{\underset{NH_2}{|}}{C}HCO_2H$$

Scheme 7 Interaction between busulfan and cysteine.

The repair enzyme, DNA nucleotidyl transferase of L-1210 leukemia cells, is inhibited by BCNU, CCNU, and 2-chloroethylisocyanate. It appears, however, that the inhibition results from carbamoylation rather than alkylation. It is known (Scheme 1) that BCNU and CCNU generate alkyl isocyanates during their decomposition. When CCNU was labeled with ^{14}C in its 2-chloroethyl group nucleic acids were labeled more than proteins. In contrast, CCNU with ^{14}C in its cyclohexyl group labeled proteins more than nucleic acids. Furthermore, 1-(2-chloroethyl)-1-nitrosourea, which cannot form an alkylisocyanate, does not inhibit the nucleotidyl transferase (Wheeler, 1974).

Alkylating Agent Design and Structure-Activity Relationships

Many ingenious approaches have been taken to the design of alkylating agents that might have improved therapeutic properties. Unfortunately, a number of them have been unfruitful or they have led to the discovery of new drugs by serendipity. In contrast to the antimetabolites, alkylating agents have not readily yielded to rational design. Most of the attempts at design have been based on real or supposed chemical and biological differences between normal and neoplastic cells. Among the real differences are greater reducing capacity and lower pH for certain cancer cells. There also are differences in certain enzyme levels and transport capabilities.

Differences in the rate and mechanism of action of alkylating agents have been explored. In principle, it might be expected that agents such as busulfan acting by the Sn2 mechanism would be more selective toward strong nucleophiles, whereas agents such as chlorambucil acting by the Sn1 mechanism would be more reactive in regions of high dielectric constant (Ross, 1962). Although this is probably so, it has made little practical difference in the antitumor activities of these compounds.

In the case of nitrogen mustards the high reactivity of mechlorethamine was moderated by replacing the methyl group with a benzene ring (Haddow et al., 1948). As discussed earlier, this replacement changes the overall reaction from second to first order. The reactivity of aryl mustards can be increased somewhat by introducing electron-releasing substituents such as NH_2, OH, and SH at the para-position (Ross, 1962). Replacement of the methyl group of mechlorethamine by hydrogen produces the compound known as nor-nitrogen mustard, noteworthy in that it is not sufficiently basic to be protonated under physiological conditions (Ross, 1962). Even the aziridinium ion formed by its cyclization loses the proton at pH 7.4 (Equation 8). Consequently, nor-nitrogen mustard should diffuse readily through cell membranes and be protonated to a greater extent in the more

$$HN(CH_2CH_2Cl)_2 \longrightarrow H\overset{\triangledown}{N}CH_2CH_2Cl \underset{H^+}{\overset{-H^+}{\rightleftharpoons}} \overset{\triangledown}{N}CH_2CH_2Cl \qquad (8)$$

acidic environment of neoplastic cells. Despite these theoretical advantages, it shows only a small increase in activity over mechlorethamine (Table 2).

The decreased reactivity of aryl nitrogen mustards allows them to be given orally, but their water solubility is too low for parenteral administration. The solubility problem was solved by introducing a carboxylic acid group into the molecule. Direct substitution of this group onto the aromatic nucleus caused strong deactivation, but when the functionalities were separated by methylene groups the activity was restored. The optimal number of methylene groups is three in benzene analogs, which produces the clinical agent chlorambucil (Figure 4; Everett et al., 1953).

Chemical reactivity of nitrogen mustards is increased by the presence of α-methyl groups in the alkylating chains (Everett & Ross, 1949) or by the replacement of chlorine with bromine; for example, $HSC_6H_4N(CH_2CH_2Br)_2$ is more active than the corresponding dichloro compound against rodent tumors (Benn, 1958). However, there are no clinical agents with any halogen other than chlorine. Increasing the distance between the nitrogen and chlorine atoms of nitrogen mustards drastically reduces the alkylating activity because an aziridinium ion cannot be formed. The mechanism of action becomes a simple Sn2 displacement of chloride by the nucleophile (Kon & Roberts, 1950).

Although monofunctional nitrogen mustards can alkylate nucleic acids, all the successful clinical agents are difunctional. This factor probably is based on the need for crosslinking of the double helix. Thus 2-chloroethylamine is inactive, whereas nor-nitrogen mustard has good antitumor activity (Burchenal et al., 1948). Studies directed toward determining the optimal distance between alkylating centers in compounds wherein both 2-chloroethyl groups were not on the same nitrogen atom revealed that the biological activity was greatest when the two centers were separated by two or three methylene groups (Kon & Roberts, 1950). This relationship also held when the two nitrogens were in a piperazine ring, as in 1,4-bis(2-chloroethyl) piperazine (Figure 4; Burchenal et al., 1949). The recently isolated antibiotic named piperazinedione (Figure 4) presents a possible exception to this generalization because the two alkylating centers are rather widely separated (Gitterman et al., 1970). It also complicates our interpretation of the alkylation process by β-chloroethylamines. Thus the diaziridinyl compound (Figure 4) that would be a likely intermediate for alkylation was synthesized and found to be less cytotoxic than the antibiotic (Brockman et al., 1976). The actual mode of action of piperazinedione remains to be elucidated.

Table 2 Antitumor Activity of Nitrogen Mustard Analogs Against L1210 Murine Leukemia[a]

Name	Structure	Optimal Dose (mg/kg)	Increase in Life Span (% T/C)
Nitrogen mustard	$CH_3N(CH_2CH_2Cl)_2$	3	19
Nor-nitrogen mustard	$HN(CH_2CH_2Cl)_2$	85	28
2-Chlorotriethylamine	$(C_2H_5)_2NCH_2CH_2Cl$	1.8	10
Glycine mustard	$HO_2CCH_2N(CH_2CH_2Cl)_2$	19	32
Melphalan (L-isomer)		60	33
medphalan (D-isomer)		23	33
merphalan (racemate)	$HO_2CCH(NH_2)CH_2\text{-}C_6H_4\text{-}N(CH_2CH_2Cl)_2$	23	40
p-Benzoic acid mustard	$HO_2C\text{-}C_6H_4\text{-}N(CH_2CH_2Cl)_2$	—	0
Chlorambucil	$HO_2C(CH_2)_3\text{-}C_6H_4\text{-}N(CH_2CH_2Cl)_2$	—	0
Tryptophan mustard	$HO_2CCH(NH_2)CH_2\text{-indole-}N(CH_2CH_2Cl)_2$	3	41
Uracil mustard	uracil-$N(CH_2CH_2Cl)_2$	3	48
Benzimidazole mustard	benzimidazole-$CH_2N(CH_2CH_2Cl)_2$	25	15
Phosphoramide mustard	$H_2N\text{-}P(=O)(OH)\text{-}N(CH_2CH_2Cl)_2$	65	47
Cyclophosphamide	cyclic-$P(=O)(NH)\text{-}N(CH_2CH_2Cl)_2$	60	145
DL-Erythritol mustard	$CH_2NHCH_2CH_2Cl$–$HCOH$–$HCOH$–$CH_2NHCH_2CH_2Cl$	23	88
D-Mannitol mustard	$CH_2NHCH_2CH_2Cl$–$HOCH$–$HOCH$–$HCOH$–$HCOH$–$CH_2NHCH_2CH_2Cl$	65	30
Chloroquine mustard	7-Cl-quinoline-$NHCH(CH_3)(CH_2)_3N(CH_2CH_2Cl)_2$	—	0

[a] Abstracted from Goldin & Wood, 1969.

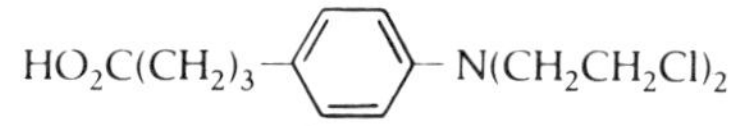

HO₂C(CH₂)₃—⟨C₆H₄⟩—N(CH₂CH₂Cl)₂
Chlorambucil

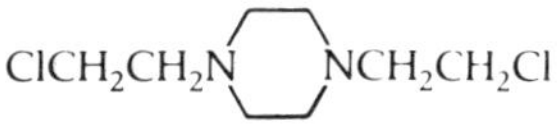

ClCH₂CH₂N⟨   ⟩NCH₂CH₂Cl
Bis(2-chloropiperazine)

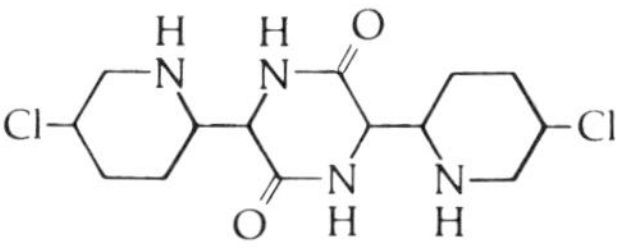

Piperazinedione

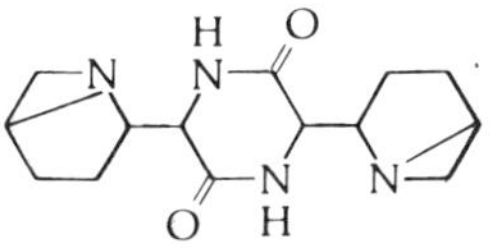

Diaziridine analog of piperazinedione

Figure 4 Separation of centers in alkylating agents.

The comparative activities of some simple nitrogen mustard derivatives against L-1210 leukemia in mice are listed in Table 2. It should be noted that none of these compounds produces significantly greater prolongation of life than nitrogen mustard itself.

Only a limited number of analogs of sulfur mustard have been prepared because this structure does not readily lend itself to modification. Bifunctionality is required, as shown by the inactivity of ethyl 2-chlorethyl sulfide compared with sulfur mustard (Figure 5; Ross, 1962). For sulfur mustard analogs in which the two 2-chloroethyl substituents are on different sulfur atoms, the maximum activity is obtained when three methylene groups separate the sulfur atoms (Gasson et al., 1948). Sulfur mustards of very short half-life in physiological solutions have been devised for intraarterial infusion, although none of them has found clinical use. A typical example is 2,3-bis[(2-bromoethyl)thio]-1-propyl 4-carboxyphenyl ether (Figure 5; Davis & Ross, 1965).

The simplest nitrosoureas, such as methylnitrosourea and ethylnitrosourea (Figure 6), are highly mutagenic and carcinogenic (Drucker et al.,

CH₃CH₂SCH₂CH₂Cl

ClCH₂CH₂SCH₂CH₂Cl
Sulfur mustard

ClCH₂CH₂S(CH₂)ₙSCH₂CH₂Cl

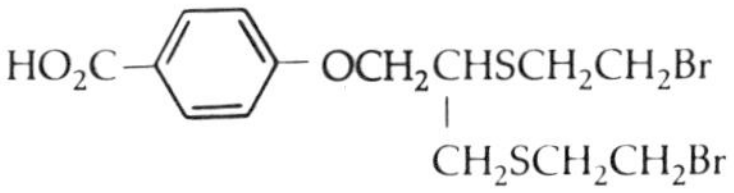

HO₂C—⟨C₆H₄⟩—OCH₂CHSCH₂CH₂Br
                               |
                        CH₂SCH₂CH₂Br

Figure 5 Analogs of sulfur mustard.

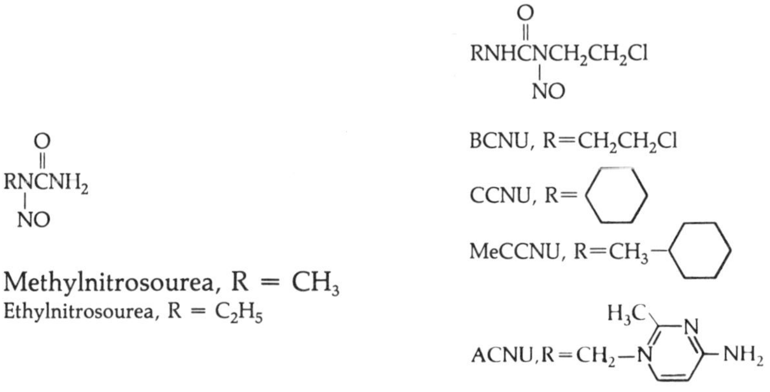

$$\text{RNHCNCH}_2\text{CH}_2\text{Cl} \text{ (C=O)}$$
$$\underset{\text{NO}}{|}$$

$$\underset{\text{NO}}{\overset{\text{O}}{\underset{|}{\text{RNCNH}_2}}}$$

BCNU, R=CH₂CH₂Cl

CCNU, R=⟨cyclohexyl⟩

MeCCNU, R=CH₃–⟨cyclohexyl⟩

Methylnitrosourea, R = CH₃
Ethylnitrosourea, R = C₂H₅

ACNU, R=CH₂–N=⟨H₃C-triazine⟩–NH₂

Figure 6 Nitrosoureas.

1967), but more complex analogs like BCNU and CCNU have much higher antitumor activity in relation to their mutagenicity (Montgomery et al., 1970). BCNU and CCNU covalently crosslink DNA under physiological conditions and the extent of this crosslinking correlates with the activity against L-1210 murine leukemia (Ludlum et al., 1975; Lown et al., 1978). The rate of crosslinking increases with pH in the range 4–10 and with the (G+C) content of the DNA. Only the 1-(2-chloroethyl)-1-nitrosoureas have substantial alkylating ability (Table 3). Other halogens, although more reactive to nucleophilic displacement, decompose by pathways that do not produce DNA alkylation. The corresponding 3-chloropropyl compounds are unreactive toward DNA and the 2-chloropropyl analogs are much less reactive than the 2-chloroethyl (Lown et al., 1978). Based on these observations, a compound that combined the cyclohexylnitrosourea feature with the functionality of a sulfur mustard (Scheme 8) was prepared. This derivative was expected to cleave to cyclohexylisocyanate and (chloroethyl)thioethyldiazo hydroxide (among other products). The latter species could alkylate DNA on decomposition of the diazo group and the resulting sulfur mustard derivative could effect the second alkylation by the usual process which involves a sulfonium ion (Scheme 8). In fact, the derivative produced 90% crosslinking of λ-DNA in 10 min under physiological conditions. It also was highly active against L-1210 leukemia (Lown et al., 1980).

The relatively high lipophilicity of certain nitrosoureas has led to their use in tumors of the central nervous system. CCNU and methyl CCNU (Figure 6) are noteworthy in this respect (Johnston et al., 1971). It also has been possible to prepare water-soluble nitrosoureas, such as ACNU (Figure 6), that are

Table 3 Antitumor Activity of Nitrosoureas Against L1210 Murine Leukemia[a]

Name	Structure	Optimal Dose (mg/kg)	Increase in Life Span (% T/C)
1-Methyl-1-nitrosourea	$H_2NCONHCH_3$, NO	52	40
1-(2-Chloroethyl)-1-nitrosourea	$H_2NCONHCH_2CH_2Cl$, NO	1.8	426
BCNU	$ClCH_2CH_2NHCONHCH_2CH_2Cl$, NO	5	429
CCNU	cyclohexyl-$NHCONHCH_2CH_2Cl$, NO	8.4	390
Streptozotocin	sugar-$NHCONHCH_3$, NO	50	31
Chlorozotocin	sugar-$NHCONHCH_2CH_2Cl$, NO	30	9/10 60-day survivors
GANU	sugar-O-$CONHCH_2CH_2Cl$, NO	8	512
5-[3-(2-Chloroethyl)-3-nitrosoureido]-1,2,3,4-cyclopentanetetrol	cyclopentanetetrol-$NHCONHCH_2CH_2Cl$, NO	16	6/6 30-day survivors

[a] Abstracted from Goldin & Wood, 1969.

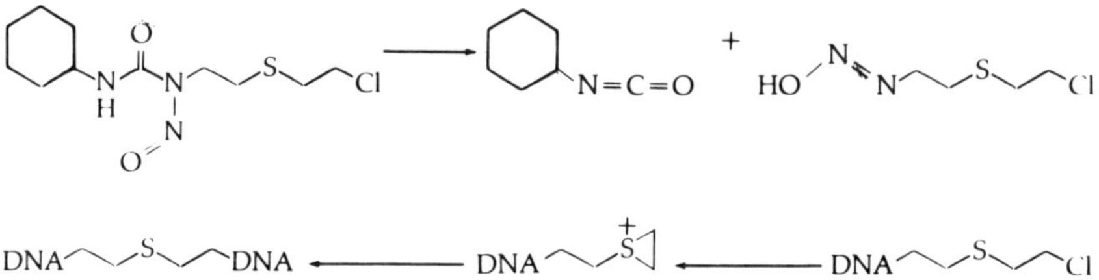

Scheme 8 Nitrosourea with sulfur mustard group.

of potential value against lung carcinoma (Saijo et al., 1978). As described subsequently, the addition of various carrier groups to the nitrosomethylurea unit has provided some highly active analogs.

Diepoxides were developed initially for use in textiles because of their ability to crosslink fibers. This property suggested possible antitumor activity if it could occur with biopolymers. A number of diepoxides have shown substantial antitumor activity, although none is used clinically (Table 4).

Table 4 Antitumor Activity of Methanesulfonates and Epoxides Against L1210 Murine Leukemia[a]

Name	Structure	Optimal Dose (mg/kg)	Increase in Life Span (% T/C)
Methyl methanesulfonate	$CH_3SO_2OCH_3$	—	0
Myleran	$CH_3SO_2O(CH_2)_4OSO_2CH_3$	—	0
Mannitol myleran	$CH_2OSO_2CH_3$—HOCH—HOCH—HCOH—HCOH—$CH_2OSO_2CH_3$	300	7
Diepoxybutane	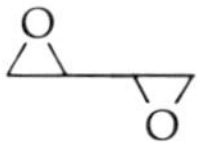 (diepoxybutane structure)	50	89[b]
Dianhydromannitol diepoxide	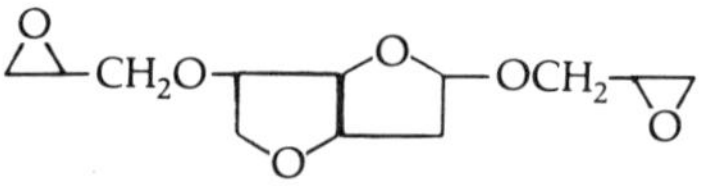 (dianhydromannitol diepoxide structure)	500	50

[a] Abstracted from Goldin & Wood, 1969.
[b] Early leukemia.

Epoxides alkylate nucleophiles by the Sn2 mechanism and this requires that the carbon atom that alkylates must be otherwise unsubstituted (Ross, 1962). Monoepoxides can be toxic but they lack antitumor activity. A series of diepoxides separated by different numbers of methylene groups was prepared. In this series the best activity was obtained when two methylene groups were present (Everett & Kon, 1950). A diepoxide containing basic nitrogen atoms (Figure 7) was designed on the premise that protonation of the nitrogens would activate the epoxides toward attack by nucleophiles (Gerzon et al., 1959).

The cytotoxicity of ethylenimine (aziridine) was first described by Ehrlich (Himmelweit, 1956). Ethylenimines react by an Sn2 mechanism similar to that of the epoxides and in appropriately substituted compounds the rates of reaction can be comparable; for example, triethylenemelamine (Figure 8) reacts with thiosulfate ion at a rate comparable to those of active diepoxides (Ross, 1950). One advantage that the ethylenimines offer for drug design is that the nitrogen atom can be substituted with a variety of functional groups, leaving the methylene groups unhindered to nucleophilic attack. Agents of clinical interest, such as triethylenemelamine and carbazilquinone (Figure 8), typically have this structural feature. In general, two ethyleneimine groups per molecule are required for antitumor activity. Compounds with three or four of these groups are not significantly more active (Table 5). A few monoaziridines are active [e.g., tetramin (Figure 8; White, 1959)], but it has been suggested that this compound can give difunctional alkylation, perhaps by bioconversion to an epoxide (Ross, 1962). Many aziridinyl benzoquinones show significant antitumor activity. Among them trenimon is especially potent (Hartlieb, 1974). By proper addition of lipophilic substituents to the aziridinyl benzoquinones antitumor activity can be extended to tumors of the

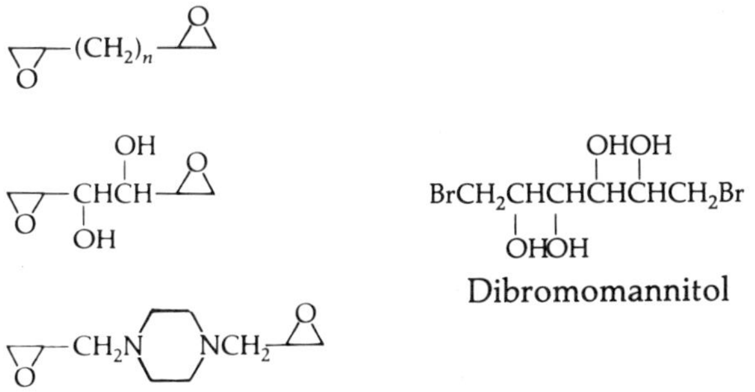

Figure 7 Diepoxides.

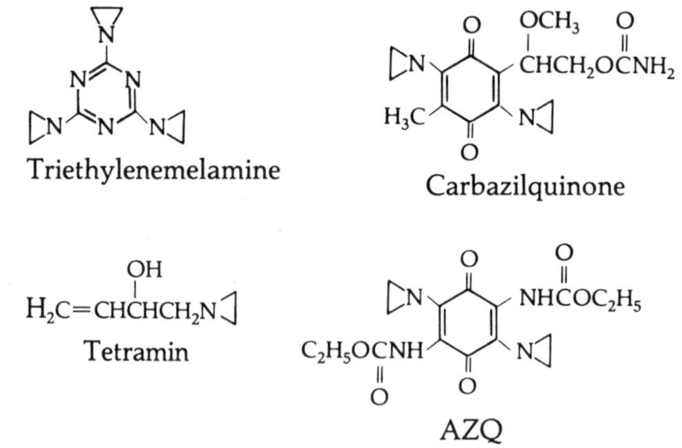

Figure 8 Aziridines.

central nervous system (Khan & Driscoll, 1976). A leading compound of this type is AZQ (Figure 8; Table 5).

Monomethanesulfonates show antitumor activity but are strongly mutagenic (Walpole, 1958). Bismethanesulfonates such as myleran (Figure 9) are about 10 times more potent as antitumor agents. In a homologous series of compounds, in which the methanesulfonyloxy groups were separated by methylene groups, the most active compounds were obtained with four or more methylene groups (Ross, 1962). A mutually deactivating effect occurred when fewer methylene groups separated the methanesulfonyloxy groups (Timmis, 1958). It was proposed earlier that bismethanesulfonates acted by way of cyclic sulfonium structures. This proposal, however, is inconsistent with the good activity of a butynyl analog (Figure 9). Methyl groups on the carbon atoms that bear oxygen enhance the rate of hydrolysis of bismethanesulfonates but not their alkylating ability. This difference probably results from a change in mechanism from Sn2 to Sn1 (Ross, 1962); for example, if one has a neighboring methyl group, unequal reactivity of the two alkylating centers offers no advantages. A beta-hydroxyl group also enhances the rate of hydrolysis at pH 7, probably because of epoxide formation. Dimethylmyleran (Figure 9) increased neutrophil depressant activity but decreased antitumor activity relative to myleran (Timmis, 1958). Mannitolmyleran (Figure 9; Table 4) retained the antitumor activity but caused no depression of neutrophils. It is not certain whether these effects are due to changes in the mechanism of alkylation or to lipophilic effects (Ross, 1962).

Table 5 Antitumor Activity of Ethylenimines Against L1210 Murine Leukemia[a]

Name	Structure	Optimal Dose (mg/kg)	Increase in Life Span (% T/C)
TEPA	[N–P(=O)–N], N (aziridinyl)	5.2	55
Thio-TEPA	[N–P(=S)–N], N (aziridinyl)	6.5	55
Uredepa	[N–P(=O)–NHCOC$_2$H$_5$], N (aziridinyl)	15	78
Trenimon	[N]-quinone-NHCOC$_2$H$_5$, [N]	0.23	20
AZQ	[N]-quinone-NHCOC$_2$H$_5$, C$_2$H$_5$OCNH, [N]	1.0	70[b]
Carbazilquinone	[N]-quinone-CH(OCH$_3$)CH$_2$OCNH$_2$, H$_3$C, [N]	0.5	100
Tetramin[c]	[NCH$_2$CH(OH)CH=CH$_2$]	14	40
1,4-bis(1-azirdinyl)-2,3-butanediol	[NCH$_2$CH(OH)CH(OH)CH$_2$N]	0.62	77
Triethylenemelamine	[N]-triazine-[N], N, aziridinyl	1.15	57

[a] Abstracted from Goldin & Wood, 1969.
[b] QD 1-9 treatment schedule. Chou et al., 1976.
[c] A mixture containing 70% of the depicted compound.

CH₃SO₂O(CH₂)₄OSO₂CH₃
Myleran

CH₃SO₂OCH₂C≡CCH₂OSO₂CH₃
Butynyl analog

CH₃SO₂OCH(CH₂)₂CHOSO₂CH₃
      | |
      CH₃   CH₃

CH₃SO₂OCH₂(CHOH)₄CH₂OSO₂CH₃
Mannitol myleran

Figure 9 Bismethanesulfonates.

Carbinolamines, also known as aminals, alkylate nucleophiles by a process in which the hydroxyl group is replaced by the nucleophile (see Equation 9). The imine resulting from loss of the elements of water from the carbinolamine might be an intermediate in this process. The pyrrolo (1,4) benzodiazepine group of antibiotics alkylate DNA by this means. This group includes anthramycin, mazethramycin, sibiromycin, tomaymycin and neothramycins A and B (Hurley, 1977; Miyamoto et al., 1977; Kunimoto et al., 1980). As shown in Figure 10, each of these antibiotics contains the same nucleus and carbinolamine function except for the neothramycins which are isolated in the anhydro form. They differ in their side chains, pyrrole-ring substituents, and benzene-ring substituents. Anthramycin and sibiromycin, the most studied compounds, showed clinical activity against gastrointestinal and

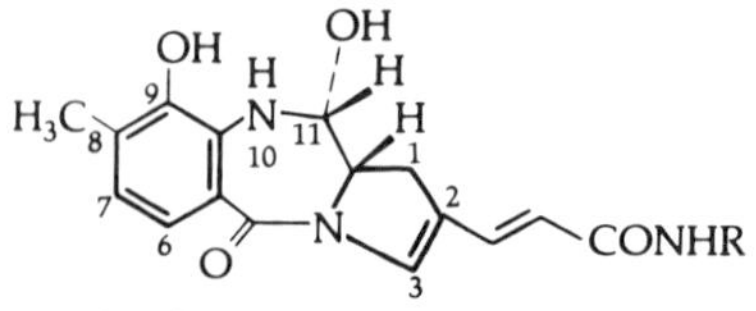

Anthramycin, R = H
Mazethramycin, R = CH₃

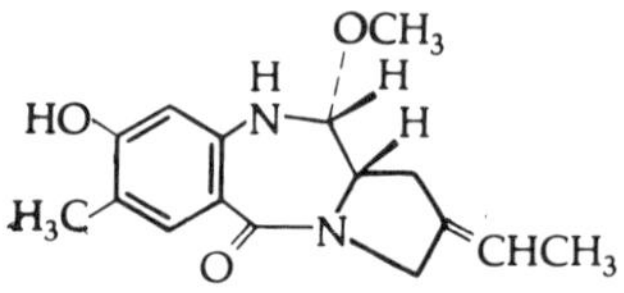

Tomaymycin

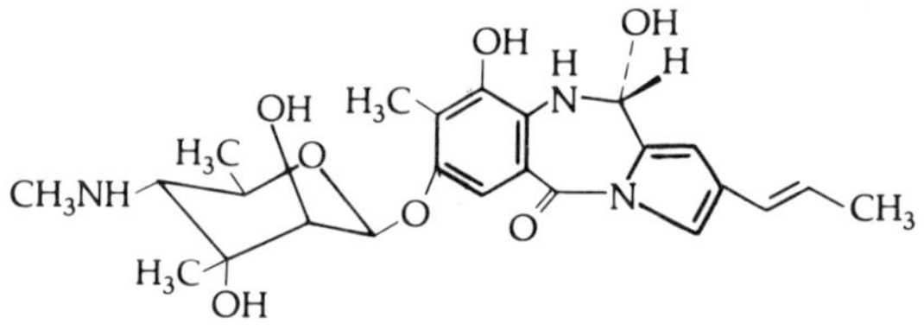

Sibiromycin

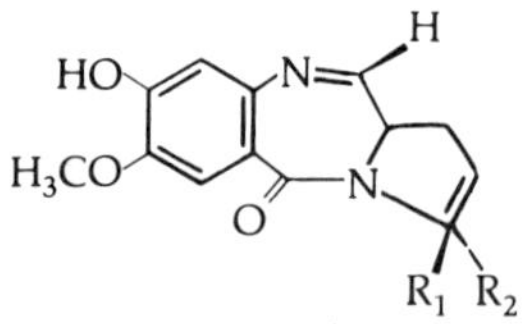

Neothramycin A: R_1 = H, R_2 = OH
             B: R_1 = OH, R_2 = H

Figure 10 Structures of the pyrrolo(1,4)benzodiazepines.

breast tumors, lymphomas, and sarcomas. Cardiotoxicity was serious, although no bone marrow toxicity was observed (Korman & Tendler, 1965.

The pyrrolo (1,4) benzodiazepines produce their antitumor effects by inhibiting DNA-directed RNA synthesis (Horwitz, 1971). This inhibition results from covalent binding with the 2-amino group of a guanine residue of the DNA (Equation 9). No crosslinking or intercalation occurs (Lown & Joshua, 1979). The covalent binding is acid-catalyzed, which suggests that it is promoted in tumor cells with low pH. Although the DNA adducts are stable in neutral and alkaline solution, they dissociate in acid. The stability order tomaymycin < anthramycin < sibiromycin reflects the relative strengths of hydrogen bonding between the antibiotics and DNA. Thus anthramycin can bind with its 9-hydroxyl group (Figure 10), whereas sibiromycin can bind with this group and its aminosugar (Hurley et al., 1977). Structure-activity relationships are being established for other analogs and derivatives. As anticipated, substituents that interfere with the fit onto DNA or the activity of the carbinolamine function destroy antitumor activity; for example, methylation of the 9-hydroxyl group, acylation of N-10, or oxidation of the 11-hydroxyl group to a carbonyl group results in inactive compounds (Horwitz et al., 1971). The minimum structure for reaction with DNA was reported to be the pyrrolo (1,4)-benzodiazepine nucleus with an 11-hydroxyl group (Lown & Joshua, 1979).

Metabolic hydroxylation of hexamethylmelamine to obtain the corresponding carbinolamine (methylol) has been identified as the means by which the relatively inert starting compound is converted into a alkylating agent. Loss of hydroxide from the methylol would generate the strongly electrophilic iminonium ion (Sanders et al., 1982; Figure 11). The vinylogous carbinolamine functionality has been suggested as one of the two alkylating centers of the pyrrole metabolite (Figure 11) of pyrrolizidine alkaloids such as heliotrine and lasiocarpine (Anderson & Corey, 1977a). This functionality also occurs in the indolohydroquinone metabolites of mitomycins (Figure 11). Although these particular structures are too unstable for prolonged existence, it has been possible to design and prepare analogs of the 5-phenyl-2,3-dihydro-6,7-bis(hydroxymethyl)-1H-pyrrolizine dicarbamate type (Anderson & Corey, 1977a) and the related 1-phenyl-2,5-dimethyl-3,4-bis(hydroxymethyl) pyrrole dicarbamate type (Anderson & Corey, 1977b; Figure 11). Examples of both types have shown activity in a variety of animal tumors.

α-Methylenelactones have been of interest in the design of potential antitumor drugs because this functionality occurs in a variety of natural products with antitumor activity (Kupchan et al., 1971). Many novel structures that contain one or more α-methylenelactone groups have been synthesized and have featured variations in substituents on the α-methylenelactone, the ring size of the lactone, and the distance between alkylating centers (Howie

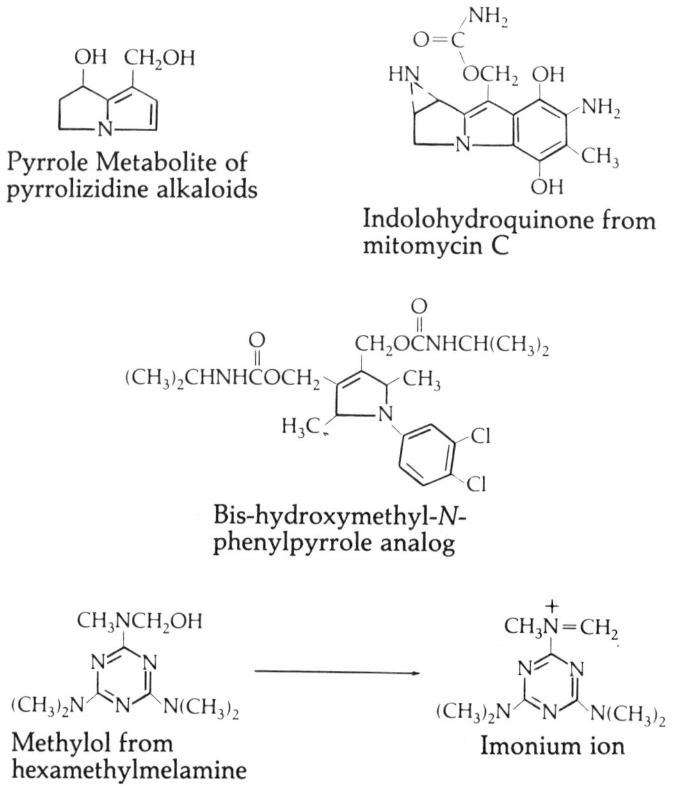

Pyrrole Metabolite of pyrrolizidine alkaloids

Indolohydroquinone from mitomycin C

Bis-hydroxymethyl-N-phenylpyrrole analog

Methylol from hexamethylmelamine

Imonium ion

Figure 11   Carbinolamine analogs.

et al., 1976; Rosowsky et al., 1974). Unfortunately, none has shown really significant activity.

From the preceding discussion it may be concluded that we know a variety of highly reactive functional groups capable of alkylating and killing cancer cells. The problem with all the agents that contain these groups is that they are toxic to normal cells, especially the rapidly dividing cells of bone marrow, the gastrointestinal tract, and certain other tissues. This lack of selective toxicity has led scientists to devise molecules that might take advantage of real or imagined differences in permeability, transport phenomena, or enzyme levels between normal and cancer cells. One design feature that has received considerable attention is the "carrier group." The theory of carrier groups holds that because cancer cells are able to compete effectively with normal cells for nutrients such as amino acids, carbohydrates, and nucleosides attachment of alkylating groups to these nutrients will result in the selective accumulation of toxic molecules in the tumor cells. Although much of the design based on carrier groups has been simplistic or even based on false premises, some important anticancer drugs, including melphalan and uracil mustard, have resulted from this research.

The role of phenylalaline in the biosynthesis of melanin pigments suggested that melanoma cells might have a high requirement for this amino acid. Thus alkylating agents that incorporate it into their structures may show selective toxicity to melanomas. Enantiomers of the structure in which the para position of the benzene ring of phenylalanine was substituted with the di(2-chloroethyl)amino function (Figure 12) were prepared and tested against various neoplasms (Bergel & Stock, 1953). Contrary to expectation, they were not active against melanomas, but they were effective in the treatment of multiple myeloma. More recently the L-enantiomer, known as melphalan, has been used against breast cancer. The corresponding D-enantiomer, named medphalan is approximately equal in activity to melphalan (Schmidt et al., 1965) and the racemic mixture, named merphalan or sarcolysin, is similarly effective (Table 2; Larionov et al., 1955). This lack of enantiomeric specificity contradicts the idea of selective active transport into cancer cells. Once it was established that melphalan was active against neoplasms a variety of derivatives was prepared with the objective of improving the therapeutic properties. Among the more interesting of these analogs are the pentapeptide and tripeptide derivatives (Bergel et al., 1959; Larionov, 1960). Related amino acid analogs that show antitumor activity include dihydroxyphenylalanine mustard (Vasil' Eva, 1970) and tryptophan mustard (Fishbein, 1964; Figure 12; Table 2).

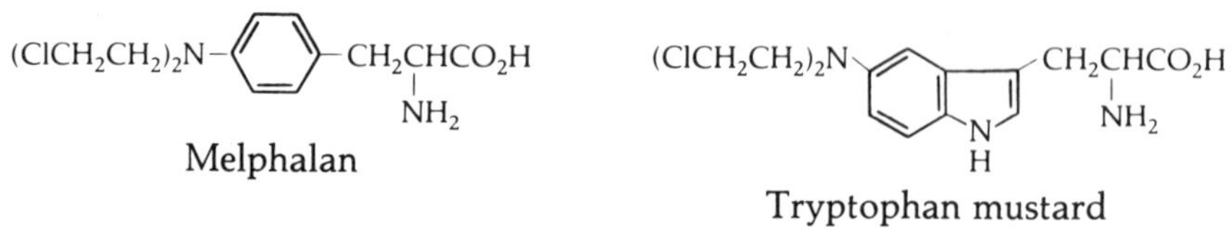

(ClCH$_2$CH$_2$)$_2$N—⟨C$_6$H$_4$⟩—CH$_2$CHCO$_2$H
        |
        NH$_2$
   Melphalan

(ClCH$_2$CH$_2$)$_2$N—[indole]—CH$_2$CHCO$_2$H
           |
          NH$_2$
       Tryptophan mustard

Figure 12 Amino acid nitrogen mustards.

Carbohydrates bearing most of the known alkylating groups have been synthesized and tested on the premise that they might be actively transported into cancer cells. However, alkylating carbohydrates, with the notable exception of the antibiotic streptozotocin, show little specificity. Nitrogen mustards corresponding to a variety of reduced sugars have been prepared; for example, D-mannitol mustard (Vargha et al., 1957) and DL-erythritol mustard (Figure 13; Table 2). The tetracetyl derivative of glucosamine mustard (Figure 13) shows antitumor activity (Schmidt et al., 1965), which indicates that active transport into the tumor cell is not a factor, at least for this compound. Dibromo derivatives such as dibromomannitol and dibromodulcitol are active against experimental tumors. Studies in vitro show that these compounds readily form the corresponding diepoxides (Figure 7; Jarman & Ross, 1967). The distance between the two epoxides is the same as that found to be optimal for the diepoxyalkanes shown in Figure 7. These diepoxides are presumed to be the alkylating species, although their existence has not been verified in vivo. Bismethanesulfonates such as D-mannitol myleran have substantial antitumor activity (Brown & Timmis, 1958; Table 4). As mentioned before, rate enhancement in the hydrolysis of the methanesulfonyl groups of these compounds suggests that epoxide intermediates are formed by participation of the neighboring hydroxyl groups. In contrast to the good activity of D-mannitol myleran against the

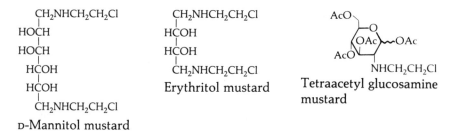

Figure 13 Carbohydrate nitrogen mustards.

Walker tumor, the corresponding L-enantiomer is inactive in this system. Thus a high degree of stereospecificity exists for this particular molecule (Brown & Timmis, 1959).

The antibiotic streptozotocin has a structure (Figure 14) in which the 2-hydroxyl group of D-glucose is replaced by the nitrosomethylurea functionality (Herr et al., 1967). This compound is taken selectively into the β-cells of the pancreas, where its cytotoxicity is expressed as diabetogenesis. However, this property makes it useful against malignant insulinomas (Kennedy, 1970). Thus streptozotocin possesses in a limited sense the kind of carrier group specificity hoped for in drug design. Perhaps not surprisingly, small modifications in the structure of streptozotocin abolish the selectivity for pancreatic β-cells. Fortunately the antitumor spectrum is enhanced at the same time that diabetogenesis is decreased. Thus the analog chlorozotocin (Figure 14; Table 3) shows attractive possibilities for cancer chemotherapy (Johnston et al., 1975). A number of newer analogs are based on the structure of chlorozotocin. Among the more interesting are the 1-β-D-glucosyl isomer (GANU; Hisamatsu & Uchida, 1977) and the 1-α-D-ribosyl analog (Fox et al., 1977). Certain of these analogs are highly active against L-1210 leukemia (Table 3).

The nitrogen mustard functionality has been attached to purines, pyrimidines, and their nucleosides (Ross, 1962). The resulting compounds have not shown the desired selective active transport into tumor cells, but this factor has not prevented some of them from having substantial antitumor activity (Table 2). Uracil mustard is an established clinical agent and a uridine derivative (Figure 15) shows promise (Belikova et al., 1967). A benz-

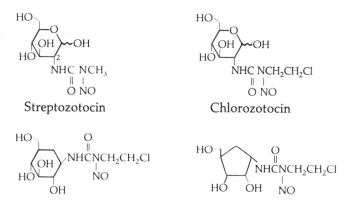

Figure 14 Streptozotocin and related analogs.

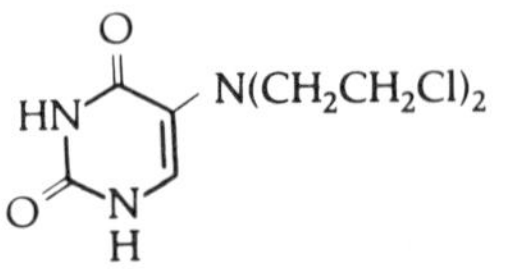

Uracil mustard

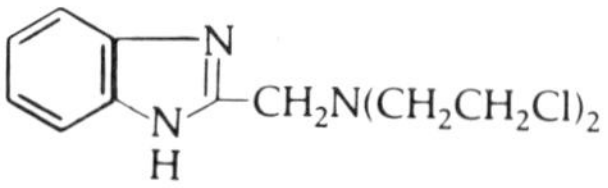

Benzimidazole mustard

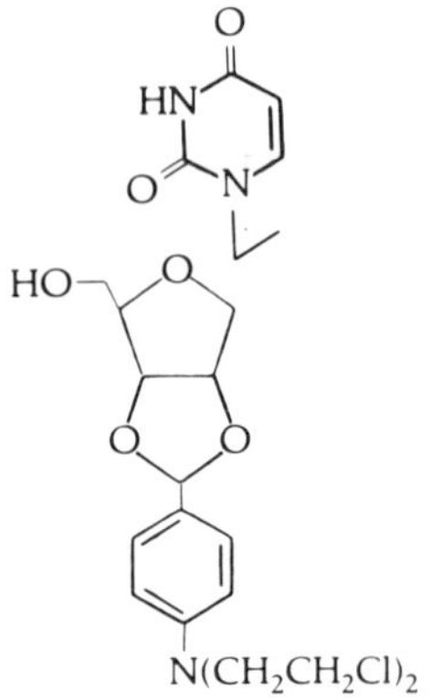

Uridine derivative

Figure 15 Purine and pyrimidine analogs.

imidazole derivative of nitrogen mustard (Figure 15), which has structural similarity to purines, also has antitumor activity (Hirschberg et al., 1958).

Antimalarial drugs have been combined with alkylating groups in the expectation that they would localize in the cell nucleus, especially of melanomatous tissues, and cause cytotoxicity. Chloroquine mustard and quinacrine mustard (Figure 16) are typical examples of these compounds. They have activity in animal tumor models (Table 2) but are not used clinically (Jones et al., 1968).

Various steroids have been utilized as potential carrier molecules for alkylating groups. Once again, the carrier principle does not appear to have succeeded, but many of the compounds have antitumor activity. The most successful alkylating functionality is p-N,N-bis(2-chloroethyl)amino phenylacetic acid, which is used to esterify hydroxyl groups of the steroids (Carrol et al., 1972). The esters of cholesterol (phenestrin, Figure 17) and estradiol have received clinical trials against breast cancer. The corresponding

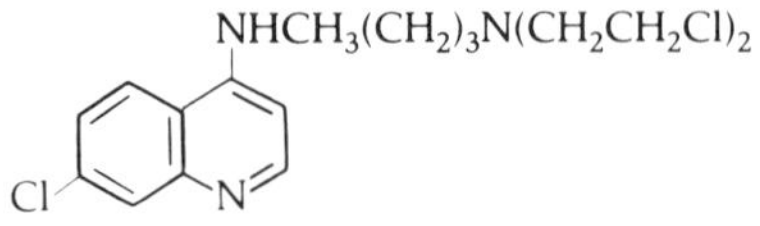

Chloroquine mustard

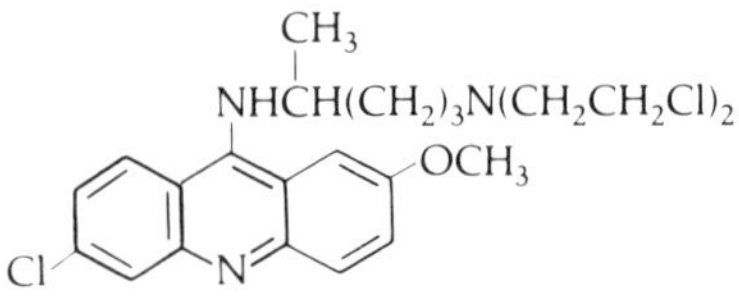

Quinacrine mustard

Figure 16 Antimalarial mustards.

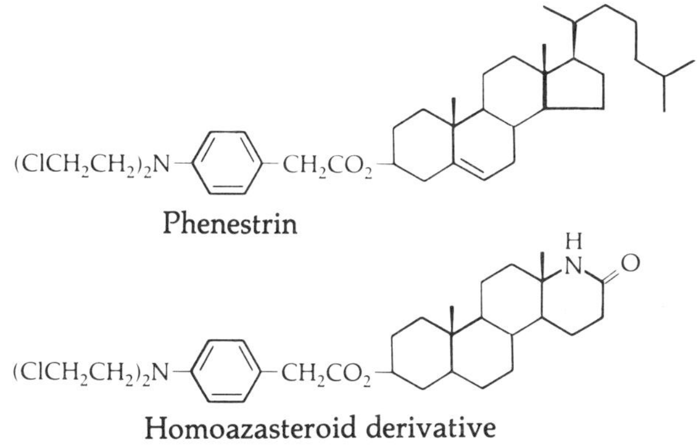

CHEMISTRY OF ANTITUMOR DRUGS 115

Phenestrin: (ClCH$_2$CH$_2$)$_2$N—C$_6$H$_4$—CH$_2$CO$_2$—[steroid]

Homoazasteroid derivative: (ClCH$_2$CH$_2$)$_2$N—C$_6$H$_4$—CH$_2$CO$_2$—[homoazasteroid]

Figure 17 Steroidal nitrogen mustards.

esters of pregnenolone, testosterone, and epiandrosterone are active as well. More recently the ester derivative of a homoazasteroid (Figure 17) has shown activity against mouse leukemias (Bukva & Gass, 1967).

As discussed above, one of the two major approaches to the design of alkylating agents has been the use of carrier groups that might promote selective uptake into cancer cells. The second major approach is based on latent activity. In this approach the alkylating agent is prepared in an inactive or less active form that can be converted into the active form by enzymic processes in the body. Selective toxicity to neoplastic cells is based on the supposition that they might contain higher levels of the activating enzymes than normal cells. The results obtained from drug design by the latent activity approach resemble those provided by the carrier group approach; that is, at least one important drug, cyclophosphamide, was discovered, but it is not activated by the mechanism for which it was designed.

Cells contain a variety of esterases, some of which appear to have elevated levels in certain tumors. A favorite alkylating agent for the study of selective esterase action has been *p*-hydroxyaniline mustard, which is a potent alkylating agent that can be converted into different types of less toxic ester. In an extensive series of carboxylic acid esters of this agent the most active antitumor agent was its *o*-methylbenzoate (Figure 18). This compound is thought to have the optimal rate of hydrolysis by esterases (Vickers et al., 1969; Bardos et al., 1969). The corresponding phosphate ester (Figure 18) was designed for use against prostatic carcinomas which have high phosphatase activity. Likewise, the sulfate ester (Figure 18) was designed

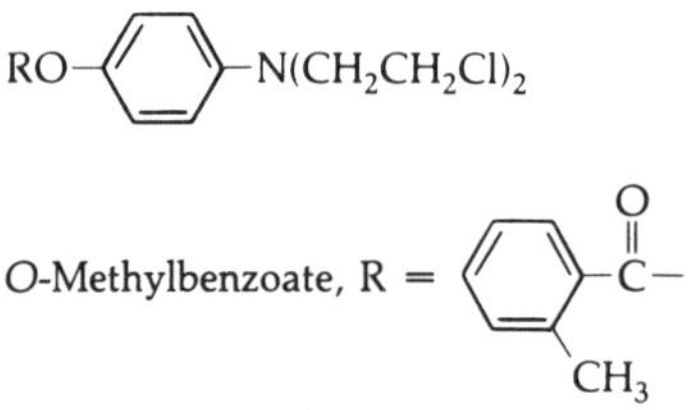

RO—⟨C₆H₄⟩—N(CH₂CH₂Cl)₂

O-Methylbenzoate, R = 2-methylbenzoyl (o-CH₃-C₆H₄-C(=O)-)

Phosphate, R = (HO)₂PO

Sulfate, R = HOSO₂

Figure 18 Esters with latent activity.

for carcinomas with high sulfatase activity. Both compounds have antitumor activity but lack the desired selectivity (Bukhari et al., 1971).

One interesting example of selective toxicity, discovered accidentally, was the high specificity of aniline mustard toward the PC5 mouse plasma cell tumor (Connors, 1969). This tumor has an extremely high level of β-glucuronidase. It is thought that aniline mustard is rapidly inactivated in the liver by conversion into the β-glucuronide of p-hydroxyaniline mustard. Normal mouse cells are not seriously damaged by this metabolite. However, the plasma cell tumor hydrolyzes it to highly toxic p-hydroxyaniline mustard (Equation 10).

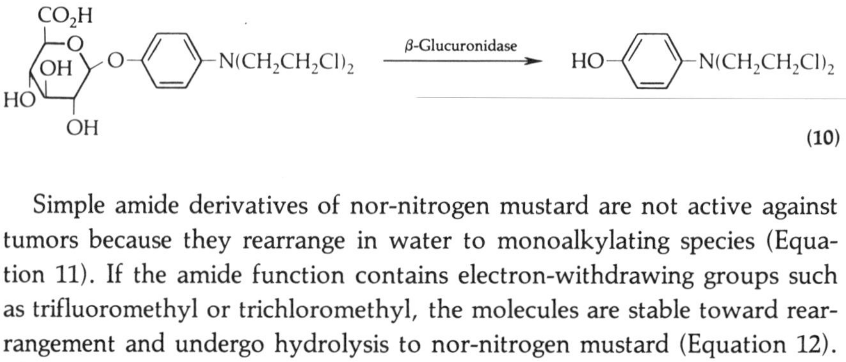

$$\text{Glucuronide-O—C}_6\text{H}_4\text{—N(CH}_2\text{CH}_2\text{Cl)}_2 \xrightarrow{\beta\text{-Glucuronidase}} \text{HO—C}_6\text{H}_4\text{—N(CH}_2\text{CH}_2\text{Cl)}_2 \quad (10)$$

Simple amide derivatives of nor-nitrogen mustard are not active against tumors because they rearrange in water to monoalkylating species (Equation 11). If the amide function contains electron-withdrawing groups such as trifluoromethyl or trichloromethyl, the molecules are stable toward rearrangement and undergo hydrolysis to nor-nitrogen mustard (Equation 12). They are active against the Walker rat carcinoma (Ross & Wilson, 1959). Sulfonyl derivatives are inactive because they resist hydrolysis. The p-acetylamino derivative of aniline mustard showed antitumor activity following hydrolysis to the p-amino analog; however, resistance developed

$$\text{RC(O)N(CH}_2\text{CH}_2\text{Cl)}_2 \xrightarrow{\text{H}_2\text{O}} \text{RC(O)OCH}_2\text{CH}_2\text{NHCH}_2\text{CH}_2\text{Cl} \quad (11)$$

$$\underset{\text{F}_3\text{CCN(CH}_2\text{CH}_2\text{Cl)}_2}{\overset{\text{O}}{\|}} \xrightarrow{\text{H}_2\text{O}} \underset{\text{F}_3\text{CCOH}}{\overset{\text{O}}{\|}} + \text{HN(CH}_2\text{CH}_2\text{Cl)}_2 \qquad (12)$$

rapidly in certain tumors. It was found that these resistant tumors had decreased peptidase activity (Danielli, 1954).

The concept of lower amidase activity in certain tumor cells led to the design of bifunctional sulfur mustards that contain amide bonds susceptible to enzyme hydrolysis. Because these compound would be more readily cleaved to the less toxic monofunctional sulfur mustards in normal cells (Equation 13), they were expected to show selective toxicity to the tumor cells. In fact, one of these compounds (Equation 13) was more active than mechlorethamine in animal studies and gave some objective responses in patients (Goodman et al., 1962).

$$\text{ClCH}_2\text{CH}_2\text{SCH}_2\overset{\overset{\text{O}}{\|}}{\text{C}}\text{NHCH}_2\text{CH}_2\text{NH}\overset{\overset{\text{O}}{\|}}{\text{C}}\text{CH}_2\text{SCH}_2\text{CH}_2\text{Cl} \xrightarrow{\text{Amidase}}$$

$$\text{ClCH}_2\text{CH}_2\text{SCH}_2\text{CO}_2\text{H} + \text{H}_2\text{NCH}_2\text{CH}_2\overset{\overset{\text{O}}{\|}}{\text{C}}\text{CH}_2\text{SCH}_2\text{CH}_2\text{Cl} \qquad (13)$$

Phosphoramide derivatives of nor-nitrogen mustards were prepared in accordance with the hypothesis that levels of phosphoramidases are higher in malignant cells. Thus preferential activation should occur in them. The earliest derivatives, such as the triamide and diamide esters shown in Figure 19, lacked significant antitumor activity (Friedman et al., 1954). The cyclic diamide ester, cyclophosphamide, showed impressive activity (Table 2; Arnold et al., 1958) and became one of the leading clinical agents. As discussed earilier (Scheme 6), cyclophosphamide is activated by hydroxylation rather than by direct amide hydrolysis.

Previously discussed examples have shown that tumors can become resistant to alkylating agents by increasing or decreasing their levels of certain hydroyltic enzymes. An ingenious method was developed to take ad-

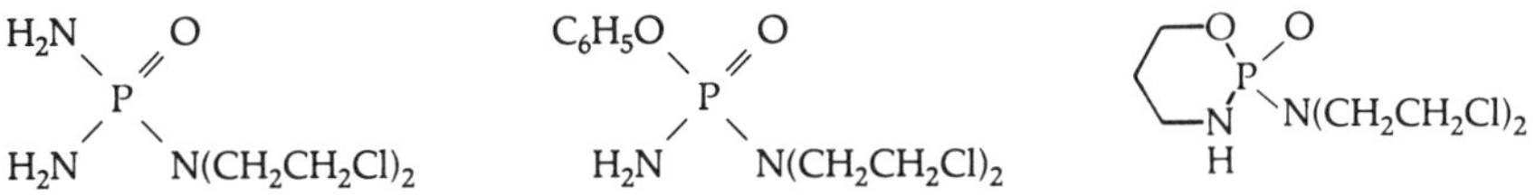

Triamide | Diamide ester | Cyclophosphamide

Figure 19 Phosphoramide mustards.

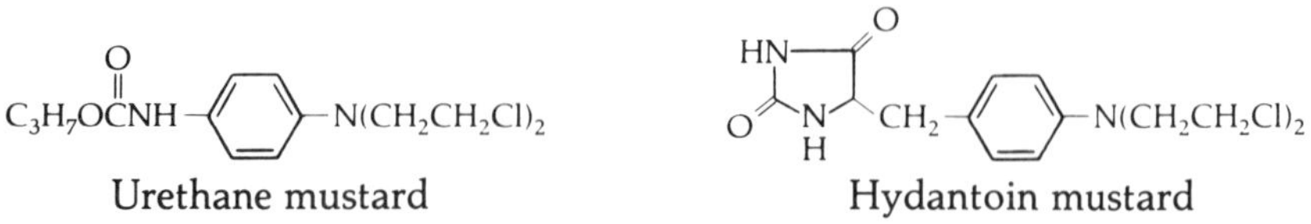

Urethane mustard: $C_3H_7OC(=O)NH\text{-}C_6H_4\text{-}N(CH_2CH_2Cl)_2$

Hydantoin mustard: hydantoin-$CH_2\text{-}C_6H_4\text{-}N(CH_2CH_2Cl)_2$

Figure 20 Urethane and hydantoin mustards.

vantage of the propensity of the Walker tumor to respond to aromatic urethanes by elaborating increased amounts of hydrolytic enzymes (Danielli, 1959). In this method tumor-bearing rats were fed the phenyl carbamate of propanol until the tumor enzyme levels were increased in relation to liver enzyme levels, and the corresponding urethane mustard (Figure 20) was given. A potent antitumor effect was produced in the rats. So far there has been no analogous system exploited successfully in human tumors.

The hydantoin mustards have proved to be one of the more disappointing classes of latent alkylating agent. They were designed according to the rationale that amino acid mustards such as melphalan are active and hydantoins are slowly hydrolyzed into amino acids (Ross & Wilson, 1959). Unfortunately, the appropriate hydantoin mustards (Figure 20) were less effective than the corresponding amino acids.

Tumor cells tend to be deficient in some of the enzymes involved in electron transport, which can cause diminished activity of oxidative systems (Ross, 1964). It is thought that the deficiency in lactic acid dehydrogenase is responsible for the lower pH of tumor cells. Following the demonstration that Walker tumor cells have a lower reduction potential than normal cells in rats (0–150 mV compared with 200–300 mV) (Cater & Phillips, 1954), a variety of compounds was designed to take advantage of anticipated selective reduction in human tumor cells; for example, benzoquinones bearing nitrogen mustard functionalities should be readily reduced to the corresponding hydroquinones, which are dihydroxyaniline mustards (Equation 14). They should be highly toxic to the cells in which this reduction occurs (Ross, 1964). Indophenol mustard (Equation 15) was designed according to the same rationale. As already discussed, quinones bearing halomethyl or

$$\text{benzoquinone-}N(CH_2CH_2Cl)_2 \rightleftharpoons \text{hydroquinone-}N(CH_2CH_2Cl)_2 \qquad (14)$$

$$O={\bigcirc}=N-{\bigcirc}-N(CH_2CH_2Cl)_2 \rightleftharpoons HO-{\bigcirc}-NH-{\bigcirc}-N(CH_2CH_2Cl)_2 \quad (15)$$

acetoxymethyl groups can form highly reactive quinonemethides on reduction (Equation 5).

The quinone reduction potential of mitomycin analogs is important for selectivity in their toxicity to tumor cells. This property is consistent with their bioreductive activation, as shown in Scheme 3. A recent study indicates a rough correlation among polarographic reduction potential, minimum effective dose in P388 murine leukemia, and leukopenia (the limiting toxicity of mitomycin C) (Iyengar et al., 1981). As shown in Table 6, the most readily reduced analogs are likely to have the greatest potency (lowest minimal effective dose) and to cause the least leukopenia in mice. The reduction potential does not correlate with the prolongation of life of the mice.

The reduction of azobenzene derivatives to anilines in the body is well known. On the basis of this knowledge, azobenzenes that bore the nitrogen mustard functionality were prepared (Ross & Warnick, 1956). One of the most effective of these compounds (Equation 16) showed clinical activity against chronic lymphocytic leukemia (Schmidt, 1960). It appears, however, that its reduction to 4-amino-3-methylaniline mustard occurs most readily in the liver.

$$\underset{}{\bigcirc}-N=N-\underset{H_3C}{\bigcirc}-N(CH_2CH_2Cl)_2 \longrightarrow H_2N-\underset{H_3C}{\bigcirc}-N(CH_2CH_2Cl)_2 \quad (16)$$

The cellular reduction of aryl nitro and nitroso groups also can be utilized in the activation of alkylating agents; for example, the active metabolite of 2,4-dinitrophenylaziridine is the corresponding 2-amino compound (Equation 17; Connors et al., 1975). For rational design the best example of a nitro group is o-nitrobenzyl chloride. Reduction of the nitro group to an amine, followed by the elimination of HCl, gives an o-quinonimine that can add nucleophiles (Equation 18; Teicher & Sartorelli, 1980). The reduction of

$$\triangleright N-\underset{O_2N}{\bigcirc}-NO_2 \longrightarrow \triangleright N-\underset{H_2N}{\bigcirc}-NO_2 \quad (17)$$

Table 6 Antitumor Activity and Quinone Reduction Potentials of Mitomycin C Analogs[a]

[Structure: mitomycin C core with X substituent, CH_2OCNH_2, OCH_3, NH, H_3C, carbonyl groups]

X	$E_{1/2}$[b] (V)	P-388 Leukemia MED[c] (mg/kg)	Leukopenia[d] (% change in WBC on day 5 at OD)
CH_3O	−0.21	0.05	−2
pyrrolidinyl (N ring)	−0.31	0.025	−11
pyridyl-NH	−0.31	≤0.2	−33
furyl-CH_2NH	−0.39	≤0.4	−54
$HC\equiv CCH_2NH$	−0.41	0.2	−69
$H_2C=\overset{CH_3}{\underset{|}{C}}CH_2NH$	−0.41	1.6	−59
H_2N	−0.45	0.2	−42

[a]Abstracted from Iyengar et al., 1981.
[b]Determined by differential pulse polarography.
[c]MED = minimum effective dose (to produce % T/C = 125). It is the median value for a group of 6 BDF female mice treated ip with the mitomycin analog on day 1 only.
[d]Mitomycin analog given at OD = optimal dose for the P-388 assay on day 0.

[Scheme: o-NO_2-C_6H_4-CH_2Cl → o-NH_2-C_6H_4-CH_2Cl → cyclohexadienone imine with $=CH_2$] (18)

the nitroso group of 4-nitroso-N,N-di(hydroxyethyl) aniline dimethanesulfonate also produces an alkylating agent (Papanastassiou et al., 1966).

Concepts of latentiation have been applied in moderating the high reactivity of compounds such as mechlorethamine and sulfur mustard. Nitromin, the N-oxide of mechlorethamine, is similar to the parent compound in antitumor activity but it is safer to handle because it has almost no vesicant

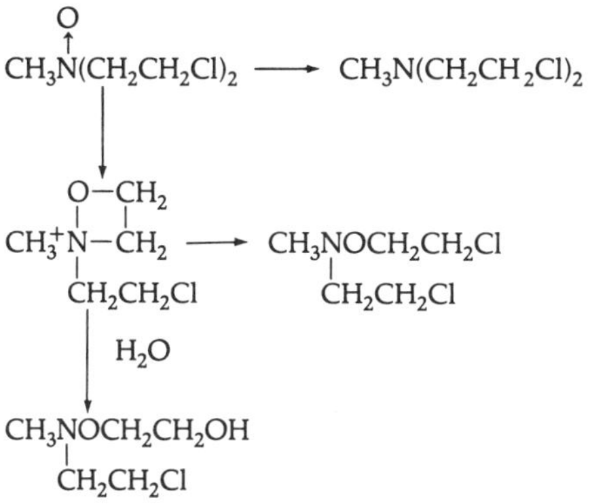

$$\text{CH}_3\overset{\uparrow}{\text{N}}(\text{CH}_2\text{CH}_2\text{Cl})_2 \longrightarrow \text{CH}_3\text{N}(\text{CH}_2\text{CH}_2\text{Cl})_2$$

$$\underset{\underset{\text{CH}_2\text{CH}_2\text{Cl}}{|}}{\overset{\overset{\text{O}-\text{CH}_2}{|\quad\;|}}{\text{CH}_3^+\text{N}-\text{CH}_2}} \longrightarrow \underset{\text{CH}_2\text{CH}_2\text{Cl}}{\overset{|}{\text{CH}_3\text{NOCH}_2\text{CH}_2\text{Cl}}}$$

$$\Big\downarrow \text{H}_2\text{O}$$

$$\underset{\text{CH}_2\text{CH}_2\text{Cl}}{\overset{|}{\text{CH}_3\text{NOCH}_2\text{CH}_2\text{OH}}}$$

Scheme 9 Reactions of nitromin.

activity (Farber et al., 1956). In addition to its desirable reduction to mechlorethamine, nitromin undergoes an intramolecular rearrangement that is detoxifying (Scheme 9; Ishidate et al., 1951). The sulfoxide corresponding to sulfur mustard is active against rodent tumors, but the corresponding sulfone is not. It undergoes decomposition into the divinyl sulfone at physiological pH (Equation 19; Ross, 1959).

$$\text{ClCH}_2\text{CH}_2\overset{\overset{\text{O}}{\uparrow}}{\underset{\downarrow}{\text{S}}}\text{CH}_2\text{CH}_2\text{Cl} \xrightarrow{-2\text{HCl}} \text{H}_2\text{C}=\text{CH}\overset{\overset{\text{O}}{\uparrow}}{\underset{\downarrow}{\text{S}}}\text{CH}=\text{CH}_2 \qquad (19)$$

In principle, the oxidizing capability of enzymes could be used to advantage in the design of latent alkylating agents. However, not many examples have appeared to date. Some important ones have been discovered accidently; that is, the oxidative metabolism of cyclophosphamide and the conversion of aniline mustard into its *p*-hydroxy derivative. A number of analogs of cyclophosphamide, which include ifosfamide and trofosfamide (Figure 21) are similarly activated by enzymic hydroxylation (Hohorat et al., 1976). The 4-trifluoromethyl derivative of cyclophosphamide was prepared in a successful attempt to increase the toxicity of the aldehydic degradation product (Scheme 4; Farmer & Cox, 1975).

Quantitative structure-activity relationships were formulated for the activity of aniline mustards against four different tumors: the leukemias L-1210 and P-388 in mice, and the solid tumors B-16 melanoma in mice and Walker 256 carcinosarcoma in rats (Panthananickal et al., 1978). For L-1210

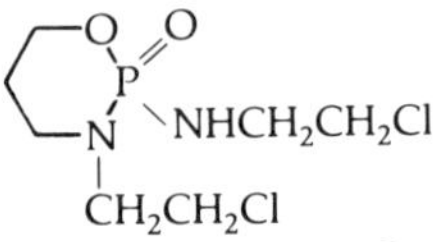

Ifosfamide: cyclic structure with P(=O)(O−)−N(CH$_2$CH$_2$Cl)−NHCH$_2$CH$_2$Cl

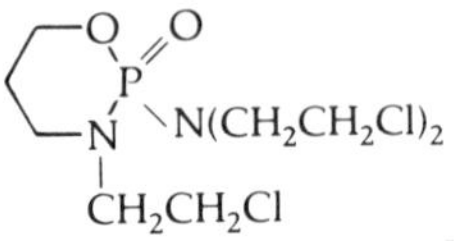

Trofosfamide: cyclic structure with P(=O)(O−)−N(CH$_2$CH$_2$Cl)−N(CH$_2$CH$_2$Cl)$_2$

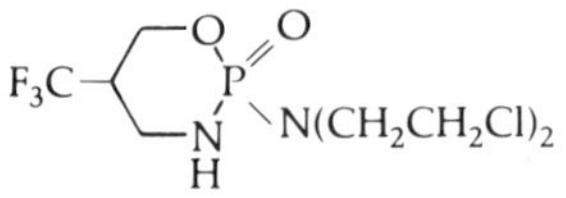

4-CF$_3$-cyclophosphamide: cyclic structure with F$_3$C− substituent, P(=O)(O−)−NH−N(CH$_2$CH$_2$Cl)$_2$

Figure 21 Analogs of cyclophosphamide.

leukemia the data were fitted by the equation log $1/C = 0.31\pi - 0.96\sigma + 0.86$ Io $+ 4.07$, wherein C is the molar concentration required to produce a 25% increase in life span over controls and the indicator variable Io is one when ortho substituents are present and zero, otherwise. The terms π and σ refer to the lipophilic and electron-withdrawing properties of the substituents, as defined by Hansch. This equation reveals that the electronic and lipophilic contributions of substituents are important, in which electron-releasing groups enhance activity, probably by increasing electron density on the nitrogen atom, and lipophilic substituents bring the overall lipophilicity of the molecule (log P_o) into the range of -1.00–0, where P_o is the octanol/water partition coefficient. The data obtained for aniline mustards that would produce an 80% increase in life span in mice with P-388 leukemia were correlated by a similar type of equation, in which hydrophilic substituents (negative coefficient of π) enhanced the activity. In contrast, the activity of neutral aniline mustards against solid tumors such as Walker 256 carcinosarcoma in rats and B-16 melanoma in mice was enhanced by more highly lipophilic substituents. Thus the equation for a 25% increase in life span for mice with B-16 melanoma was log $(1/C) = -2.06\sigma - 0.15\pi - 0.13\pi^2 + 4.13$. This equation gives a shape parabolic in its dependence on π; the maximum value corresponds to a log P_O of 2.33. Of course, the electron-releasing power of the substituent is the most important effect in this equation. In the series of aniline mustards used to derive these equations one compound, phenylalanine mustard, was significantly more active than predicted. Its enhanced activity was attributed to active transport into the cancer cells (Panthananickal et al., 1979).

It also was possible to correlate the dose levels required for a series of nitrosoureas to prolong by 25% the life span of mice with Lewis lung carcinomas with the lipophilicity of the compound (Montgomery et al., 1974). The equation $\log (1/C) = -0.082 (\log P)^2 + 0.14 \log P + 1.12$ indicated that the ideal log P was 0.83; the high activity occurred in the range of -0.20–1.34. This result, and the observation that nitrosoureas with negative log P values were less toxic, suggested that slightly hydrophilic compounds would be important as future analogs.

In the formulation of quantitative structure-activity relationships among 1-aryl-3,3-dialkyltriazines with L-1210 leukemia in mice it was found that sets of congeners for substituted phenyl-, pyrazolyl-, and imidazolyltriazines had the same ideal lipophilicity of $\log P_o = 1$ (Hatheway et al., 1978). Electron-releasing substituents increased potency and ortho substituents decreased it. A parallel correlation of the toxicity of 1-aryl-3,3-dialkyltriazines with their lipophilicity revealed that the optimum value for toxicity was $\log P_o = O$. The lipophilicity term had a small coefficient and changes in lipophilicity had little effect on toxicity or activity compared with changes in the electron-releasing power of substituents. This result suggested that strongly-electron releasing groups would be desirable in analogs. Unfortunately, compounds with substituents more electron-releasing than 4-OCH_3 are too unstable for biological studies. It was suggested that the practical inability to separate toxicity from potency in the 1-aryl-3,3-dialkyltriazines should discourage further work on these compounds as antitumor agents (Hansch et al., 1978).

An independent study of the use of cluster analysis to derive quantitative structure-activity relationships for 1-phenyl-3,3-diaryltriazines revealed that the electronic nature of substituents accounted for 85% of the variance in the antitumor activity (Sarcoma 180), whereas the hydrophobicity was not a significant factor in determining activity (Dunn et al., 1976). The same correlation was made with the toxicity of these triazines.

Synthesis of Alkylating Agents

In this section only the synthesis of clinically used alkylating agents is described. The chemistry involved in their synthesis is representative of that in the groups of analogs related to them. For mechlorethamine, the prototype nitrogen mustard, the synthesis begins with the reaction of methylamine with two equivalents of ethylene oxide. Treatment of the resulting diethanolamine derivative with thionyl chloride yields the product (Abrams et al., 1949; Equation 20). Chlorambucil and uracil mustard are prepared in

$$CH_3NH_2 + 2 \triangle\!\!\!\!\!\!O \longrightarrow CH_3N(CH_2CH_2OH)_2 \xrightarrow{SOCl_2} CH_3N(CH_2CH_2Cl)_2 \quad (20)$$

the same manner from from *p*-aminophenylbutyric acid (Bergel & Stock, 1954) and 5-aminouracil (Lyttle & Petering, 1958), respectively. Phenylalanine mustard (melphalan) is obtained from the phthalimido derivative of *p*-aminophenylalanine, which is treated with ethylene oxide and then phosphorus oxychloride, followed by acid hydrolysis of the phthalimido group (Bergel & Stock, 1954) (Equation 21). The synthesis of cyclophosphamide involves the treatment of dichloro-*N*-(dichloroethyl) phosphoramide with 3-propanolamine (Arnold & Bourseaux 1958; Equation 22). As anticipated, the synthesis of busulfan is accomplished by treating 1,4-butanediol with methanesulfonyl chloride in pyridine (Timmis, 1959; Equation 23).

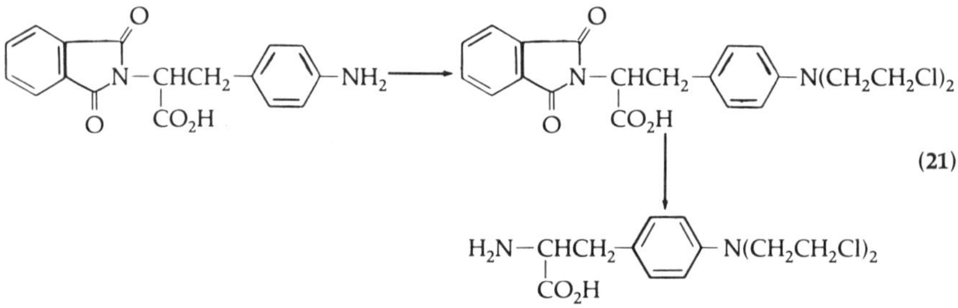

(21)

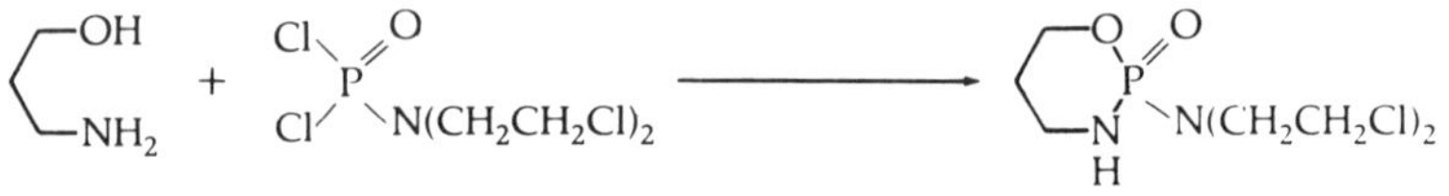

(22)

$$2CH_3SO_2Cl + HO(CH_2)_4OH \longrightarrow CH_3SO_2O(CH_2)_4OSO_2CH_3 \quad (23)$$

The chemistry of nitrosoureas has been carefully studied to provide good yields of the commercial antitumor agents. Carmustine (BCNU) is prepared in a straightforward manner by condensing 2-chloroethylamine with 2-chloroethyl isocyanate, followed by nitrosation with sodium nitrite in formic acid (Johnston et al., 1963; Equation 24). Lomustine (CCNU) and semustine (Methyl CCNU) are prepared in a similar manner from cyclohexylamine and *trans*-4-methylcyclohexylamine, respectively (Johnston et al.,

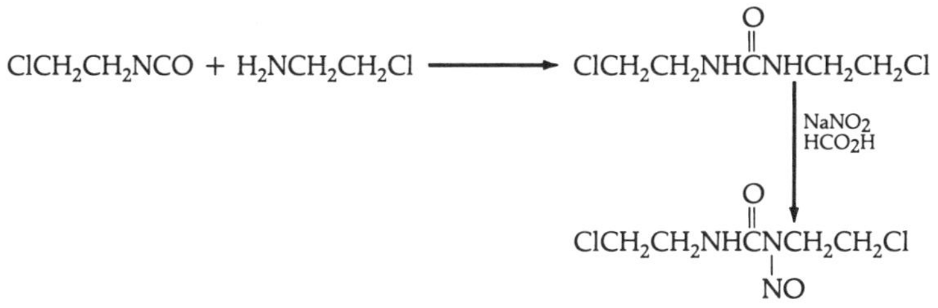
ClCH2CH2NCO + H2NCH2CH2Cl
ClCH2CH2NHCNHCH2CH2Cl
NaNO2
HCO2H
ClCH2CH2NHCNCH2CH2Cl
NO

(24)

1977). For the preparation of the methylcyclohexylamine the *p*-toluenesulfonate of *cis*-4-methylcyclohexanol is treated with sodium azide and the resulting trans azide is reduced catalytically (Equation 25). Although streptozotocin is a naturally occurring antibiotic, it is more conveniently prepared from 2-amino-2-deoxy-D-glucose by treatment with methyl isocyanate and dinitrogen trioxide (Hessler & Johnke, 1970; Equation 26). In the related preparation of chlorozoticin it was found that the best condition for nitrosation consisted of dinitrogen tetroxide in concentrated hydrochloric acid (Johnston et al., 1975; Equation 27).

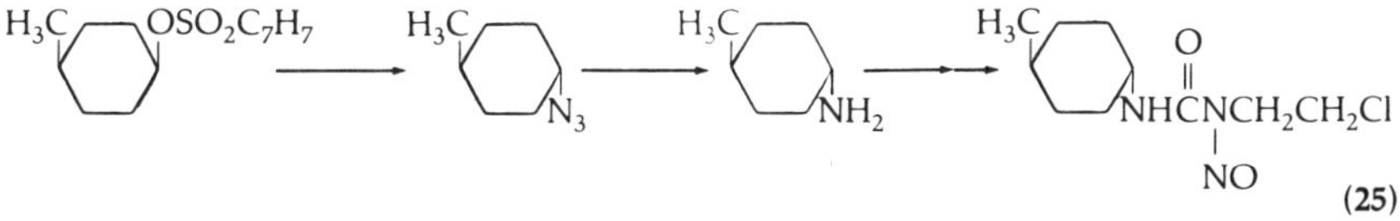
H3C
OSO2C7H7
N3
NH2
NHCNCH2CH2Cl
NO

(25)

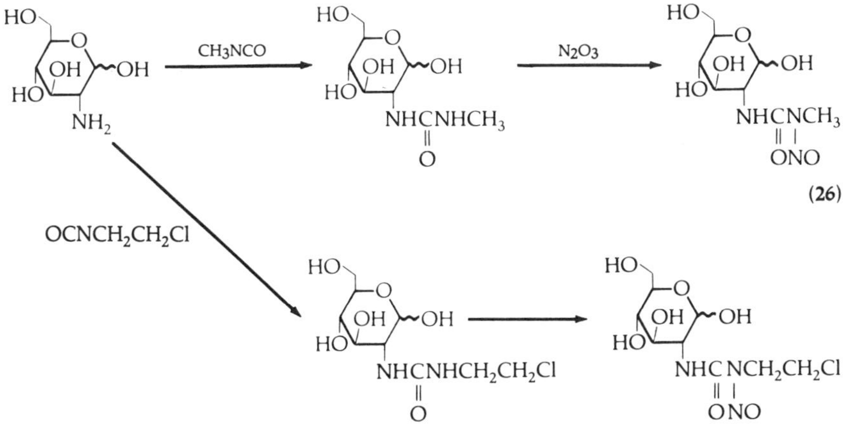
CH3NCO
N2O3
NHCNHCH3
NHCNCH3
ONO
OCNCH2CH2Cl
NHCNHCH2CH2Cl
NHCNCH2CH2Cl

(26)

(27)

Triethylenemelanine is prepared from ethylenimine and cyanuric chloride (Wystrach et al., 1955; Equation 28), whereas thio-TEPA (triethylenethiophosphoramide) is prepared from thiophosphoryl chloride and ethylenimine in the presence of triethylamine (Kuh & Seeger, 1954; Equation 29). The interesting synthesis of procarbazine begins by condensing diethyl azodicarboxylate (DEAD) with the isopropyl amide of 4-methylbenzoic acid. The product is methylated on a hydrazine nitrogen by treatment with sodium hydride and methyl iodide. Finally, the two carbethoxy groups are removed by refluxing hydrochloric acid (Hoffmann-La Roche, 1962; Equation 30). Dacarbazine is prepared from the diazonium salt of 5-aminoimidazole-4-carboxamide which reacts with dimethylamine to give the desired triazine (Shealy et al., 1962; Equation 31).

ANTIMETABOLITES

The Biochemistry of Antimetabolite Action

In the preceding sections the interaction between alkylating agents and DNA was described. It was concluded that the anticancer effects of these agents was based primarily on interference with the function of DNA. Another major group of antitumor agents, the antimetabolites, acts by mechanisms that complement the mechanisms of the alkylators, that is, antimetabolites prevent the biosynthesis of nucleotides and their incorporation into DNA. Antimetabolites also can be incorporated into nucleic acids in place of the normal nucleotides. The result is "fraudulent" copies that malfunction.

Antimetabolites are defined as compounds that prevent the biosynthesis or use of normal cellular metabolites. Often they are closely related in structure to the antagonized metabolites. This is especially true when they combine with an enzyme active site as if they were the substrate or a cofactor. They are more likely to resemble the end product of a biosynthetic pathway when they bind to the allosteric regulatory site of an enzyme that plays a key role in a pathway under feedback control. Sometimes the mode of action of an antimetabolite is highly complex, for example, 6-mercaptopurine and its transformation products (anabolites) inhibit at least 30 enzymes (Montgomery, 1970a). This makes it difficult to establish the inhibitions that are most important for the antitumor effect.

The biosynthesis of purine nucleotides is given in an abbreviated form in Scheme 10. It is apparent that all the compounds in this scheme are 5-phosphoribosyl derivatives of the purine intermediates. Consequently purine analogs are most effective as antimetabolites when they have been converted to their 5-phosphoribosyl anabolites. The first step in Scheme 10, conversion of 5-phosphoribosylpyrophosphate into 5-phosphoribosylamine, is a key one for purine biosynthesis. It is subject to feedback regulation by the final products, adenylate and guanylate (Montgomery, 1970b). Analogs of adenine and guanine that are converted into their ribouncleotides can be effective in limiting the *de novo* biosynthesis by binding to the regulatory site on the rate-limiting enzyme, phosphoribosylamine synthetase. Two compounds, 6-mercaptopurine (6-MP) and 6-thioguanine (Figure 22) are effective for this purpose (Elion & Hitchings, 1965). 6-Mercaptopurine enters cells and is converted into its ribonucleotide, 6-thioinosinate (Figure 22), by the enzyme hypoxanthine-guanine phosphoribosyltransferase (Lukens & Herrington, 1957). This ribonucleotide is an active enzyme inhibitor. It is not possible to con-

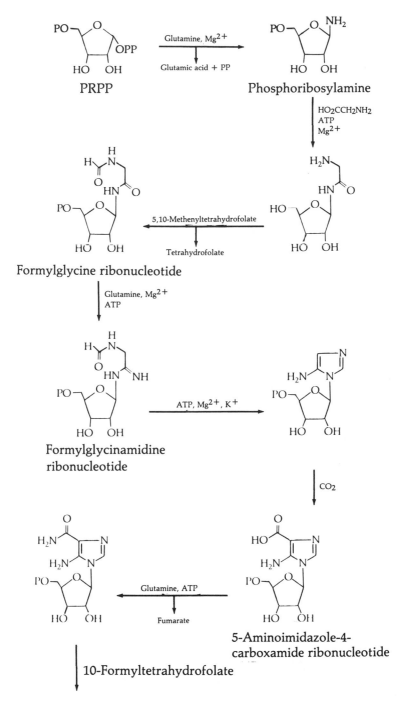

Scheme 10 Biosynthesis of purine nucleotides (simplified).

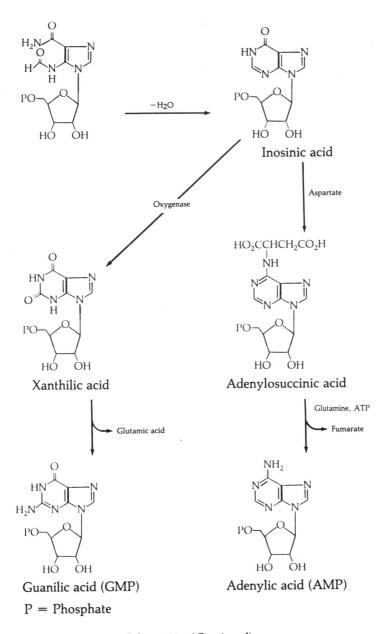

Scheme 10 (Continued).

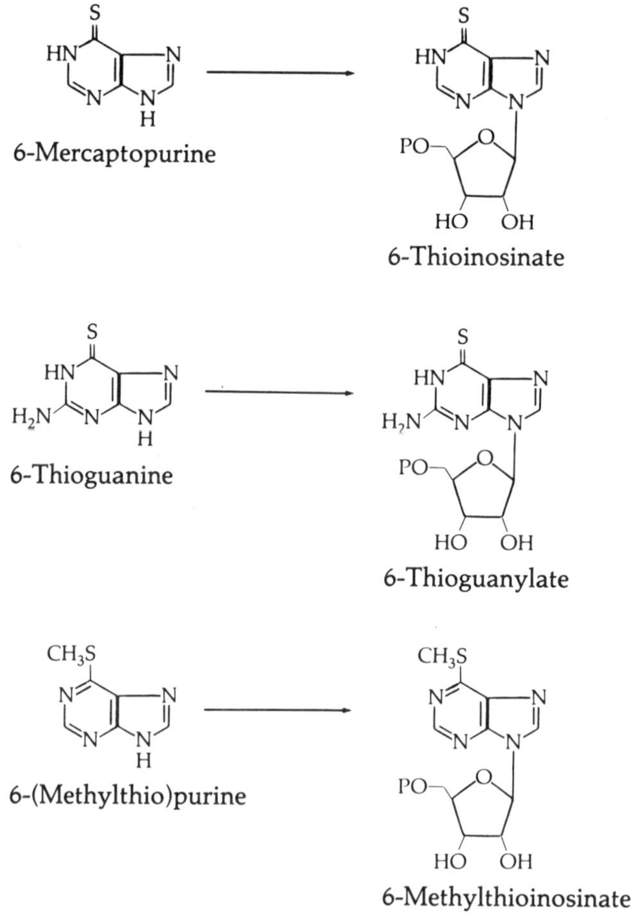

Figure 22 6-Mercaptopurines.

duct therapy directly with 6-thioinosinate because this compound does not enter cells readily. A parallel process is observed for 6-thioguanine, which is converted into 6-thioguanylate by the same phosphoribosyltransferase (Law, 1958). The mode of action of 6-thioinosinate and 6-thioguanylate is complicated because they are further anabolized to their diphosphate and triphosphate derivatives, which are also enzyme inhibitors (Caldwell, 1969; Moore & LePage, 1958). Furthermore, 6-thioinosinate acts as a substrate for a methyl transferase that converts it into 6-methylthioinosinate (Figure 22), which is also an antimetabolite (Remy, 1963). The conversion of 5-phosphoribosylpyrophosphate into 5-phosphoribosylamine involves the transfer of an amino group from glutamine with the formation of glutamate. This process can be inhibited by glutamine antagonists such as 6-diazo-5-oxo-L-norleucine (Figure 23; Hartman, 1963). However, these agents exert their main influence on purine biosynthesis at the step in which formylglycinamidine ribonucleotide is formed (Bennett et al., 1956).

5-Phosphoribosylamine is converted into its glycinyl derivative by a reaction that involves glycine, ATP, and magnesium ion (Scheme 10). This derivative is then formylated on the glycine amino group by a process in which a formyl group is transferred from 5,10-methenyltetrahydrofolate. As described below, dihydrofolate reductase inhibitors like amethopterin (Methotrexate) prevent the formation of 5,10-methenyltetrahydrofolate, which lim-

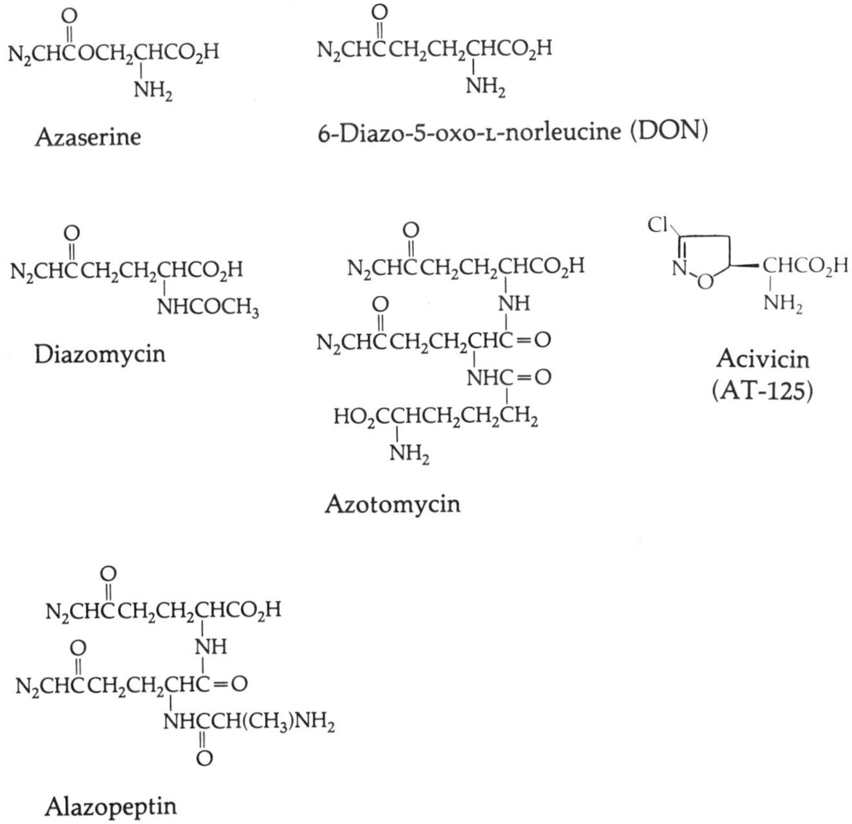

$$N_2CHCOCH_2CHCO_2H \quad | \quad NH_2$$

Azaserine

$$N_2CHCCH_2CH_2CHCO_2H \quad | \quad NH_2$$

6-Diazo-5-oxo-L-norleucine (DON)

$$N_2CHCCH_2CH_2CHCO_2H \quad | \quad NHCOCH_3$$

Diazomycin

$$N_2CHCCH_2CH_2CHCO_2H \quad | \quad NH \\ N_2CHCCH_2CH_2CHC=O \quad | \quad NHC=O \\ HO_2CCHCH_2CH_2CH_2 \quad | \quad NH_2$$

Azotomycin

Acivicin (AT-125)

$$N_2CHCCH_2CH_2CHCO_2H \\ O \quad NH \\ N_2CHCCH_2CH_2CHC=O \\ NHCCH(CH_3)NH_2 \\ \| \\ O$$

Alazopeptin

Figure 23 Glutamine antagonist antibiotics.

its the *de novo* synthesis of purines. The next step in purine biosynthesis is the conversion of formylglycine ribonucleotide into the corresponding amidine. This reaction requires glutamine, ATP, magnesium ion, and an enzyme that is strongly inhibited by DON, azaserine, and other glutamine antagonists (Hartman, 1963). Their inhibition is competitive, but it becomes irreversible when incubated in the absence of glutamine (Levenberg et al., 1957). Thus a study of incubation of azaserine-^{14}C with the enzyme, followed by digestion with proteolytic enzymes and acid hydrolysis, produced S-carboxymethylcystine-^{14}C (Equation 32). This study demonstrated that the azaserine had reacted covalently with a sulfhydryl group of a cysteine residue of the enzyme (Dawid et al., 1963).

$$\underset{\underset{NH_2}{|}}{HO_2CHCH_2O\overset{\overset{O}{\|}}{C}CHN_2} + \underset{\underset{NHR}{|}}{HSCH_2CHCONHR} \longrightarrow \underset{\underset{NH_2}{|}}{HO_2CHCH_2O\overset{\overset{O}{\|}}{C}CH_2SCHCONHR} \quad (32)$$

$$\downarrow \text{hydrolysis}$$

$$\underset{\underset{NH_2}{|}}{HO\overset{\overset{O}{\|}}{C}CH_2SCHCO_2H}$$

From formylglycinamidine ribonucleotide the biosynthesis of purines progresses through the cyclization to an aminoimidazole, which is carboxylated and converted into an amide, as illustrated in Scheme 10. The next step involves formylation of the amino group by transfer of the formyl group from 10-formyltetrahydrofolate. This step is limited by the availability of the 10-formyltetrahydrofolate, which can be decreased by dihydrofolate reductase inhibitors (Mead, 1975). In a subsequent biosynthetic step, the pyrimidine ring of the purine structure is closed and inosinate results.

The bioconversion of inosinate into adenylate and guanylate (Scheme 10) is inhibited at three points by the ribonucleotides of 6-mercaptopurine and 6-thioguanine (Elion, 1967; Meich et al., 1969). Inosinate is converted into adenylosuccinate by the incorporation of aspartate and adenylate is formed by the elimination of fumarate. The enzyme for the second step is inhibited strongly by alanosine (Gale et al., 1968; see Equation 39). The final step in guanylate biosynthesis in Scheme 10 is the transfer of an amino group from glutamine; it is also inhibited by the glutamine antagonists, especially acivicin (Jarjaram et al., 1975; Figure 23).

Scheme 11 shows, in abbreviated form, the biosynthesis of pyrimidine nucleotides. This scheme differs from that for purines in that the phosphori-

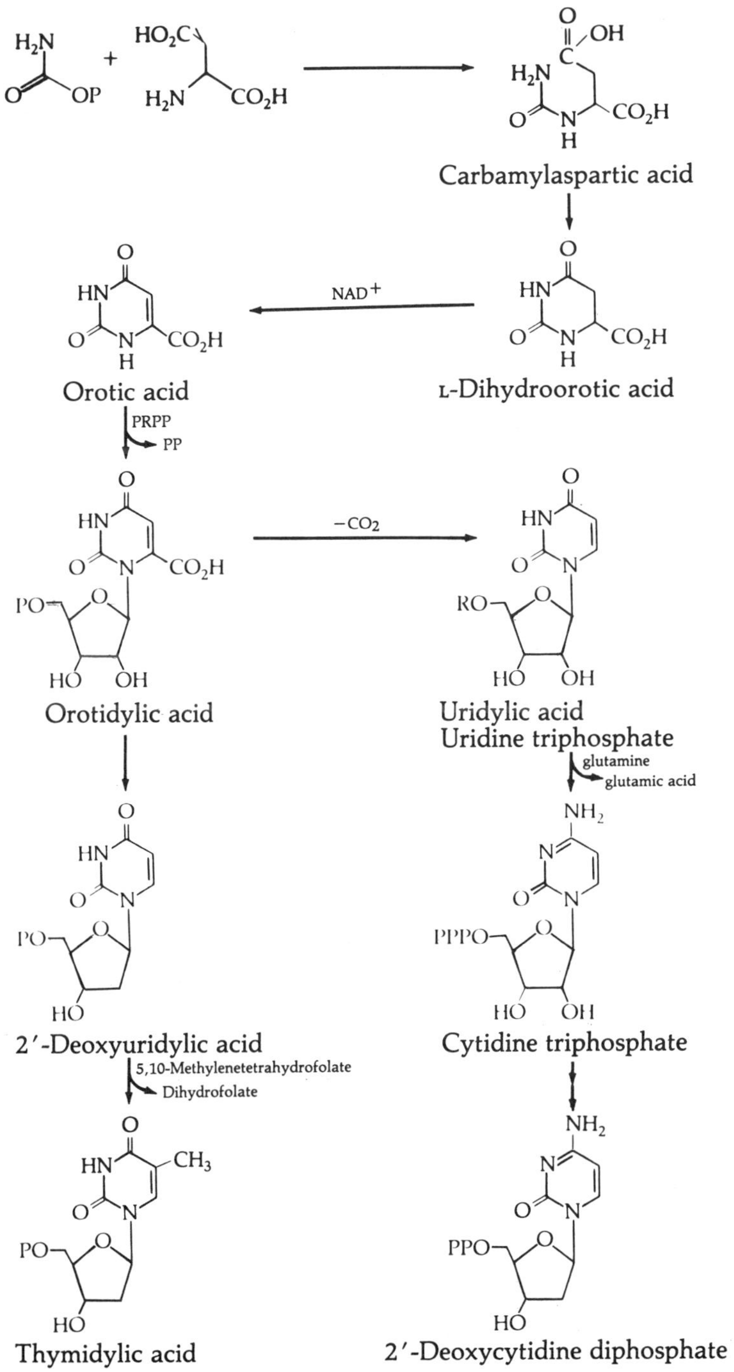

Scheme 11 Biosynthesis of pyrimidine nucleotides (simplified).

bosyl portion is not incorporated until orotic acid, a pyrimidine compound, is fully formed. Orotidylic acid, the resulting nucleotide, is the first compound in Scheme 11 whose further biotransformation has been effectively inhibited by antimetabolites. Thus the decarboxylase that converts orotidylate to uridylate is inhibited by the nucleotides formed by anabolism of 5-azacytidine, 6-azauridine, and 5-aminouridine (Habermann & Sorm, 1958; Vesely et al., 1968a; Smith et al., 1966; Figure 24). It has been reported, however, that 5-aminouridylate has some inhibitory effect on the incorporation of carbamoylaspartate into pyrimidine (Eidinoff et al., 1959). The conversion of uridine triphosphate into cytidine triphosphate requires glutamine and is inhibited by antagonists such as DON and azaserine (Eidinoff et al., 1958). As described above, this inhibition is not the most important function of these antagonists.

The most important step in pyrimidine nucleotide biosynthesis that presents a target for anticancer agents is the formation of thymidylate from 2'-deoxyuridylate and 5,10-methylenetetrahydrofolate. A mechanism for this transformation (Scheme 12) involves a covalently bound intermediate formed from the substrate, folate derivative, and the enzyme thymidylate synthetase.

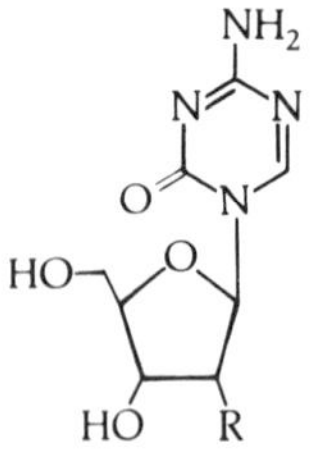

R = OH, 5-azacytidine
R = H, 5-aza-2'-deoxycytidine

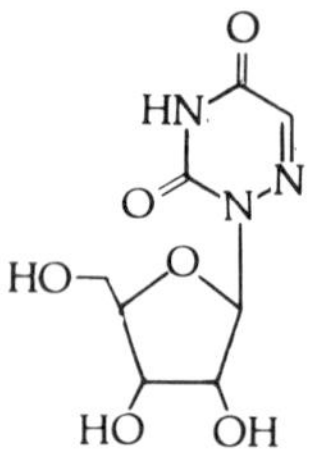

6-Azauridine

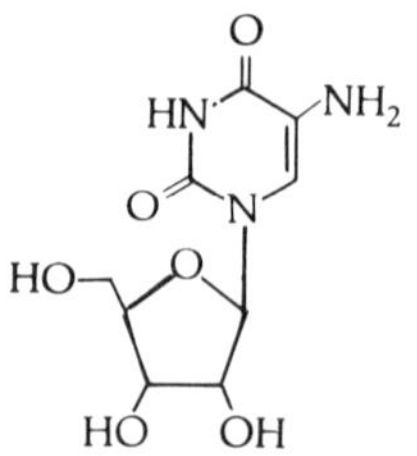

5-Aminouridine

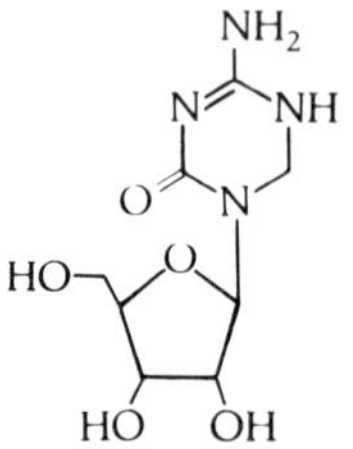

5,6-Dihydro-5-azacytidine

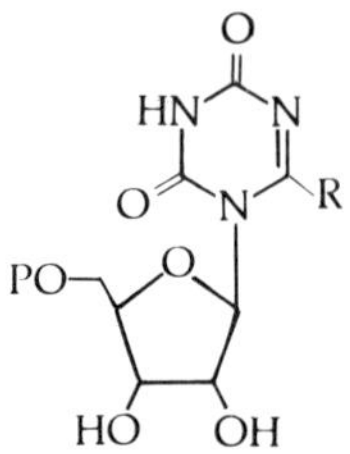

R = H, 5-azauridylate
R = CO_2H, 5-azaorotidylate

Figure 24 Azapyrimidines.

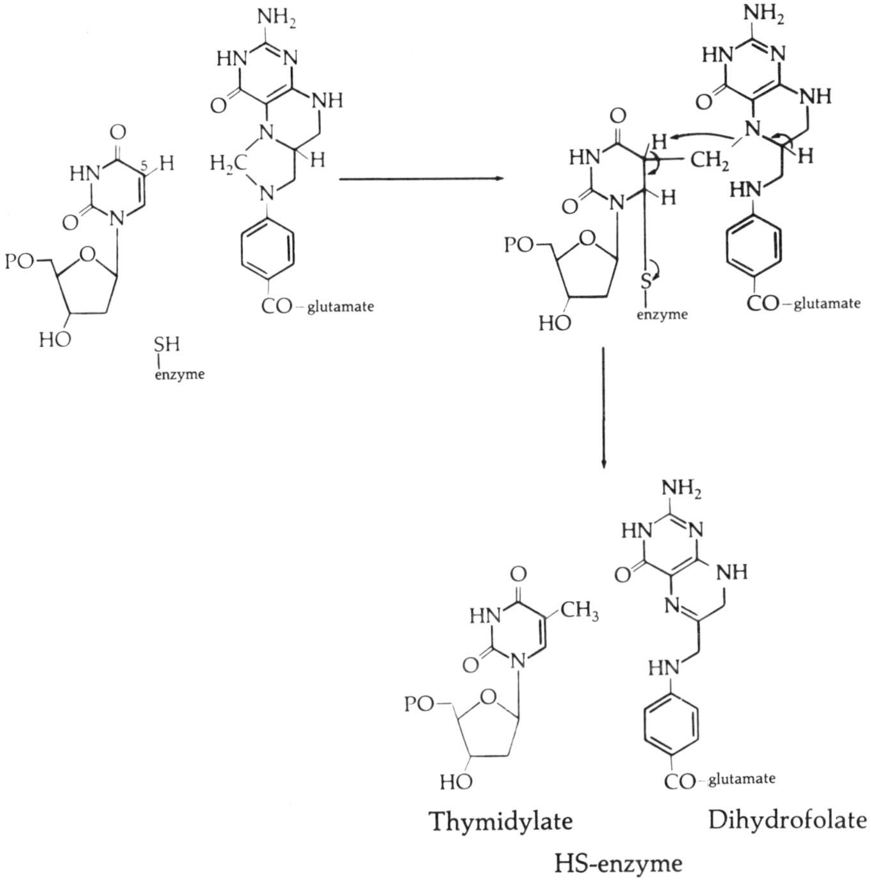

Scheme 12 Proposed formation of thymidylate from 2′-deoxyuridylate.

This intermediate dissociates into thymidylate, dihydrofolate, and the enzyme. An essential action in this scheme is the movement of the 5-hydrogen of uridine onto the methylene group to form the 5-methyl group of thymidylate. This action cannot take place if the 5-hydrogen is replaced by a fluorine atom because the resulting covalent intermediate is stable and an inhibition of the K_{cat} type results (Danenberg & Heidelberger, 1976). In this manner the 2′-deoxyribonucleotide of 5-fluorouracil acts as a powerful antagonist of thymidylate synthetase. 5-Fluorouracil (5-FU, Figure 25) can be administered as the simple pyrimidine for cancer chemotherapy because it is readily converted into its 2′-deoxyribonucleotide in cells (Harbers et al., 1959).

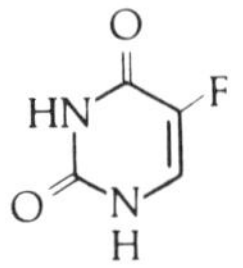

5-Fluorouracil

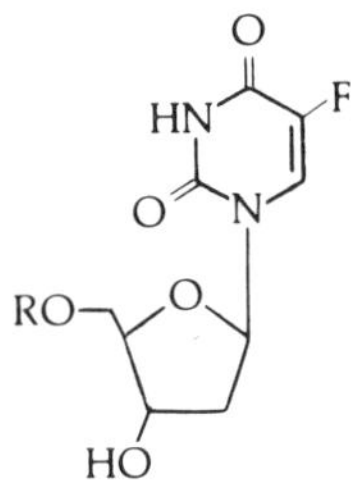

R = H, 5-FUdR
R = PO$_3$H^{2-}, 5-F-2'-deoxyuridylate

Figure 25 5-Fluorouracil and anabolites.

Alternatively, it can be given as the 2'-deoxyribonucleoside (5-FUdR) which is readily phosphorylated by thymidine kinase (Kessel & Wodinsky, 1970). Other important inhibitors of thymidylate synthetase are trifluorothymidine and 5-mercapto-2'-deoxyuridine. The biosynthesis of thymidylate is also decreased by dihydrofolic reductase inhibitors, such as amethopterin, which limit the amount of 5,10-methenyltetrahydrofolate that is available (Mead, 1975). This step appears to be more responsible for the general toxicity of amethopterin than for its selective toxicity to cancer cells because it is possible to reduce amethopterin toxicity by "rescue therapy" with thymidine while maintaining the antitumor effect. Scheme 13 illustrates the interconversion of various forms of folic acid. It is evident that the inhibition of dihydrofolic reductase can limit the species needed for the transfer of one-carbon units in nucleic acid biosynthesis and that this type of limitation can be overcome by supplying exogenous thymine or N^5-formyltetrahydrofolate (known as leucovorin and citrovorum factor) in "rescue therapy" protocols.

In addition to inhibiting the biosynthesis of ribonucleotides, antimetabolites are able to interfere with the conversion of ribonucleotides into the corresponding 2'-deoxyribonucleotides. This process is catalyzed by the enzyme ribonucleotide reductase, which has two iron-containing subunits (Moore, 1969). Compounds like hydroxyurea, which complex with the iron, are able to inhibit ribonucleotide reductase. In regard to the antimetabolites, the triphosphate of cytosine arabinoside (Scheme 14), which is epimeric at the 2'-position with the triphosphate of cytosine, decreases the formation of 2'-deoxycytosine phosphates (Chu & Fischer, 1962). Although this was once thought to be its main mode of action, recent studies have shown that it is no more effective than various deoxyribonucleosides in reducing cellular levels

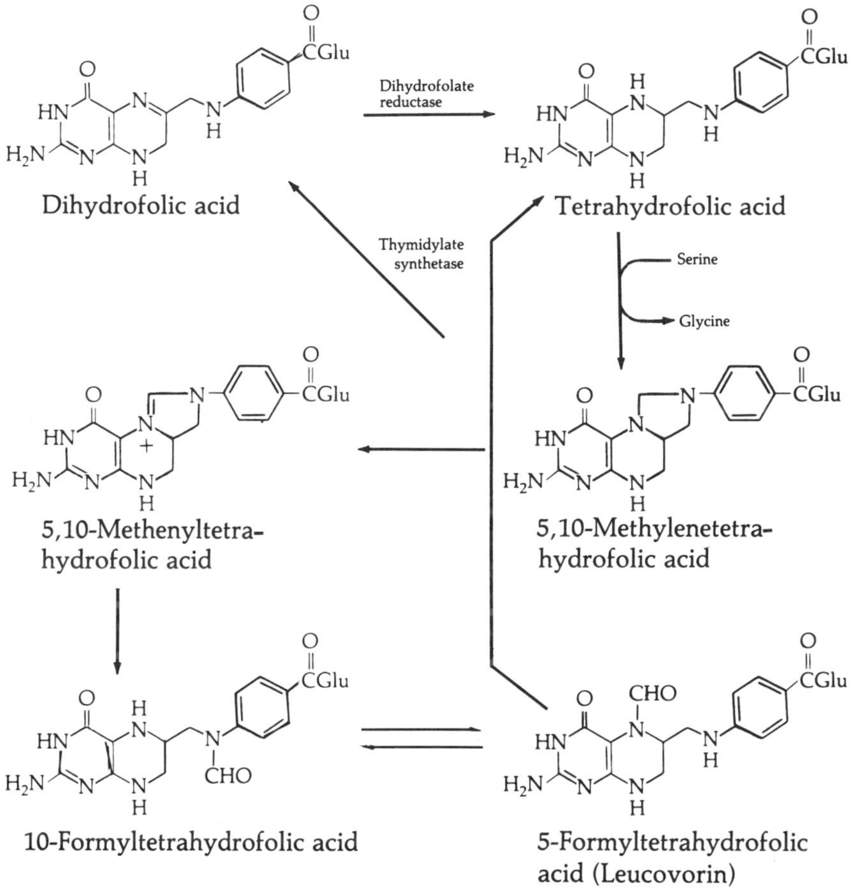

Scheme 13 Interconversions of folic-acid derivatives.

of 2'-deoxycytidylate (Inagaki et al., 1969). Cytosine arabinoside is readily transported into cells and phosphorylated by deoxycytidine kinase (Momparler & Fischer, 1968). Evidently this enzyme is not affected by the presence of the 2'-hydroxyl group in an unnatural configuration.

Another stage of DNA synthesis at which cytosine arabinoside interferes is in the operation of DNA-dependent DNA polymerase. The phosphate esters of cytosine arabinoside are competitive inhibitors of this enzyme (Kimball & Wilson, 1968). Adenine arabinoside, the analogous antimetabolite based on adenosine, inhibits DNA polymerase in a similar manner (Furth & Cohen, 1968).

Incorporation of the anabolites of certain purine and pyrimidine antimetabolites into DNA has been observed, for example, with 6-mercaptopurine, 6-thioguanine, 5-fluorouracil, and cytosine arabinoside (Scannel & Hitchings, 1966; Champe & Benzer, 1962; Momparler, 1972). It is thought that this results in "fraudulent" DNA which malfunctions in its role of directing cell processes. The relative importance to antineoplastic activity of this effect, as opposed to other antimetabolite effects such as the inhibition of DNA synthesis, is uncertain at this time.

Although a variety of powerful mechanisms for the inhibition of DNA synthesis and function by antimetabolites has been described, the effects of these agents frequently are less than anticipated because antineoplastic cells have methods of resisting the antimetabolites. These methods include using salvage pathways for purines and pyrimidines, decreasing the levels of phosphorylating enzymes to prevent the anabolism of antimetabolites, increasing the enzymes for *de novo* nucleic acid synthesis, and increasing the enzymes that degrade antimetabolites.

Salvage pathways are especially important to pyrimidines. Tumor cells scavenge products of nucleic acid degradation from the bloodstream and use them for the synthesis of their own nucleotides, for example, the ability to scavenge thymidine is significant for tumor cells with blocked thymidine synthetase. Uracil and uridine also are readily taken in by tumor cells, where they are converted into uridylic acid by uridine phosphorylase and uridine kinase (Matsushita & Fanburg, 1970).

A significant mode of resistance to 6-mercaptopurine and 6-thioguanine is the decrease in hypoxanthine-guanine phosphoribosyltransferase levels in tumor cells (Brockman, 1963). Pyrimidine antagonists that lose their antitumor effectiveness because of the low levels of the appropriate phosphorylases and kinases include cytosine arabinoside, 5-azauridine, 5-azacytidine, 6-azauricil, and 5-aza-2'-deoxycytidine (Kessel, 1967; Pasternak et al., 1961).

Tumor-cell resistance to amethoperin includes an increase in the number of gene copies for dihydrofolate reductase and a decrease in the active transport of the antimetabolite into the cells.

The most important degradative process for the deactivation of purines and pyrimidines in tumor cells is the conversion of amino substituents into the corresponding hydroxyl substituents. Thus cytosine arabinoside is converted into uridine arabinoside by cytidine deaminase (Scheme 14; Steuart & Burke, 1971). Similarly, adenine arabinoside (Figure 27) is transformed into hypoxanthine arabinoside, a less active compound, by adenosine deaminase

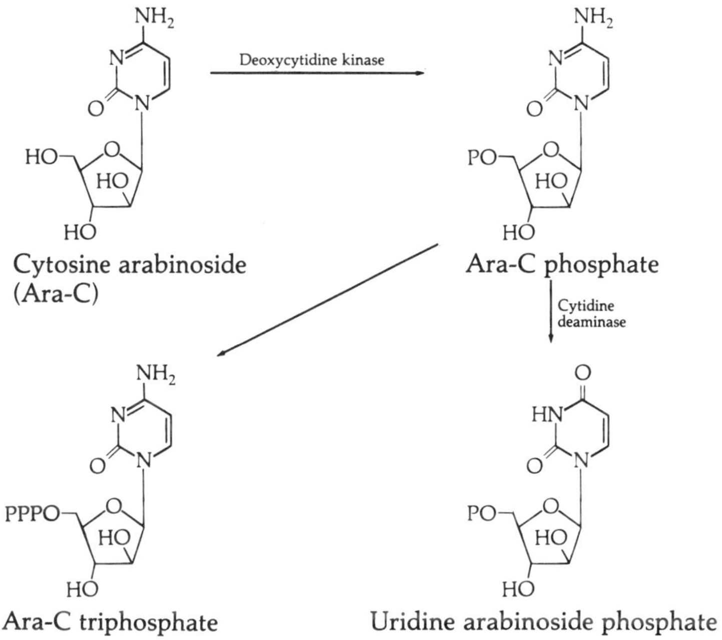

Scheme 14 Biotransformations of cytosine arabinoside.

(Brink & LePage, 1964). Because of the serious limitations that deamination has placed on chemotherapy, intensive efforts have been made to develop inhibitors of the deaminases. These efforts are discussed in a following section.

Another type of degradative transformation occurs in the conversion of 6-mercaptopurine into 6-thioxanthine by guanase, thence into 6-thiouric acid by xanthine oxidase (Atkinson et al., 1965; Equation 33). Allopurinol,

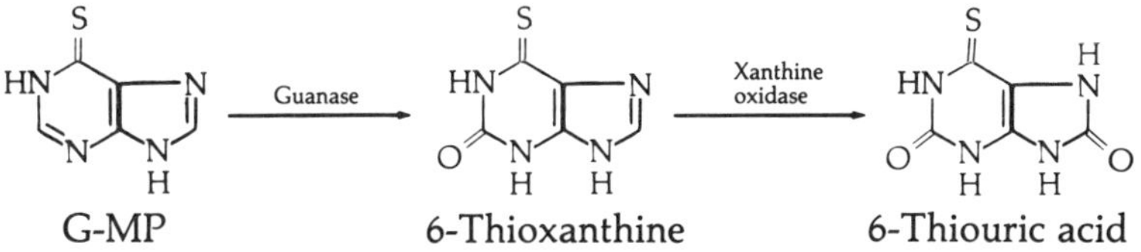

an inhibitor of xanthine oxidase, increases the potency of 6-mercaptopurine significantly but causes a corresponding increase in its toxicity (Levine et al., 1969).

The Design and Discovery of Antimetabolites

Three different types of structural variant of purines and their nucleosides have been designed as antimetabolites: those in which carbon and nitrogen atoms are interchanged in the nucleus (azapurines and deazapurines), those with novel substituents on the purine nucleus, and those with abnormal sugars attached to the purine. It is significant that these variants have been found in nature. In a number of examples rationally designed compounds were prepared before they were discovered in nature.

The 8-azapurines (v-triazolo[4,5-d]pyrimidines), which include 8-azaadenine, 8-azahypoxanthine, and 8-azaguanine (Figure 26) were among the first purine antimetabolites prepared (Roblin et al., 1945). Among these compounds 8-azaguanine showed antitumor activity in mice (Kidder et al., 1949).

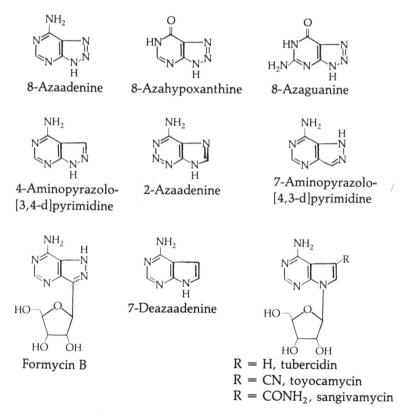

Figure 26 Azapurines and deazapurines.

It was introduced into clinical studies, but abandoned in favor of newer agents such as 6-mercaptopurine. The structure of 8-azaguanine is acceptable to the enzyme hypoxathine-guanine phosphoribosyltransferase, which converts it into the ribonucleotide (Way & Parks, 1958). This intermediate is converted into the triphosphate by guanylate kinase (Way et al., 1959). The triphosphate is incorporated into m-RNA and t-RNA, which results in inhibition of protein synthesis. The isomeric 2-azapurines (imidazo[4,5-d]-v-triazines), such as 2-azaadenine (Figure 26), are highly cytotoxic but lack good antitumor activity (Montgomery et al., 1970).

Pyrazolopyrimidines are compounds in which one of the nitrogen atoms has been interchanged with a carbon atom in the imidazole ring of purines. A subgroup contains the pyrazolo[3,4-d]pyrimidines. The 4-amino derivative of this nucleus (Figure 26), an isomer of adenine, showed significant activity against animal tumors, but it failed as a clinical agent because of hepatotoxicity (Robins, 1956; Falco & Hutchings, 1956). The corresponding 4-chloro analog reacts readily with amines to produce a variety of active 4-N-alkyl derivatives, but none is used clinically (Noell & Robins, 1958). The other subgroup is the pyrazolo[4,3-d]pyrimidine. A 7-amino derivative of this nucleus was synthesized and found to be inactive against tumors (Robins et al., 1956). The corresponding 3-ribonucleoside (Figure 26), an antibiotic named formycin B, is highly active (Hori et al., 1964) and phosphorylated readily in mammalian cells. Apparently the inactivity of 7-aminopyrazolo[4,3-d]pyrimidine is due to the inability of cells to convert it into a ribonucleotide (Montgomery et al., 1970b).

Deazapurines lack one or more of the nitrogen atoms in the nucleus. Examples have been prepared in which each of four nitrogen atoms is missing. However, only the 7-deazapurine system (pyrrolo[2,3-d]pyrimidine) is associated with antineoplastic activity. 7-Deazaadenine and its 6-oxo and 2,6-diamino analogs were not biologically active, presumably because they are not converted into their ribonucleotides in cells (Davoll, 1960). In contrast, a group of antibiotics, including tubercidin (7-deazaadenosine), toyocamycin, and sangivamycin (Figure 26), has the ribosyl group (Anzai et al., 1957; Nishimura et al., 1956; Rao & Renn, 1963). These compounds are readily phosphorylated and show antitumor activity in animal models.

Considerable success has been obtained in replacing the 6-substituent of purines with other groups. The most important therapeutic agent of this type is 6-mercaptopurine (Figure 22; Elion et al., 1952). Its biochemical transformations and mode of action were discussed in the preceding section. A homolog, 6-(methylthio) purine, is used as its ribonucleoside (Figure 22), which is phosphorylated by adenosine kinase. This property confers activity against tumor cells that have become resistant to 6-mercaptopurine by their lack of hypoxanthine-guanine phosphoribosyltransferase (Montgomery et al., 1961). Many other substituents have been attached to the 6-thio functionality by alkylating the 6-thiolate anion or by the action of thiolate ions on 6-chloropurine. Imuran, the 1-methyl-5-nitroimidazolyl derivative, a compound widely used in organ transplants because of its immunosuppressive activity, is prepared by the latter method (Hitchings & Elion, 1962; Equation 34). It is a prodrug that forms 6-mercaptopurine in the body. A family of compounds related to 4-(purin-6-ylthio) butyric acid has received considerable attention in experimental tumor models (Cerny et al., 1967). The selenium isostere of 6-mercaptopurine is equal to the parent compound in therapeutic potential (Mautner, 1956).

$$(34)$$

Imuran

Among the other 6-substituted purines 6-chloropurine and its ribonucleoside show potent anticancer activity (Brown & Weliky, 1953). Furthermore, they are intermediates in the chemical synthesis of other active compounds. Thus displacement with hydroxylamine gives N-hydroxyadenosine (Equation 35), a compound that is active against mouse leukemias and serves as an inhibitor of adenosine deaminase (Chang, 1965). The corresponding hydrazine analog and certain-N-alkyladenosines and N-aralkyladenosines also show antitumor activity (Montgomery & Holum, 1957).

$$(35)$$

6-Thioguanine has important clinical activity and its biochemical transformations and mode of action mostly parallel those of 6-mercaptopurine (Elion & Hitchings, 1967). It has been converted into a variety of S-alkyl derivatives which include Guaneran, the 2-amino analog of Imuran (Hitchings & Elion, 1962).

2-Substituted adenines (Figure 27) are important in cancer research. 2-Aminoadenine (2,6-diaminopurine) was the first adenine analog to be prepared and the first purine to show antitumor activity in experimental systems (Taylor et al., 1959). Among the 2-halopurines, 2-fluoroadenosine is highly active against tumors and resistant to adenosine deaminase (Montgomery & Hewson, 1967). It is too toxic for clinical use; however, the corresponding 9-β-D-arabinofuranosyl derivative (F-ara-A) has activity against L1210 leukemia in mice comparable to that of ara-A given in combination with a deaminase inhibitor (Brockman et al., 1977). 2-Methoxyadenosine, known as spongosine (Figure 27), was isolated from a Caribbean sponge (Bergmann & Stempien, 1957). This compound and other 2-substituted adenosines have biological activity, but they are not useful antitumor agents. A series of 8-substituted adenosines was prepared, but only the 8-bromo compound showed antitumor activity (Ikehara et al., 1967).

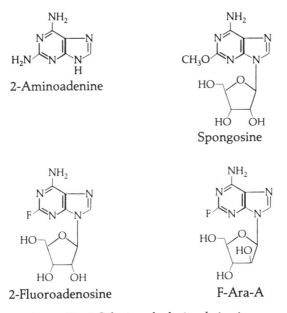

Figure 27 2-Substituted adenine derivatives.

Variations in the ribose moiety of purine nucleosides have been effective in the development of antimetabolites with antitumor activity (Figure 28). The 9-β-D-arabinofuranosyl derivative of adenine (ara-A) was prepared first by chemical synthesis (Lee et al., 1960) and isolated subsequently from *Streptomyces antibioticus* (Parke-Davis & Co., 1967). It is anabolized to the triphosphate, whereupon it becomes an inhibitor of DNA polymerase (Furth & Cohen, 1968). It is a powerful antiviral agent, but its activity against cancer is severely limited by adenosine deaminase inactivation (Brink & LePage, 1964). Recent studies of the combination of ara-A with a deaminase inhibitor are promising (see below). The corresponding arabinofuranosyl derivative of 6-mercaptopurine (ara-6-MP) is unusual in that it is active without conversion into a nucleotide (LePage et al., 1969). It appears to be an inhibitor of cytidylate reductase (LePage, 1971).

Another analog of adenosine, 9-β-D-xylofuranosyladenine, is epimeric with adenosine at C-3' but is less active than ara-A (Lee et al., 1963). The related 3'-deoxyadenosine, known as cordycepin, was the first nucleoside antibiotic to be discovered (Cunningham et al., 1951). Puromycin contains in its structure a 3-amino-3-deoxy-D-ribose unit linked to N,N-dimethyladenine and substituted with the p-methoxyphenylalanyl group (Figure 28). It is a powerful inhibitor of protein synthesis but shows no significant antitumor activity (Porter et al., 1952). The synthetic analog 3'-deoxy-3'-N-methyl-6, 6-N,N-dimethyladenosine, however, showed antitumor activity in experimental systems (Baker et al., 1955). The simpler analog 3'-amino-3'-deoxyadenosine, first synthesized and later isolated from a *Helminthosporum*, also had antitumor activity (Ammann & Safferman, 1958).

Two additional antitumor antibiotics, psicofuranine and decoynine, consist of adenine with novel hexoses linked to N-9 (Schroeder & Hoeksema, 1959; Hoeksema et al., 1964; Yuntsen et al., 1954).

Activities of various purine antimetabolites against L1210 leukemia in mice are listed in Table 7.

The periodate oxidation of ribonucleosides yields products in which the 2',3'-diol is cleaved to the corresponding dialdehyde; for example, inosine is converted into its dialdehyde, a compound active against experimental tumors, by this method (Equation 36). The corresponding dialdehyde from 5'-deoxyinosine had more potent activity (Corey & Parker, 1979).

Research efforts directed toward the design and development of pyrimidine nucleoside antimetabolites have taken a course parallel to that of the purine nucleosides. The same general types of structural variation include in-

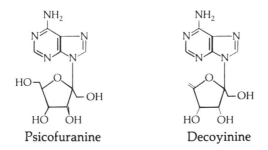

Figure 28 Adenosine analogs with novel sugars.

Table 7 Antitumor Activity of Purine Antimetabolites Against L1210 Murine Leukemia[a]

Name	Synonyms	Dose (mg/kg)	Increase in Life Span (% T/C)
6-Mercaptopurine	6-MP	160	57
6-Mercaptopurine, 9-β-D-ribofuranosyl		160	55
6-(Methylthio)purine, 9-β-D-ribofuranosyl		20	60
6-(Allythio)purine, 9-β-D-ribofuranosyl		150	75
6-Mercaptopurine, 9-β-D-arabinofuranosyl		400	63
3′-Amino-3′-deoxy-N,N-dimethyladenosine		50	32
6-Thioguanine	6-TG	1	60
6-Thioguanine, 9-β-D-ribofuranosyl		1	64
6-Thioguanine, 9-β-D-3-deoxyribofuranosyl		200	89
6-Thioguanine, 9-β-D-2-deoxyribofuranosyl		2	99
6-Thioguanine, 9-α-D-2-deoxyribofuranosyl		15	95
Toyocamycin		0.2	25
Sangivamycin		1.5	50
Formycin		20	22

[a] Abstracted from Goldin et al., 1968.

(36)

terchange of nitrogen and carbon atoms in the nucleus, variation of substituents on the pyrimidine ring, and the use of other sugars in place of D-ribose. Important results have been obtained in certain examples of each of these analogs. One example, 5-fluorouracil, represents a landmark in the rational design of anticancer drugs.

Among the nuclear modifications of pyrimidine antimetabolites, the 5- and 6-aza derivatives (Figure 24) have the most significant antitumor activity. 5-Azacytidine, an accepted clinical agent, represents another instance in which an antimetabolite was first synthesized (Sorm et al., 1964) and later isolated from a microorganism (Hanka et al., 1966). The mode of action of 5-azacytidine has two different aspects: it is incorporated into soluble RNA to produce inhibition of protein synthesis (Doskocil et al., 1967) and interferes with the *de novo* synthesis of pyrimidine nucleotides by inhibiting orotidylate decarboxylase (Vesely et al., 1968b). The corresponding 2'-deoxyribonucleotide, 5-aza-2'-deoxycytidine, has antitumor activity based on its inhibition of the incorporation of 2'-deoxyuridine into uridylic acid (Vesely et al., 1969). Another active analog is 5,6-dihydro-5-azacytidine. This compound is less potent than 5-azacytidine but produces equal prolongation of life for mice with L1210 leukemia at a 33-fold higher dose. The dihydro analog is more stable than 5-azacytidine to hydrolytic decomposition but it is also a better substrate for cytidine deaminase (Futterman et al., 1978). The ribonucleotides of 5-azauracil and 5-azaorotic acid (Figure 24) are competitive inhibitors of orotidylate pyrophosphorylase, the enzyme that converts orotic acid into orotidylic acid. Thus they inhibit pyrimidine biosynthesis (Hakala et al., 1956). 5-Azauracil decomposes to *N*-formyl biuret (Cihak et al., 1964) which inhibits the cyclization of ureidosuccinic acid (*N*-carbamylaspartic acid) to dihydroorotic acid. This effect may contribute to the modest antitumor activity of 5-azauracil (Skoda, 1963).

6-Azauracil is a toxic compound with poor antitumor activity. Its ribonucleoside, 6-azauridine, however, is more active and has fewer side effects (Handschumacher et al., 1962). This nucleoside is readily phosphorylated by uridine kinase, whereupon it becomes a specific competitive inhibitor of orotidylate decarboxylase, an important enzyme in pyrimidine biosynthesis (Handschumacher & Pasternak, 1958). The use of 6-azauridine against experimental tumors was complicated by its poor oral absorption. This problem was solved by using the corresponding 2',3',5'-O-triacetyl derivative, a latent form (prodrug) with good oral absorption (Slavik et al., 1971). 6-Azacytidine is anabolized to 6-azacytidylate, which is one-tenth as potent as 6-azauridylate in inhibiting orotidylate decarboxylase (Handschumacher et al., 1963). It is deaminated to 6-azauracil but this conversion provides no increase in antitumor activity.

3-Deazauridine (Figure 29), presently the most important pyrimidine nucleoside lacking one of the nuclear nitrogens, has been used with some success in acute myelogenous leukemia patients who no longer respond to ara-C (Wang & Bloch, 1972). the 5-β-D-ribofuranoside of uracil, known as pseudouridine (Figure 29) occurs in all species of *t*-RNA that have been sequenced (Phillips, 1969). It has no antitumor activity. However, the closely related antibiotic, oxazinomycin (minimycin), is active in a number of tumor systems (Haneishi et al., 1972; Kusakabe et al., 1972).

The design of 5-fluorouracil was based on the observation that certain tumors used uracil rather than orotic acid, the normal precursor, for pyrim-

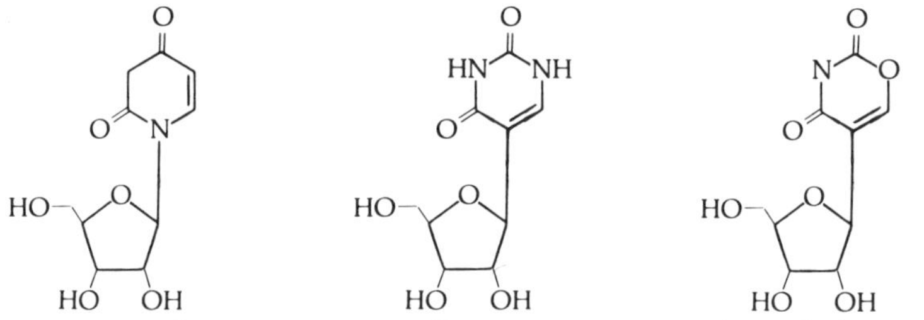

3-Deazauridine **Pseudouridine** **Oxazinomycin**

Figure 29 Deazapyrimidine nucleosides.

idine nucleotide biosynthesis (Heidelberger, 1975). A substituent at the 5-position was chosen in the expectation that it would block the conversion of uridylate into thymidylate and decrease DNA synthesis. This substituent would have to be compatible with anabolism to the active ribonucleotide and able to increase the strength with which this species binds to thymidylate synthetase. The fluorine atom fulfilled these requirements admirably and 5-fluorouracil (5-FU) became the most important agent for treating solid tumors (Heidelberger, 1975). Anabolism of 5-FU into its active form, 5-fluoro-2'-deoxyuridylate (FdUMP; Figure 25), occurs by a route that involves reaction with 2-deoxyribose-1-phosphate to give FUdR (Sköld, 1958), followed by phosphorylation to FdUMP by thymidine kinase (Harbers et al., 1959). An alternative route is based on reaction with ribose-1-phosphate, phosphorylation to the diphosphate, and reduction by ribonucleotide reductase (Kent & Heidelberger, 1972). The direct conversion of 5-FU into 5-fluorouridylate by a phosphoribosyltransferase operates in certain tumors (Reyes, 1969). It is possible to conduct cancer chemotherapy by administering 5-fluorouracil-2'-deoxyribonucleoside (FUdR), which is converted into the nucleotide by thymidine kinase (Heidelberger, 1975).

Inhibition of thymidylate synthetase is considered to be the main lethal effect of 5-fluorouracil-2'-deoxyribonucleotide (Cohen et al., 1958). This inhibitor binds competitively to the enzyme at first and its binding strength is promoted by the increased acidity conferred by the fluorine atom. The normal enzymic processes illustrated in Scheme 13 begin with a sulfhydryl group of cysteine adding to the 6-position of the pyrimidine. The 5-position then binds to the methylene group of 5,10-methylenetehydrofolate. When 2'-deoxyuridylate is the substrate its 5-hydrogen is transferred to this methylene group to provide the methyl group of thymidylate. The fluorine-carbon bond is stable, however, and this transfer does not occur with 5-fluorouracil-2'-deoxyribonucleotide. A terminal product with covalently bonded enzyme, substrate, and cofactor results (Santi & McHenry, 1972).

The therapeutic importance of 5-FU, coupled with its serious side effects, has stimulated an intensive search for better derivatives and analogs. A tetrahydrofuranyl derivative, known as ftorafur, has shown equal activity with reduced toxicity against gastrointestinal tract tumors (Hiller et al., 1967; Saito et al., 1974). It is now known that ftorafur is metabolized to 5-FU; hence it must be considered a prodrug form (Benevenuto et al., 1978). Other latent forms of 5-FU are 5-fluoropyrimidin-4-one (Johns et al., 1966) and 5-bromo-5,6-dihydro-5-fluoro-6-methoxy-2-deoxyuridine (Duschinsky et al., 1967; Figure 30). Further variations of the 5-FU structure which provides inhibitors of thymidylate synthetase are the β-D-arabinofuranoside of 5-fluorocytidine (ara-FC, Heidelberger, 1975) and its 2,5'-anhydro derivative (Alberto et al., 1978; Figure 32). 5'-Fluoro-5'-deoxythymidine was active against the synthetase obtained from the Ehrlich ascites carcinoma (Langen et al., 1969).

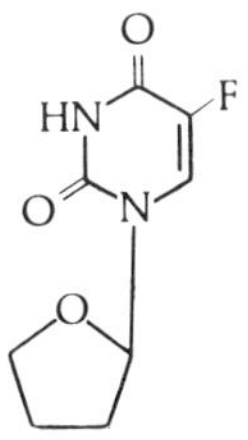

Ftorafur

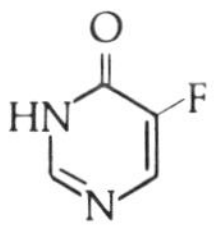

5-Fluoropyrimidin-4-one

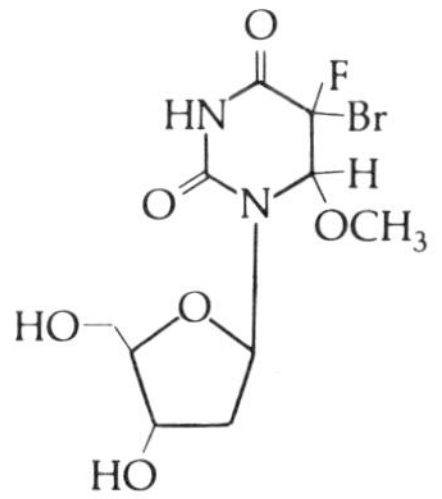

Figure 30 Latent forms of 5-FU.

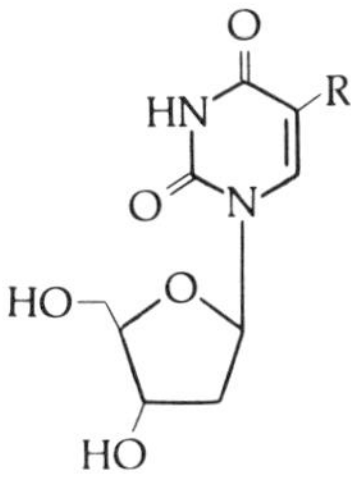

R = I, 5-IUdR
R = SH
R = CH$_3$S   **Figure 31** Analogs of 5-FUdR.

Another rationally designed antimetabolite is trifluorothymidine (Figure 32). This compound was conceived as an inhibitor of thymidylate synthetase (Heidelberger, 1975). It must be given as the 2'-deoxyriboside because mammalian cells are unable to form nucleosides from thymine and its analogs. Trifluorothymidine is readily phosphorylated by thymidine kinase and the nucleotide binds initially in a reversible manner to the enzyme. After preincubation, however, it becomes covalently bound (Reyes & Heidelberger, 1965) as a result of the extraordinary lability of the trifluoromethyl group on this molecule. It reacts with nucleophiles such as the amino group of glycine at neutral pH. With glycine, an amide is formed (Scheme 15; Heidelberger et al., 1964). Kinetic studies indicate that the reaction involves an initial nucleophilic attack at C-6, followed by loss of HF to give a difluoromethylene group. Addition of glycine to this group and hydrolysis of the two remaining fluorine atoms then gives the amide (Santi & Sakai, 1971).

Other 5-substituted uracil derivatives have been synthesized and evaluated as antineoplastic agents. The 5-chloro-, 5-bromo-, and 5-iodouracils and their ribosides show little antitumor activity (Prusoff & Goz, 1975). The corresponding 2'-deoxyribonucleosides are biologically active and one of

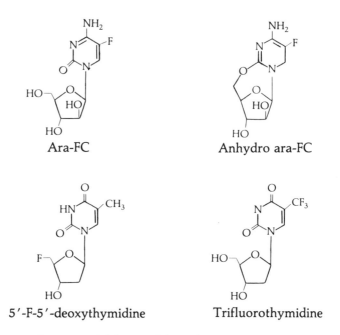

Figure 32 Inhibitors of thymidylate synthetase.

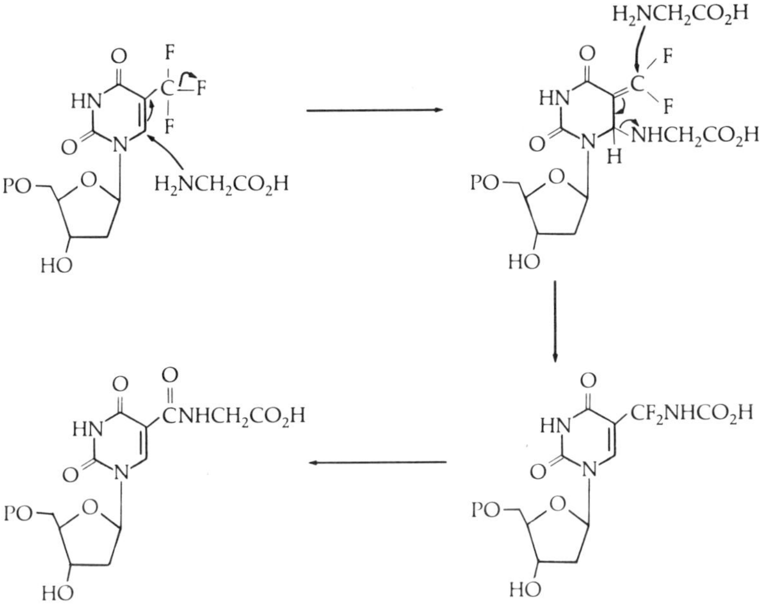

Scheme 15 Reaction of trifluorothymidylate with glycine.

them, 5-iodo-2'-deoxyuridine, is used against viral infections (Kaufman, 1965). 5-aminouridine has antitumor activity (Visser, 1955). It is anabolized to the phosphate, which inhibits orotidylate decarboxylase (Smith et al., 1966). Treatment of 5-aminouridine with nitrous acid produces a 5-diazo derivative in which the 5'-oxygen is added to C-6 to form a new ring (Thurber & Townsend, 1971). Dimethylamine converts this derivative into a 5-(dimethyltriazenyl) derivative (Equation 37) with significant antitumor activity (Golden et al., 1968). 5-Mercapto-2'-deoxyuridine (Figure 31), which is active against basal cell carcinomas in man (Schwartz et al., 1970), is converted into its 5'-phosphate by thymidine kinase, after which it inhibits

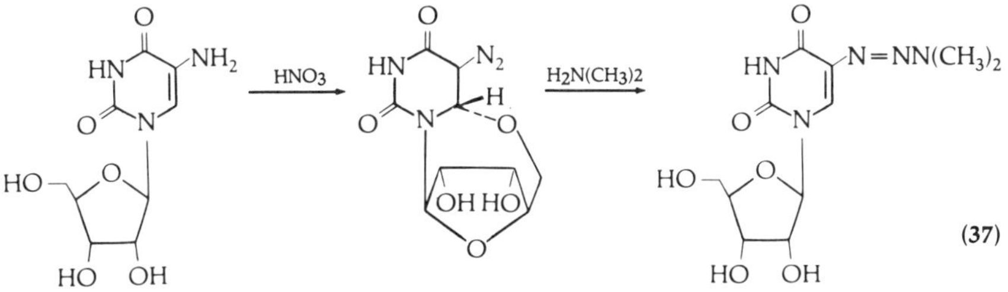

(37)

thymidylate synthetase (Kalman & Bardos, 1970). One problem with this compound is its oxidation to the corresponding disulfide, which is not a substrate for thymidylate synthetase. 5-Methylthio-2'-deoxyuridylate does act as a substrate for this enzyme (Kalman & Bardos, 1970).

The one really significant modification that has been made in the sugar moiety of pyrimidine nucleosides is the substitution of D-arabinose for D-ribose in cytosine (Creasey, 1975). The resulting drug, cytosine arabinoside (ara-C), is especially useful for inducing remissions in acute leukemia in adults (Bodey et al., 1969). It represents yet another example of an antimetabolite prepared first by synthesis and later isolated from a microorganism (Bergman & Feeney, 1951). Ara-C is anabolized to the triphosphate, which inhibits DNA polymerase (Furth & Cohen, 1968; Graham & Whitmore, 1970). The conversion of ara-C monophosphate into the inactive uridine analog by cytidine deaminase (Scheme 14), combined with diminished cytidine kinase activity, severely limits the duration of its effectiveness against leukemias (Steuart & Burke, 1971; Kessel, 1967). This difficulty has stimulated attempts to modify ara-C in ways that diminish its susceptibility to deamination. Thus the N-oxide (Figure 33) is resistant to cytidine deaminase and is slowly converted into ara-C by reduction (Panzica et al., 1971). Another slow-release derivative of ara-C is the 5'-O-palmitate (Wechter, et al., 1975). The interesting structure 2,2'-O-cyclocytidine also resists the deaminase and is converted slowly into ara-C (Kanai, et al., 1970).

In another approach to overcoming deamination of nucleosides compounds that are specific inhibitors of the deaminases are designed or discovered. One inhibitor is tetrahydrouridine (Hanze, 1967; Figure 34), a compound that requires preincubation with the enzyme for maximum effectiveness and produces noncompetitive inhibition of cytidine deaminase. It can be

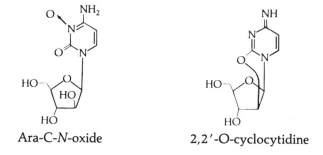

Ara-C-N-oxide   2,2'-O-cyclocytidine

Figure 33 Cytidine-deaminase-resistant derivatives of ara-C.

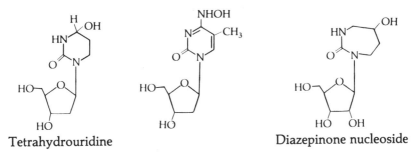

Figure 34 Inhibitors of cytidine deaminase.

considered a "transition-state-analog" inhibitor (Cohen & Wolfenden, 1971). A hydroxylamino analog, 4-hydroxylamino-5-methylpyrimidin-2-one-2'-deoxyribonucloside, acts as a competitive inhibitor and undergoes direct dehydroxylamination (Camiener, 1967; Trimble & Maley, 1971).

A useful inhibitor of adenosine deaminase, EHNA (Figure 35), was derived from the rational study of 9-substituted adenines. This study showed that for 2-hydroxy-3-alkyl substituents (in relation to the carbon attached to adenine) the erythro configuration was more active than the threo and maximum activity was obtained with an n-nonyl chain, which indicates that the enzyme has a large lipophilic region appropriately situated (Scheaffer & Schwender, 1974). Coformycin, an unusual nucleoside of microbial origin, has no antibacterial or antitumor activity but it potentiates the activity of ara-A by inhibiting adensoine deaminase (Nakamura et al., 1974). The corresponding 2'-deoxyribonucleoside (Figure 35) and isocoformycin, a syn-

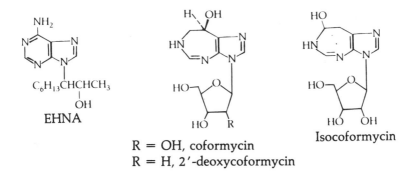

Figure 35 Inhibitors of adenosine deaminase.

thetic analog, are also potent inhibitors of this enzyme (LePage et al., 1976; Shimazaki et al., 1979). More recently a related diazepinone nucleoside (Figure 34) was found to be highly potent in inhibiting cytidine deaminase (Marquez et al., 1980).

The relative activities of selected pyrimidine antimetabolites against L-1210 mouse leukemia are listed in Table 8.

The development of antimetabolites based on folic acid represents a prime example of frustration in the effort to improve the clinical antitumor activity of an early lead compound. Following the structure elucidation of folic acid (Angier et al., 1946), a systematic series of variations on its structure was undertaken. The next year its 4-amino analog, aminopterin, was synthesized and the clinical activity of this analog was discovered one year later (Seeger et al., 1947; Farber et al., 1948). In 1949 amethopterin (Methotrexate), the N^{10}-methyl derivative of aminopterin was synthesized (Seeger et al., 1949). It soon supplanted aminopterin because of its decreased toxicity. Since that time no better antifolate has emerged for cancer chemotherapy, although antibacterial and antimalarial agents based on folate antagonism have been prepared. The failure to develop a better anticancer agent has occurred despite many variations on the structure. Furthermore, the mode of action of folic acid has been largely elucidated and this advance has allowed

Table 8 Antitumor Activity of Pyrimidine Antimetabolites Against L1210 Murine Leukemia[a]

Name	Synonyms	Dose (mg/kg)	Increase in Life Span (% T/C)
5-Diazouridine		300	45
5-Fluorouracil	5-FU	35	60
2'-Deoxy-5-fluorouridine	FUdR, Floxuridine	80	55
2'-Deoxy-5-iodouridine		600	40
2'-Deoxy-5-(trifluoromethyl)-uridine		50	75
Cytosine, 1-β-D-arabinofuranosyl	Ara-C	20	133
Cytosine, 1-β-D-arabinofuranosyl, N,O,O,O-tetraacetyl		1350	195
5-Fluorocytosine, 1-β-D-arabinofuranosyl	F-Ara-C	24	156
2'-Deoxy-5-fluorocytidine		67	35
6-Azauridine		200	35
6-Azauridine triacetate		300	44
5-Azacytidine		5.25	156

[a]Abstracted from Goldin et al., 1968.

the rational design of antifolates (Mead, 1975). Of course, it is still possible that a better drug, or at least an alternative to amethopterin, will emerge. Two antifolates, 5-methyltetrahydrohomofolate and a substituted 4,6-diaminodihydrotriazine, are listed as "compounds in development" by the National Cancer Institute.

The structure of folic acid (Figure 36) consists of a pteridine nucleus substituted with a 2-amino group, 4-hydroxyl group (keto tautomer), and 9-methylene group attached to the nitrogen atom of p-aminobenzoate. The carbonyl group of this benzoate is linked to the amino group of glutamatic acid.

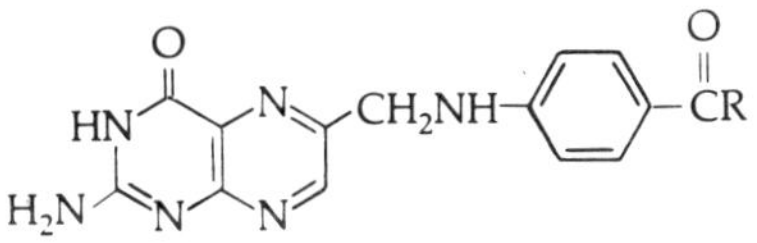

R = Glutamate, folic acid
R = Aspartate

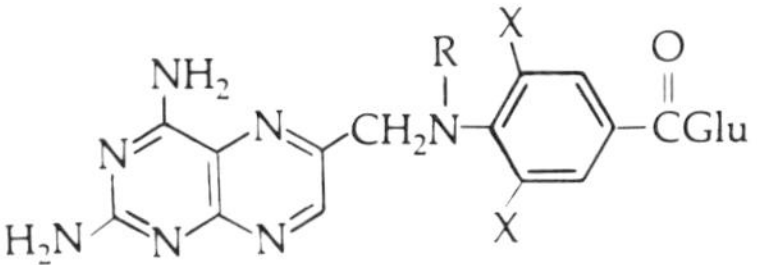

R = X = H, aminopterin
R = CH_3, X = H, amethopterin (MTX)
R = CH_3, X = Cl, dichloroamethopterin

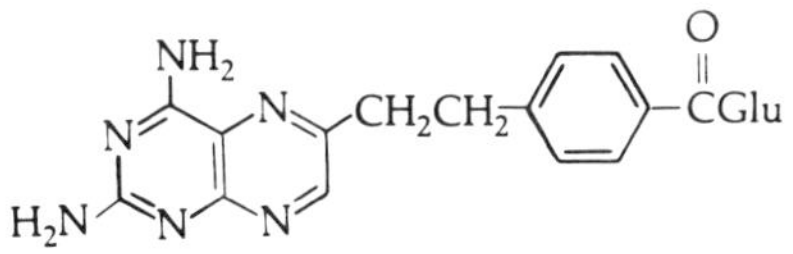

10-Deazaaminopterin

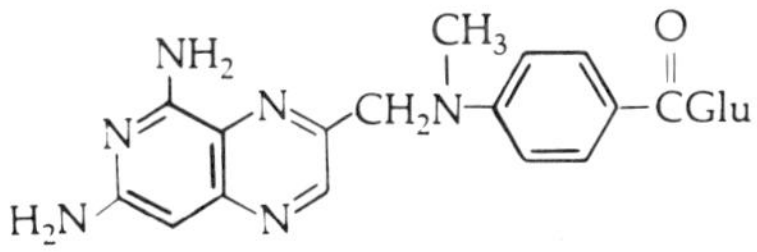

3-Deazaamethopterin

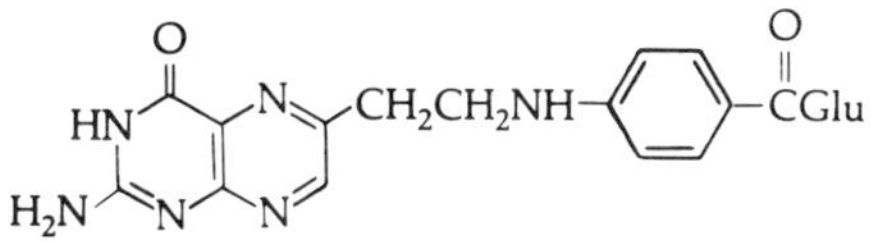

Homofolate

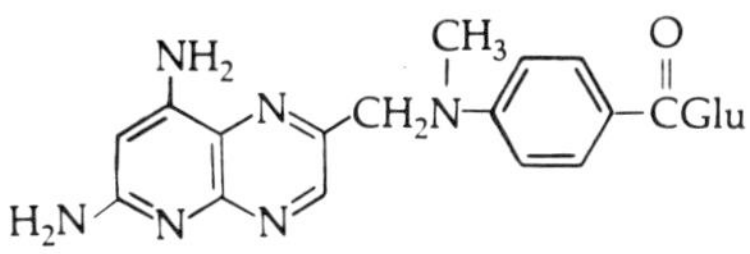

1-Deazaamethopterin

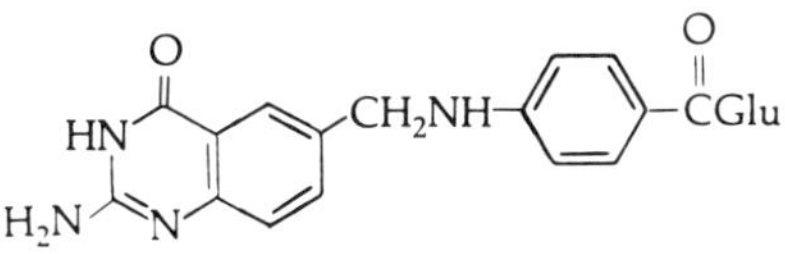

5,8-Deazafolate

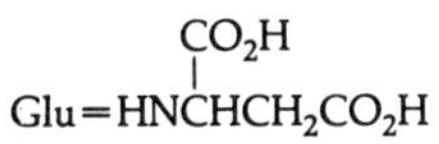

Glu = HNCHCH$_2$CO$_2$H
      |
      CO$_2$H

Figure 36 Folic acid and antimetabolites.

All three segments have been varied in analog structures. The glutamic acid has been replaced by asparate and other amino acids, by di-, tri-, and polyglutamate, and by mixed dipeptides (Hutchings et al., 1947; Boothe et al., 1949; Baugh & Krumdieck, 1971). None of these replacements enhanced the antitumor activity. A variety of substituents has been placed on the benzene ring of the p-aminobenzoate segment. The substitution pattern was a single substituent at 3' or disubstitution at the 2',3' and 3',5' positions (Cosulich et al., 1953). Halogens were used most frequently as the substituents, and one compound of this type, 3',5'-dichloroamethoptorin (Angier et al., 1959; Figure 36) showed a better therapeutic ratio than amethopterin against experimental tumors in mice (Goldin et al., 1959). Unfortunately this advantage was not found in the subsequent clinical trial. It was observed later that the 3',5'-dichloro analog is metabolized to a 7-hydroxy derivative in mice more readily than amethopterin, but there is little difference in metabolic hydroxylation in humans (Johns & Valerino, 1971). This observation probably explains the decreased toxicity and consequent therapeutic advantage in mice. Methylation of the amino nitrogen (N^{10}) of the p-aminobenzoic and segment of pteroylglutamic acid produced increased antitumor activity and it represents one of the two structural changes made in going from folic acid to amethopterin (Cosulich & Smith, 1948). Other substituents on N^{10} were less effective than methyl. Replacement of N^{10} of aminopterin with a carbon atom gave an analog (10-deazaaminopterin, Figure 36) with a broader spectrum of activity against experimental tumors (Sirotnak et al., 1978). However, a 9,10-double bond decreased activity (Struck et al., 1971). Substitution of pyridine for benzene, as in 2' and 3'-azofolates, or replacement of p-aminobenzoate by sulfanilamide also decreased activity (Roberts & Shealy, 1971; Fahrenbach et al., 1954).

The most important change in the pteridine segment was the replacement of the 4-hydroxyl group with an amino group which increased the binding strength to dihydrofolic reductase and provided powerful inhibitors such as aminopterin and amethopterin (Seeger, et al., 1947). The basicity of the amino group, which results in a protonated species at neutral pH, is responsible for the increased binding. Thus at pH 6 amethopterin bonds stoichiometrically with dihydrofolic reductase ($K_I = 10^{-10} M$) to give a degree of inhibition termed "psudoirreversible" (Bertino et al., 1964). At higher pH, however, the binding is weaker and competitive with the substrate. 4-Alkoxy-and 4-alkylamino folates did not show enhanced activities (Roth et al., 1950; Roth et al., 1951b). The latter folate should be highly basic but the compounds apparently suffer from steric hindrance to binding. Introduction of alkyl

CHEMISTRY OF ANTITUMOR DRUGS 157

groups at position 7 or on the 2-amino group also reduced activity (Boothe et al., 1952; Roth et al., 1951a). Replacement of nitrogen by carbon in the pteridine nucleus was done in two ways: changing the pyrimidine ring to pyridine as in 1-deazaamethopterin and 3-deazaamethopterin (Elliott et al., 1971; Figure 36) and changing the pyrazine ring to benzene as in 5,8-dideazafolate (Davoll & Johnson, 1970). Both types of change reduced activity. The 2,6-diaminopurine congeners of aminopterin and amethopterin were synthesized but they did not show useful activity (Weinstock et al., 1970).

Another change in folate structure that does look important is the homologation of the 9-methylene group to ethylene (De Graw et al., 1965). Homofolic acid (Figure 36), the resulting analog, is a substrate for dihydrofolic reductase rather than an inhibitor. However, tetrahydrohomofolate, its reduction product, is a specific inhibitor of thymidylate synthetase (Goodman et al., 1964). Tetrahydrohomofolate and dihydrohomofolate were investigated for activity against tumors that had become resistant to amethopterin because they produced large amounts of dihydrofolic reductase. Activity was found against these tumors, but only borderline inhibition was found against amethopterin-sensitive tumors (Mead et al., 1966; Mishra & Mead, 1972). The 5-methyl derivative of tetrahydrohomofolic acid (Figure 37) was then prepared (Knott & Taunton-Rigby, 1971) and found to have activity against both sensitive and resistant strains of L-1210 leukemia (Mishra & Mead, 1972). It is also more stable than tetrahydrohomofolate, which improves its chances of becoming a clinical agent. Other reduced analogs of folic acid, including tetrahydroaminopterin (Kisliuk, 1960), tetrahydroamethopterin (Figure 37; Mead et al., 1961) and 5-methyltetrahydroamethopterin (Slavik & Zakrzewski, 1967), have been synthesized but none was as active as amethopterin. Table 9 shows the relative activities of folic acid antagonists against L-1210 leukemia in mice.

In contrast to dihydrofolate reductase inhibitors, which resemble the substrate closely in structure, a large number of fairly potent inhibitors differ

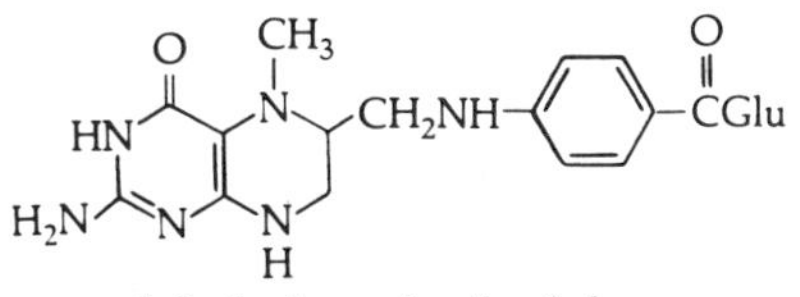

5-Methyltetrahydrofolate

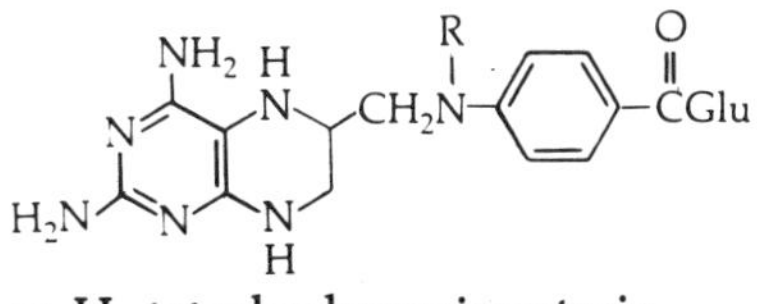

R = H, tetrahydroaminopterin
R = CH_3, tetrahydroamethopterin

Figure 37 Analogs of tetrahydrofolate.

Table 9 Antitumor Activity of Folic Acid Antagonists Against L1210 Murine Leukemia[a]

Name	Synonyms	Optimal Dose (mg/kg)	Increase in Life Span (% T/C)
Aminopterin		0.18	45
Amethopterin	Methotrexate	0.8	160
3′,5′-Dichloroamethopterin		40	>700
3′-Bromo-5′-chloroamethopterin		60	>695
3′-Chloroamethopterin		1.25	205
3′-Bromoamethopterin		1.5	178
N^{10}-Ethylaminopterin		1.25	93
9-Methylamethopterin	Adenopterin	0.7	136
4-Aminopteroylaspartic acid	Amino-AN-FOL	30	82
4-Aminopteroylalanine		18	54
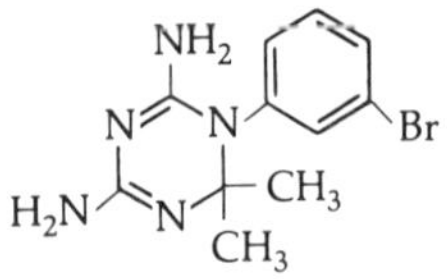 (phenyl-dimethyl triazine with Br, structure shown)		25	50

[a]Abstracted from Goldin et al., 1968.

significantly from dihydrofolate. They include the 2,4-diaminopyrimidines and the 4,6-diaminodihydrotriazines. These inhibitors display significant species specificity toward the reductases; for example, trimethoprim (Figure 38) is highly selective for the bacterial enzyme (60,000 times more effective on an *E. coli* enzyme than a pigeon liver enzyme), whereas pyrimethamine and cycloguanil (Figure 36) are highly selective for the enzyme from *Plasmodium* (Hitchings & Burchall, 1965). Observations of this type led Baker to advance the hypothesis that in each of the various types of dihydrofolate reductase the active site was the same but there were structural differences in areas adjacent to the active site. Compounds like amethopterin do not show species specificity because they bind completely within the active site. In contrast, compounds such as trimethoprim and cycloguanyl bind partly to the adjacent area (Baker, 1969). Baker then assumed that there are differences between normal and cancer cells in areas adjacent to the active site of their dihydrofolate reductases and prepared many compounds designed to probe these differences. One compound, the phenylbutyl derivative of a 4,6-diaminodihydrotriazine (Figure 38), showed a 100-fold difference in selectivity between a normal rat liver enzyme and a rat tumor enzyme (Baker, 1968). To magnify this selectivity the inhibitor structure was modified further to include

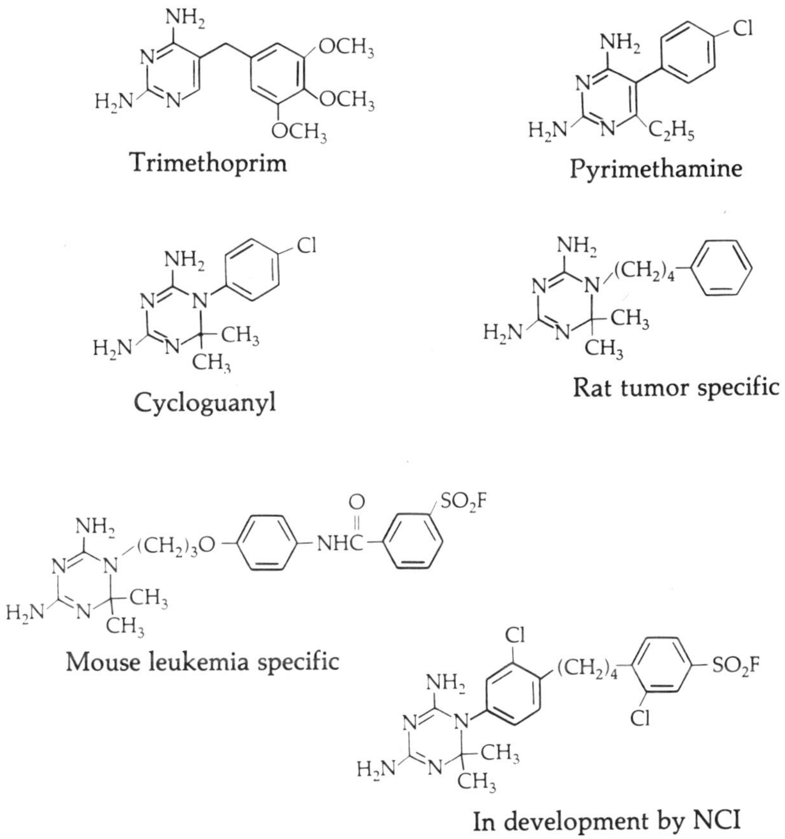

Figure 38 Species specific inhibitors of dihydrofolic reductase.

fluorosulfonyl, a functional group that may bind covalently with hydroxyl groups on the adjacent site. The resulting compound (Figure 38) showed a 2000-fold difference in selectivity between the isolated enzymes from mouse liver and mouse leukemia. However, it failed to penetrate the intact cells (Baker, 1969). More recently an analog with two chlorine atoms (Figure 38) showed good antitumor activity.

Little progress has been made in developing improved analogs of compounds such as azaserine and DON which are glutamine antagonists (Bennett, 1975). Three additional antibiotics that contained the DON structure in conjugated form were identified (Figure 23; Rao et al., 1960; DeVoe et al., 1957). They appear to exert their antitumor effects only after cleavage to

DON (Bennett, 1975). Synthetic analogs such as alanylserine diazoacetic ester and 5-diazo-4-oxonorvaline are much less active than the parent compounds because their diazomethyl group is not properly positioned for reaction with a sulfhydryl group in phosphoribosylformylglycinamidine synthetase or other glutamine-requiring enzymes (Baker, 1967).

Acivicin (AT-125) is an antibiotic isolated from *Streptomyces svicens* (Martin et al., 1973). It functions as an irreversible inhibitor of glutamine amidotransferases, including those involved in purine and pyrimidine biosynthesis (Jarjaram et al., 1975). It also inhibits asparagine synthetase, the enzyme that transfers an amino group from glutamine to aspartic acid, a property that has led to its experimental use in combination with asparaginase (Cooney et al., 1974). The structure of acivicin (Equation 38) shows similarity to DON and azaserine and resembles them in its two-step inhibition of amidotransferases. Thus initially it forms a reversible covalent complex that is followed by irreversible covalent bond formation (Tso et al., 1980). A cysteine residue at the enzyme active site participates in the covalent attachment, presumably by an alkylation reaction in which it replaces the chlorine atom of acivicin (Equation 38). Phase I clinical studies on acivicin have revealed the development of CNS toxicity. It is thought that this toxicity results from a structural resemblance to ibotenic acid (Figure 39), a known CNS toxin found in mushrooms.

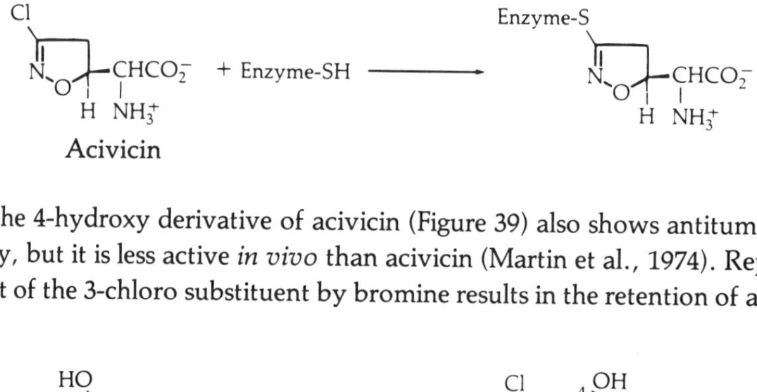

$$\text{Acivicin} + \text{Enzyme-SH} \longrightarrow \text{Enzyme-S-product} \tag{38}$$

The 4-hydroxy derivative of acivicin (Figure 39) also shows antitumor activity, but it is less active *in vivo* than acivicin (Martin et al., 1974). Replacement of the 3-chloro substituent by bromine results in the retention of antitu-

Figure 39 Structures related to acivicin.

mor activity at the same level of potency. Studies are presently underway to attempt the separation of antitumor activity from CNS toxicity in acivicin analogs.

Another rather remarkable antimetabolite is alanosine, an antitumor antibiotic isolated from *Streptomyces alanosinicus* (Murthy et al., 1966). This compound antagonizes aspartic acid by blocking its transport and incorporation into adenylosuccinate (see Scheme 10; Gale et al., 1968). Thus adenosine monophosphate cannot be formed. The isolation of an alanosine metabolite which is a condensation product between alanosine and 5-amino-4-imidazolecarboxylic acid ribonucleotide (Equation 39), suggests that this metabolite is the actual inhibitor of adenylosuccinate synthetase (Tyagi & Cooney, 1980).

$$\text{Alanosine} + \text{AICAR} \longrightarrow \text{product} \quad (39)$$

A further mode of antimetabolite action is found in the inhibitors of ribonucleoside diphosphate reductase. These compounds prevent the conversion of ribonucleotides into the corresponding 2'-deoxyribonucleotides, thereby blocking the biosynthesis of DNA. The reductase is composed of two subunits, one of which contains iron (Krakoff et al., 1968). Antitumor compounds that inhibit this enzyme do so by interfering with the iron-containing subunit. They are typically small organic molecules with the ability to chelate iron strongly. Examples include hydroxyurea (Stearns et al., 1963) and guanazole (Brockman, et al., 1970a; Figure 40). A variety of thiosemicarbazones has antitumor activity but toxicity has prevented clinical use. 5-Hydroxypyridine-2-carboxaldehyde thiosemicarbazone (Figure 40) and related heterocyclic compounds are powerful chelating agents for iron and other transition metals and a direct correlation between their chelating ability and antitumor activity has been claimed (Brockman et al., 1970b; Michaud & Sartorelli, 1968). Bis(thiosemicarbazones) of α-ketoaldehydes and

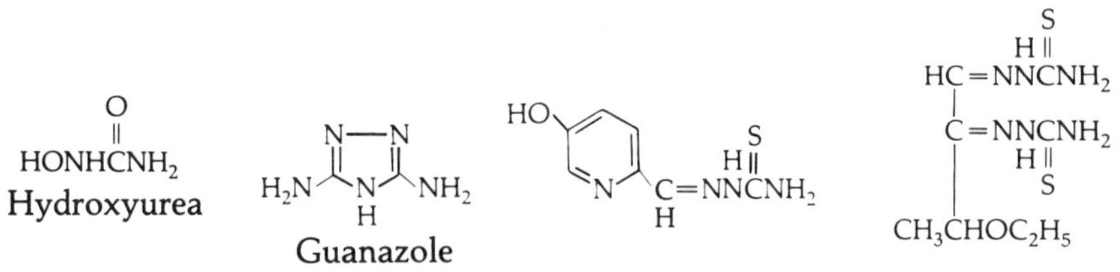

Figure 40 Inhibitors of ribonucleoside diphosphate reductase.

α-diketones also have antineoplastic activity. The most thoroughly studied compounds of this type are the derivatives of methylglyoxal and 3-ethoxy-2-oxobutyraldehyde (Figure 40; French & Friedlander, 1958) which form strong chelates with metals. It has not been established that this property is responsible for their cytotoxicity (Booth & Sartorelli, 1967).

A different type of antimetabolite effect is provided by the enzyme L-asparaginase, which creates a deficiency in the amino acid L-asparagine. Discovery of the antitumor activity of this enzyme resulted from experiments in which guinea pig serum was injected into mice and rats with certain transplanted tumors (Kidd, 1953). When regressions of these tumors occurred, the causative agent in the serum was L-asparaginase. Apparently there is a qualitative difference in the L-asparagine requirement of normal and certain tumor cells or at least in their ability to respond to the deficiency (McCoy et al., 1959). The limited spectrum of neoplasms inhibited by L-asparaginase (certain leukemias) indicates that these differences are not a general phenomenon. Commercial production and clinical use were made possible by the discovery that *E. coli* produces a form of this enzyme with antineoplastic activity (Mashburn & Wriston, 1964). Clearance of the enzyme from plasma results from an immunological reaction. Patients who become sensitive to this reaction can be treated with the enzyme obtained from *Erwinia carotovora* (Hrushesky, 1976).

Synthesis of Antimetabolites

In this section, as in the case of the alkylating agents, only the syntheses of clinically used agents are described. The following references are provided for the reader who wishes to consult more extensive treatments of purines and pyrimidines and their nucleosides: Walker, R. T., 1979; Goodman, L., 1974; Ohno, M., 1980; Zorbach & Tipson, 1968; Townsend & Tipson, 1978.

6-Mercaptopurine is prepared by heating hypoxanthine with phosphorus pentasulfide. The yield and reliability of this reaction is imporved by running it in refluxing pyridine (Beaman & Robins, 1961; Equation 40). This method is also used for the synthesis of 6-thioguanine from guanine (Elion & Hitchings, 1955). A nucleophilic displacement by the sulfur atom of 6-mercaptopurine on the chlorine atom of 5-chloro-1-methyl-4-nitroimidazole produces azathioprine (Hitchings & Elion, 1962; Equation 40).

The original synthesis of 5-fluorouracil involved the condensation of 5-ethylisothiouronium bromide with the potassium salt (enolate) of 2-fluoro-2-formylacetate (Duschinsky et al., 1957; Equation 41). More recently, preparation of 5-fluorouracil has been accomplished by direct fluorination of uracil with F_3COF (Earl & Townsend, 1972). For the preparation of 5-fluorouracil-2'-deoxyribonucleoside the monomercuri derivative of 5-fluorouracil is condensed with 3,5-di-O-(4-methylbenzoyl)-2-deoxyribosyl-1-chloride and the resulting intermediate is deprotected by alkaline hydrolysis (Hoffer et al., 1959; Equation 42).

The preparation of cytosine arabinoside involved the conversion of the 4-carbonyl group of uracil arabinoside into an amino group by way of the intermediate 4-thiocarbonyl group. Thus treatment with phosphorus pentasulfide in pyridine, followed by heating with ammonia at 98–105° in a bomb, furnished the desired product (Hunter, 1963; Equation 43). 5-Azacytidine

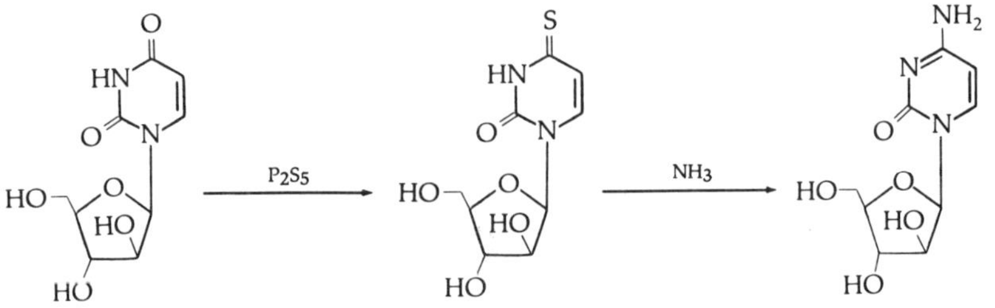

was prepared by a sequence beginning with peracetylribosyl-1-isocyanate (Scheme 16). This compound was condensed with 2-methylisourea and the product was treated with methyl orthoformate to give the peracetylriboside of 5-azauridine. Deacetylation and amination then completed the synthesis (Piskala & Sorm, 1964).

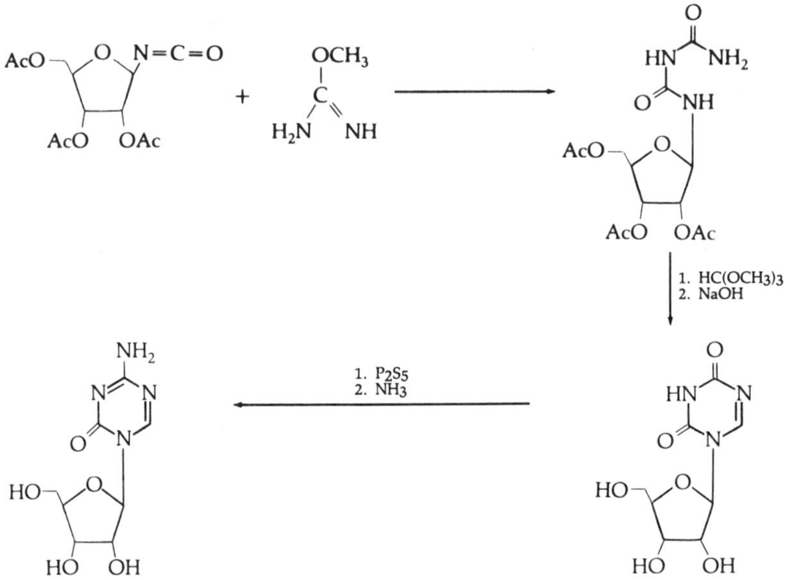

Scheme 16 Synthesis of 5-azacytidine.

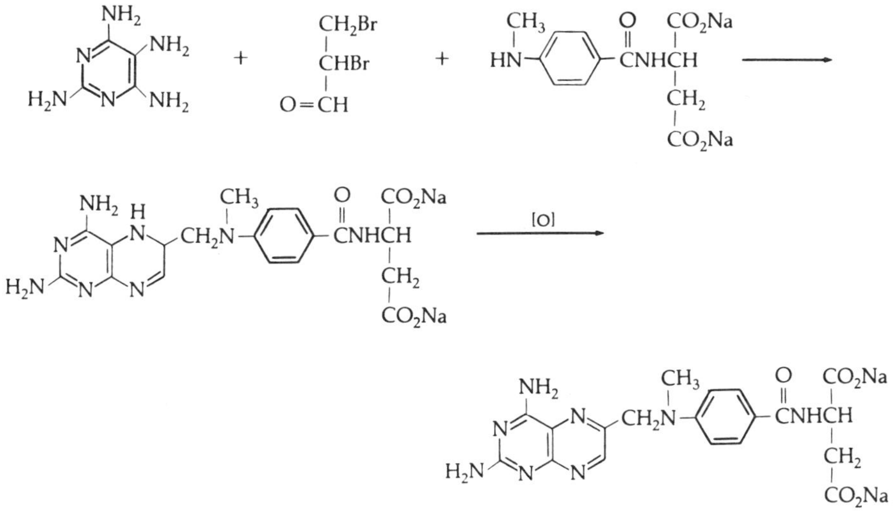

Scheme 17 Synthesis of amethopterin.

Amethopterin has a complex pteridine structure, but its preparation was accomplished in two steps from 2,4,5,6-tetraaminopyrimidine. A combination of this compound, the disodium salt of N-(4-methylaminobenzoyl)glutamic acid, and 2,3-dibromopropionaldehyde was condensed in the presence of iodine and potassium iodide. This mixture was filtered, heated with lime water, and acidified to yield the product (Seeger et al., 1949; Scheme 17).

INTERCALATION OF ANTITUMOR DRUGS INTO DNA

The Intercalation Process

Some of the most important antitumor drugs, for example, adriamycin and bleomycin, owe at least part of their cytotoxic effect to DNA intercalation. Intercalation is a rather remarkable binding process that occurs between double helical DNA and molecules that have relatively flat polycylic areas. In this process the deoxyribose-phosphate backbone of the helix unwinds partially to form a separation between adjacent base pairs. The intercalating agent then moves into this separation, as shown in Figure 41. In ethidium (Figure 42) the separation distance occupied is 3.4 Å and the helix unwinding

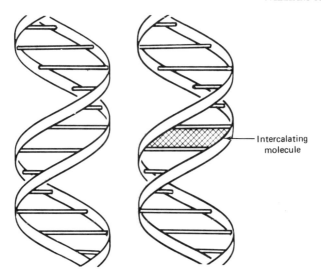

Figure 41 Intercalation into double helical DNA.

angle is 12° (Fuller & Waring, 1964). The intercalated portion of the drug lies perpendicular to the helix axis and binds to the base pairs above and below by van der Waals and other noncovalent forces. Additional binding can occur between the nonintercalated portion of the drug and the DNA outer surface. Thus daunorubicin (Figure 52) is thought to have a specific hydrogen bond between the 4'-hydroxyl group of its sugar moiety and a phosphate group on the DNA and an electrostatic attraction between its 3'-ammonium group and DNA phosphate anion (Pigram et al., 1971). Furthermore, the binding of antinomycin D (Figure 49) and double helical DNA involves intercalation of the phenoxazone chromophore between adjacent G-C pairs, with the two pentapeptide lactone functionalities lying in the minor groove in opposite directions and making numerous van der Waals interactions with the DNA. A specific hydrogen bond between the 2-amino groups of guanines and the carbonyl

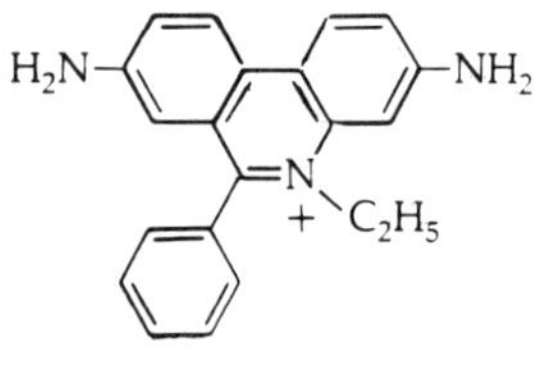

Figure 42 Ethidium.

groups of threonines in the pentapeptide rings also occurs (Sobell & Jain, 1972).

Intercalation of drugs and other molecules causes pronounced changes in the properties of DNA (Lerman, 1961). Most of these changes reflect the unwinding, lengthening, and stiffening of the double helical structure. Thus the viscosity increases, the buoyant density decreases, and the thermal denaturation temperature increases. Changes also occur in the intercalating agent, as shown by the shift in the visible light absorption maximum of ethidium from 480 to 520 nm (Waring, 1965) and the strong enhancement of its fluorescence quantum yield (Le Pecq & Paoletti, 1967). These changes are used as indicators of the occurrence of intercalation. Other indicators are shifts in the NMR spectra for the base pairs and drug, flow dichroism studies that show the plane of the drug perpendicular to the helical axis, and X-ray diffraction patterns that show the DNA lengthened but not thickened. Recent studies of the effects of intercalation on the supercoiling of closed circular duplex DNA have greatly improved our understanding of this binding process (Crawford & Waring, 1967); for example, when negatively superhelical circular DNA is treated with an intercalating agent such as ethidium the supercoiling is diminished until a relaxed, untwisted circular state is obtained. The addition of more ethidium causes the circles to develop reversed supercoils. These changes in the DNA structure can be followed by changes in the sedimentation coefficient, which falls to a minimum for the relaxed, untwisted state. In this manner the helix unwinding angle can be determined.

The intercalation of drugs and other molecules effects the function of DNA. Cytotoxicity results if the normal operation of DNA as a template is seriously impaired. One way in which this impairment is expressed is the inhibition of polymerases. Thus ethidium and daunorubicin inhibit DNA-dependent DNA polymerase I and DNA-dependent RNA polymerase from *E. coli* (Waring, 1975; Di Marco et al., 1965), whereas actinomycin D and aureolic acid inhibit the DNA-dependent RNA polymerase selectively (Reich et al., 1961; Behr & Hartmann, 1965). These inhibitions depend on binding the drugs to primer DNA rather than direct action on the enzyme or competition with cofactors. In a representative experiment inhibition of the RNA polymerase from *E. coli* by actinomycin D could not be reversed by increasing the concentration of enzyme, cofactors, or precursors, but it was overcome by adding more DNA (Kersten & Kersten, 1962). It is thought that intercalation prevents the progression of the enzyme along the DNA template (Hyman & Davidson, 1970). The mode of action of actinomycin D is actually more complex than the experiments outlined indicate. Thus recent studies have shown that it does not inhibit the transcription of ribosomal RNA directly. Instead, it acts at a nucleoplasmic site to inhibit the transcription of messenger RNA which codes for a protein-initiation factor for nucleolar RNA polymerase I (Lindell et al., 1978; Tsurugi et al., 1972).

Drugs that effect fundamental biological mechanisms like DNA replication and transcription are active across a broad spectrum of biological systems. The potent intercalating agents show activity against DNA viruses, bacteria, fungi, and mammalian cells, although there are many species differences; for example, actinomycin strongly inhibits DNA-dependent RNA polymerase from *E. coli*, but it is inactive against the intact cells because it fails to penetrate them (Hartmann & Coy, 1962). The lethal action of intercalating agents is usually expressed at the stage of mitosis, where cell division fails to occur. There is severe damage to the nucleolus and chromatin (Di Marco et al., 1967). In bacteria threshold concentrations permit cell growth without cell division that results in long, filamentous forms (Kersten & Kersten, 1962).

Antitumor Drugs that Intercalate into DNA

The acridines are one family of chemical structures that show the ability to intercalate into double helical DNA. Members of this family, including acridine, proflavine, and acridine orange (Figure 43), were used in the pioneering investigation of intercalation (Lerman, 1961). Although 9-aminoacridine is an intercalator, it lacks antitumor activity. Modifications of its structure, however, have led to a number of compounds that show good activity against rodent tumors. One of these compounds, the methanesulfon-*m*-anisidide derivative (*m*-AMSA, Figure 44), is undergoing extensive clinical trials. It was developed in the following way. On the basis of the literature report of antitumor activity for 9-(4-dimethylamino)anilinoacridine (Goldin, et al., 1966) and on their own earlier studies, Cain and coworkers pre-

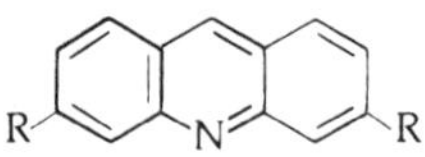

Acridine, R = H
Proflavine, R = NH_2
Acridine orange, R = $N(CH_3)_2$

Figure 43 Intercalating acridines.

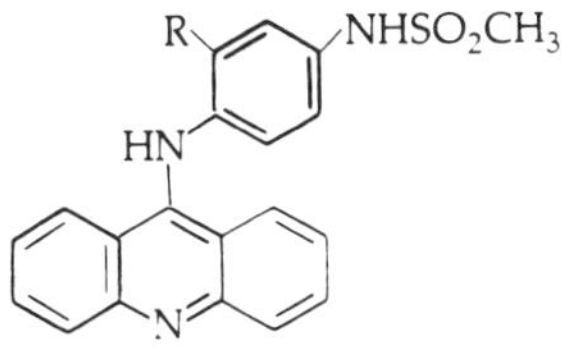

Figure 44 9-Anilinoacridines.

4-Methanesulfonamido derivative, R = H
m-AMSA, R = OCH$_3$

pared a variety of analogs substituted in the aniline ring. The 3,4-diamino analog had good activity but was unstable to air oxidation. Reasoning that a point electron donor site was needed on the anilino ring, they decided that a sulfonamide group, which generated a small proportion of negatively charged species at physiological pH, might serve the same purpose and be more stable. The resulting 4-methanesulfonamido derivative (Figure 44) was equal in activity to the diamine (Atwell et al., 1972). With this derivative as the lead, a new series of analogs was prepared. Electron-releasing substituents such as methyl, amino, and methoxy in the aniline ring produced active analogs, one of which, a compound known as m-AMSA (Figure 44), showed clearly superior antitumor activity. Its synthesis is illustrated in Equation 44. Hydrophobic substituents such as halogen in the acridine nucleus also increased activity. A quantitative structure-activity analysis was made by a modified Free-Wilson approach in which all comparisons were made at isolipophilicity (Cain, et al., 1975).

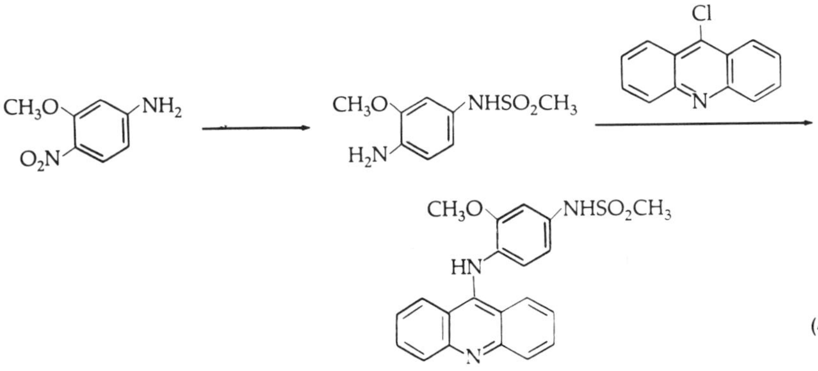

(44)

In recent years the use of bisintercalating agents has been explored. The basic premise of these agents is that a single molecule that can intercalate into DNA at two different sites will provide strong binding and inhibition of the template function. Quinoxaline antibiotics such as echinomycin are naturally occurring bis-intercalators (Wakelin & Waring, 1976). Among the synthetic compounds, a series of bis(9-aminoacridines) linked by their amino groups with alkyl chains, showed significant inhibition of P-388 murine leukemia (Canellakis et al., 1976). The compound in which the alkyl chain is six methylene units (diacridinyl hexanediamine, Figure 45) is undergoing study by the National Cancer Institute. A related series of compounds had the acridine nitrogen methylated to give positively charged intercalating groups (Figure 45). The acridinium rings were substituted by 2-methoxy and 6-chloro groups in some cases. It was found that compounds linked by 6, 8, 10, or 12 methylene units were bis-intercalators, whereas those with 4-methylene units could give only monointercalation. These compounds inhibited the DNA-dependent DNA polymerase from *E. coli* by interfering with template function. The bisacridinium compounds inhibited synthesis more efficiently than the corresponding monoacridines (Lown et al., 1978b).

A number of antimalarial drugs, including quinacrine and chloroquine, intercalate into DNA, although few have potent antitumor activity. One of the most active antitumor compounds of this type is hycanthone (Figure 46), whose structure suggests that is should be able to make an electrostatic interaction with a phosphate anion on the DNA and to intercalate between the base pairs.

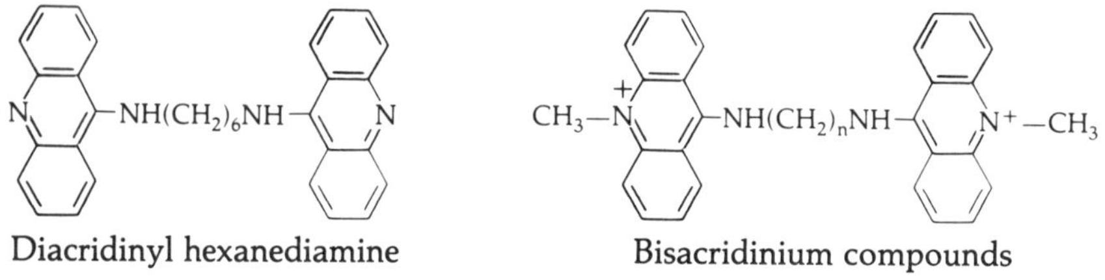

Diacridinyl hexanediamine

Bisacridinium compounds

Figure 45 Bisintercalating acridine derivatives.

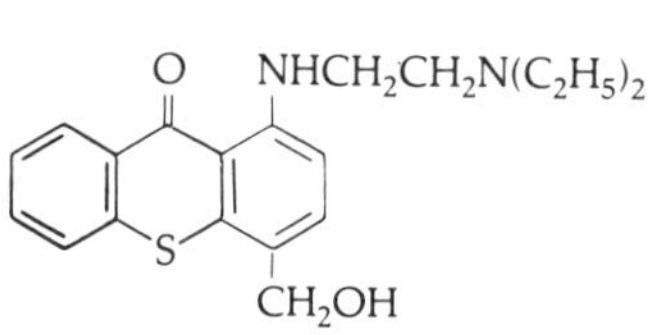

Figure 46 Hycanthone.

As mentioned above, phenanthridinium compounds such as ethiduim are effective intercalators. Ethidium is not a practical anticancer drug, although it can inhibit tumor cells in combination with azaserine (Kandaswamy & Henderson, 1962). Its main use is in veterinary medicine for the treatment of trypanosomiasis. Naturally occurring alkoxybenzo[c]phenanthridine alkaloids such as nitidine and 6-methoxy-5,6-dihydronitidine (Figure 47) and their synthetic analogs are active against experimental tumors (Zee-Cheng & Cheng, 1973). Coralyne, a related dibenzo[1,g]quinolizinium alkaloid and certain of its synthetic analogs (Figure 47) also have antileukemia activity in mice (Zee-Cheng et al., 1974).

The alkaloid ellipticine (Figure 48) exerts its antitumor effect by way of intercalation but only after microsomal oxidation to its 9-hydroxyl derivative (Paoletti et al., 1978). This derivative is further oxidized in a two-electron process to a quinone imine (Equation 45) that is thought to provide covalent alkylation of DNA. Ellipticine has received considerable attention as a potential clinical agent, but its poor solubility and toxicity have delayed its development (Rahman et al., 1978). A variety of derivatives and analogs have been prepared with particular emphasis on alkylation of the indole nitrogen or its replacement with oxygen and sulfur. The 9-methoxy

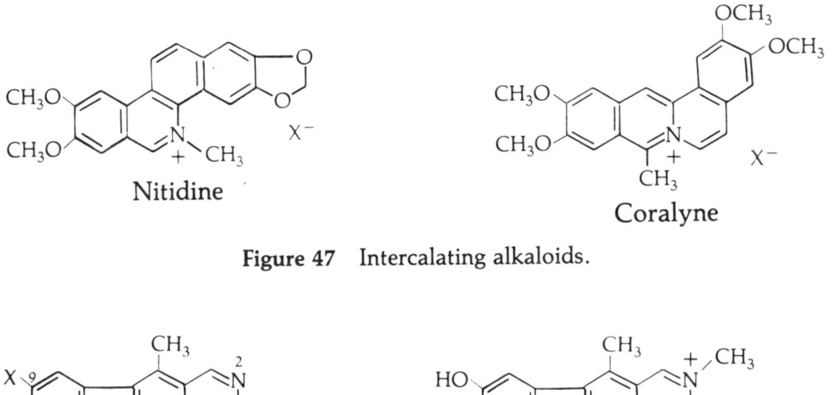

Nitidine

Coralyne

Figure 47 Intercalating alkaloids.

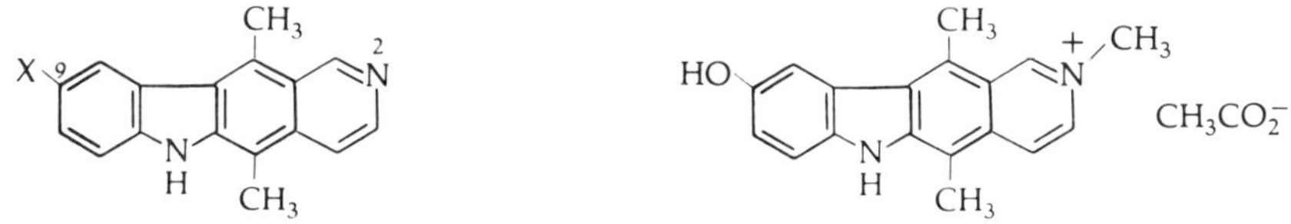

Ellipticine, R = H
9-Hydroxyellipticine, R = OH
9-Methoxyellipticine, R = OCH_3

9-Hydroxy-2-methylellipticinium acetate

Figure 48 Ellipticine analogs.

analog had a brief clinical trial in Europe (Mathe et al., 1970). Recent studies in France have demonstrated that the 9-hydroxy metabolite is as good or better than ellipticine in DNA binding and antitumor activity in mice (Paoletti et al., 1979). This metabolite proved to be inactive in humans, but it was found that the inactivation was caused by rapid glucuronide formation (Van-Bac et al., 1980). A way around this difficulty was found by methylating the pyridine nitrogen. The resulting compound, 9-hydroxy-2-methylellipticinium acetate (Figure 48) has antitumor activity in humans and is in Phase II clinical trial (Van-Bac et al., 1980). This improvement in activity results from quaternary salt formation, which inhibits glucuronide formation, but not oxidation to a quinone imine. Quarternization also prevents 9-hydroxylation, as shown by the inactivity of 2-methylellipticinium acetate (Paoletti et al., 1979).

(45)

Actinomycin C_1 (D) was the first antitumor antibiotic isolated (Waksman & Wodruff, 1940), although its activity against tumors was not established until later (Farber, 1958). In the meantime, actinomycin C_3 was isolated (Brockmann & Grubhofer, 1950) and shown to be effective against Hodgkin's disease (Schulte, 1952). As already noted, compounds in the actinomycin family have complex structures composed of a chromophoric portion, 2-amino-4,5-dimethyl-3-phenoxazone-1,9-dicarboxylic acid, known as actinocin, and two pentapeptide lactone rings (Figure 49). All naturally occurring actinomycins have the same chromophore, but they differ in the amino acid composition of the pentapeptide lactones. These lactones can be identical (isoactinomycines) or different (anisoactinomycins) in the same molecule. The lactone ring attached to the 9-carboxyl group is denoted α, and the other, as β. Actinomycin D is an isoactinomycin with the following sequences of amino acids, starting from the chromophore: L-threonine, D-valine, L-proline, sarcosine and L-N-methylvaline. The lactone ring is formed between the hydroxyl group of threonine and the carboxyl group of L-N-methylvaline. The structure of actinomycin C_3 is the same, except that it has D-alloisoleucine instead of D-valine in each lactone ring. A rational naming system for actinomycins is based on two parent compounds: actino-

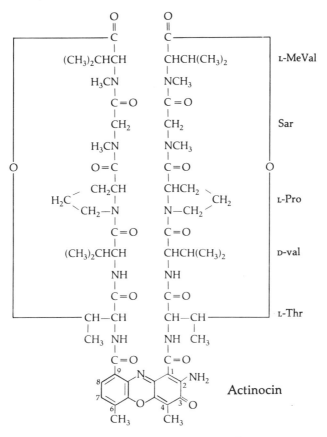

Figure 49 Actinomycin D (C_1).

mycin D takes the abbreviated name Val_2-AM and actinomycin C_3 becomes a Ile_2-AM. Other actinomycins are named according to how they differ in amino acid compositions from these two; standard IUPAC abbreviations representing the amino acids (Meienhofer & Atherton, 1973). Thus actinomycin V (X_2), which differs from actinomycin D by having L-4-oxoproline instead of L-proline as the third amino acid in the β-chain, is designated [βOpr³]-Val_2-AM. About 17 naturally occurring actinomycins have been identified, but the two parent compounds remain the only ones used clinically.

The serious toxicity of actinomycins has stimulated efforts to prepare analogs with improved therapeutic properties. Three different approaches have been taken toward this goal: directed biosynthesis, partial synthesis, and total synthesis. Directed biosynthesis can change the distribution of known constituents in a mixture; for example, the normal fermentation of *S. antibioticus* produces mainly actinomycin D (Val$_2$-AM) with only traces of analogs [Sar^3]-Val$_2$-AM and [Sar^3]$_2$-Val$_2$-AM in which one or both prolines are replaced by sarcosine. However, addition of sarcosine to the fermentation greatly enhances the formation of the trace analogs (Katz & Goss, 1959). Novel actinomycins can also be produced by directed biosynthesis by the addition of unnatural substrates. This approach is limited by substrate specificity of the enzyme complex that synthesizes actinomycins, but all new structures made have shown biological activity. The most fruitful kind of new substrates have been the proline analogs. They include the 4-methyl, chloro, and bromo derivatives of proline (Katz et al., 1977; Mauger, 1980), the thiazolidine-4-carboxylic acid isostere (Nishimura & Bowers, 1967), and compounds such as pipecolic acid and azetidine-2-carboxylic acid that have different ring sizes (Figure 50). Pipecolic acid gave a mixture of six products, some of which contained 4-hydroxypipecolic acid or 4-oxypipecolic acid (Katz, 1974). One of the products obtained from azetidine-2-carboxylic acid [Aze^3]-Val$_2$-AM, known as azetomycin I (Formica & Apple, 1976), has received attention because it was reported to be superior to actinomycin D in antitumor activity in one study (Rose et al., 1978).

Actinomycin analogs with variations in the chromophore and pentapeptide lactones have been prepared by partial synthesis. In the chromophore replacement or modification of the 2-amino group led to significant new structures (Scheme 18). Thus treatment of an actinomycin with aqueous acid yields the 2-hydroxy derivative (Brockmann & Franck, 1954), which can be converted into the corresponding 2-chloro derivative with thionyl chloride (Brockmann et al., 1958). Displacement of chloride by primary or secondary amines has produced about 20 new analogs (Brockmann et al., 1967; Moore et al., 1975). It has also been possible to prepare the 2-benzylamino analog directly from actinomycin D by treatment with dimethylsulfonium methylide, followed by benzyl bromide (Chaykovsky et al., 1977). The 7-position of actinomycins is reactive to electrophilic substitution. It is possible to chlorinate

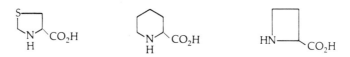

Figure 50 Amino acid analogs of proline used in actinomycin biosynthesis.

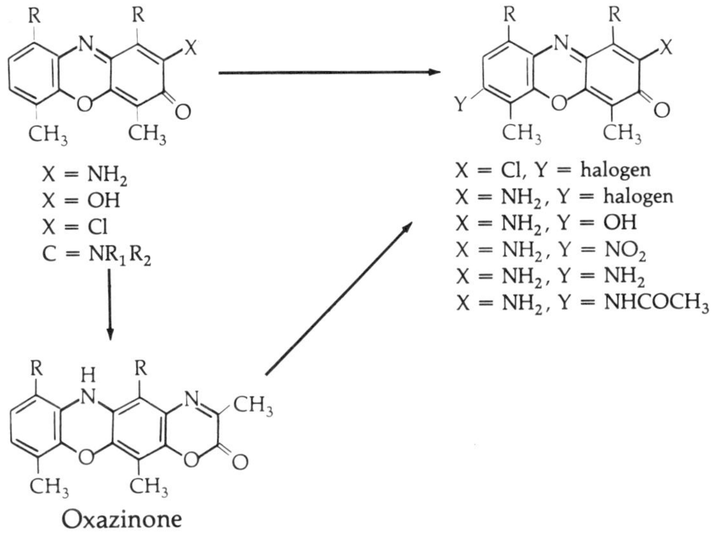

Scheme 18 Synthesis of actinomycin analogs.

or brominate the 7-position of 2-chloro-2-deaminoactinomycins directly, and the products can be converted into 7-haloactinomycins by treatment with ammonia (Scheme 19; Brockmann et al., 1966). α-Ketoacids react with reduced actinomycins to produce oxazinones. These derivatives have antitumor activity following hydrolysis in vivo to the parent compounds (Sengupta et al., 1979). Oxazinone derivatives are useful as protecting groups in the preparation of 7-substituted actinomycins. Thus their treatment with an oxidizing agent such as 2,3-dichloro-5,6-dicyanoquinone gives the quinone imine, which can be converted into a 7-hydroxyactinomycin by hydrolysis and air oxidation (Scheme 18; Sengupta et al., 1975). Nitration of the oxazinone, followed by hydrolysis and air oxidation, affords the 7-nitroactinomycin (Sengupta et al., 1971). This product also can be obtained by direct nitration of the actinomycin (Sengupta et al., 1975). Catalytic reduction of the nitro derivative followed by air oxidation produces a 7-aminoactinomycin (Sengupta et al., 1971). It is possible to acylate selectively the 7-amino group of this type of compound (Muller, 1962).

Chemical transformations of the lactone rings have been limited mainly to hydroxyl or keto groups present in certain actinomycins. Reduction of the carbonyl group in the 4-oxoproline residue of [Opr^3]-Val$_2$-AM with aluminum isopropoxide gave [Hyp^3]-Val$_2$-AM, which has hydroxyproline, whereas cat-

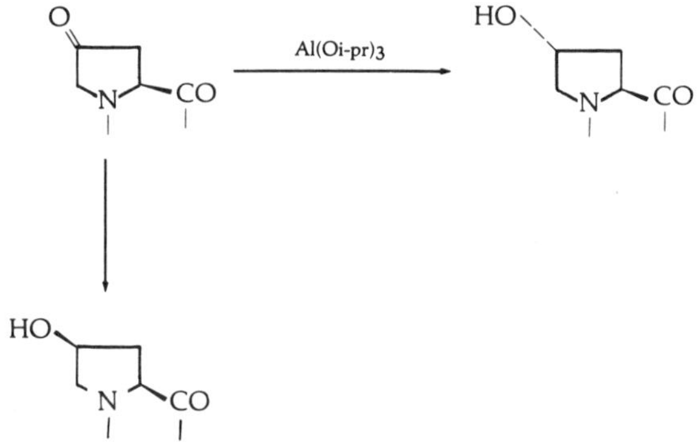

Scheme 19 Reduction of 4-oxoproline in an actinomycin.

alytic reduction gave [aHyp^3]-Val$_2$-AM with allohydroxyproline (Brockmann & Manegold, 1960; Scheme 19). The aluminum isopropoxide reduction of actinomycin Z_1, which has a 5-methyl-4-oxoproline residue, was less specific. It yielded both epimers (Brockmann & Manegold, 1962). The hydroxyl groups of these products and other actinomycins that contain hydroxyproline or γ-hydroxythreonine residues have been converted into O-acetyl derivatives which have diminished biological activity (Muller, 1962).

The lactone rings of actinomycins can be opened by alkaline hydrolysis (Brockmann & Manegold, 1964) or microbial transformation (Perlman et al., 1966); however, the resulting actinomycinic acids (Figure 51) are inactive, as are the derived methyl esters and di-O-acetates (Brockmann & Franck, 1956).

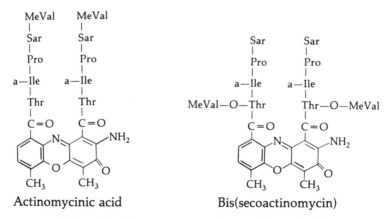

Figure 51 Cleavage products from actinomycin C_3

Cold, concentrated hydrochloric acid cleaves actinomycins selectively between the sarcosine and L-N-methylvaline residues to form inactive bis(secoactinomycins) (Figure 51; Brockmann & Sunderkotter, 1960).

Total synthesis of actinomycins has provided new functional groups on the chromophore and in the pentapeptide lactones. The methyl groups at positions 4 and 6 on the chromophore have been replaced by hydrogen, ethyl, *t*-butyl, bromine, and methoxy (Brockmann & Seela, 1971). None of the resulting analogs was so active as actinomycin D. For the synthesis of these analogs the route generally involved oxidative coupling of the monomeric benzoylpentapeptide, followed by lactonization (Scheme 20). Variations in the actinomycin amino acids have included serine or α, β-diaminopropionic acid (to give a lactam ultimately) at site 1 (Brockmann & Lackner, 1964; Meienhofer & Patel, 1971), D-alanine or D-leucine at site 2 (Brockmann & Lackner, 1968), glycine at site 4 (Brockmann & Lackner, 1968), and valine or sarcosine at site 5. Hexapeptide lactones (Mauger, 1980) and tetrapeptide lactones (Vlasov et al., 1979) also have been made.

No significant summary of the quantitative structure-antitumor activity relationship has been published for the actinomycins, but a qualitative comparison of a variety of natural and semisynthetic actinomycins in three differ-

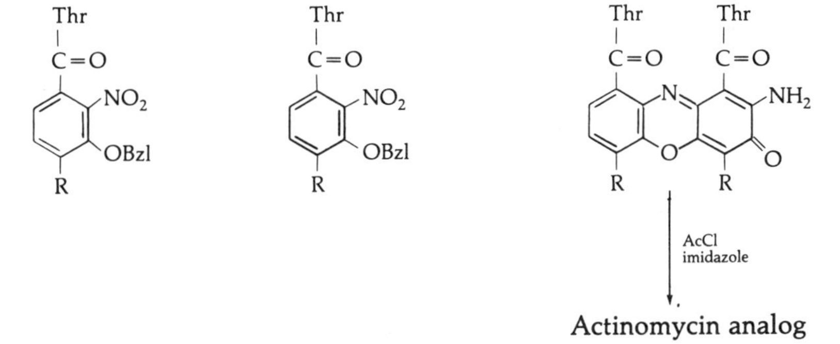

Scheme 20 Synthesis of actinomycin analogs.

ent experimental tumor systems is available in the literature (Meienhofer & Atherton, 1977).

The chemistry of actinomycins has been covered in a number of authoritative reviews (Brockmann, 1960; Katz, 1967; Meienhofer & Atherton, 1973; Hollstein, 1974; Remers, 1979; Mauger, 1980).

The anthracyclines are a large and most important family of antitumor antibiotics, among which doxorubicin (Adriamycin) is the leading drug in present cancer chemotherapy. Other anthracyclines, which include daunorubicin, carminomycin, AD-32, rubidazone, and aclacinomycin, are receiving clinical trials. There are so many other promising leads (marcellomycin, 4'-deoxydoxorubicin, 4'-epidoxorubicin, and 4'-O-tetrahydropyranyldoxorubicin) that it is difficult to choose among them for future clinical candidates.

The name anthracycline is derived from the structural feature common to members of the family: they have anthraquinone chromophores contained within linear hydrocarbon skeletons related to the tetracyclines (Figure 52). Anthracyclines occur naturally as glycosides and aglycones, known as anthracyclinones. They bear varying numbers of phenolic hydroxyl groups on their benzenoid rings and differ in the substitution patterns in the alicyclic ring. These differences determine the names of anthracycline subgroups. Within each subgroup differences can occur in the number and structures of the sugar units attached to the aglycone. No comprehensive naming system other than the rather complex systematic one used in *Chemical Abstracts* has been devised. Therefore the literature on anthracyclines abounds in trivial names based on microorganisms from which they were isolated or on their reddish colors; for example, carminomycin (Figure 52) was isolated from *Ac-*

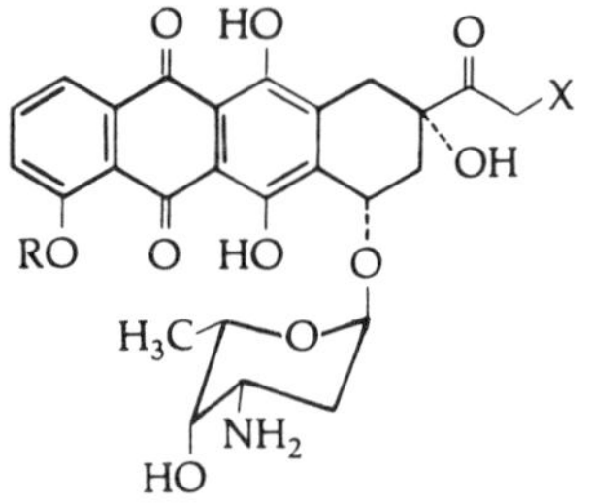

Carminomycin: R = H, X = H
Daunorubicin: R = CH_3, X = H
Doxorubicin: R = CH_3, X = OH

Figure 52 Type-I anthracyclines.

tinomadura carminata (Gause et al., 1973) and cinerubin A (Figure 53) is one of the numerous anthracyclines isolated from *Streptomyces* species such as *galilaeus* and *antibioticus* (Ettlinger et al., 1959).

The anthracyclines have been divided into two classes based on their selective effects on the inhibition of nucleic acid synthesis. Type I anthracyclines, exemplified by doxorubicin and carminomycin (Figure 52), inhibit DNA, whole cell RNA, and nucleolar RNA synthesis at approximately comparable concentrations, whereas Type II anthracyclines, exemplified by aclacinomycin and marcellomycin (Figure 53), inhibit whole cellular RNA synthesis at six- or seven-fold lower concentrations and nucleolar preribosomal synthesis at 170–1250-fold lower concentrations than those required for the inhibition

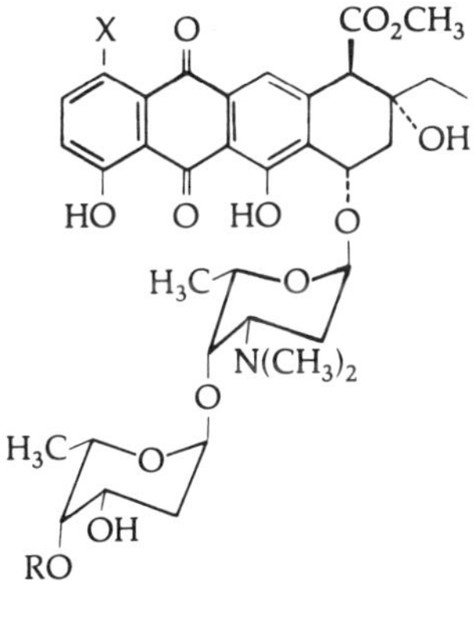

Aclacinomycin A: X = H, R =

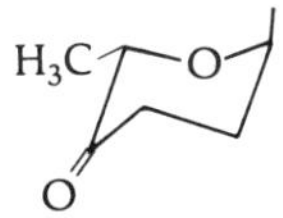

Cinerubin A: X = OH, R =

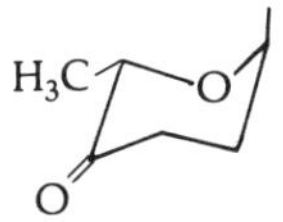

Marcellomycin: X = OH, R =

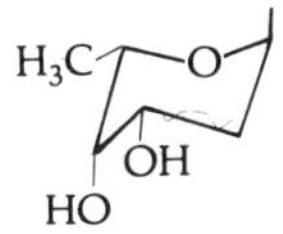

Rudolfomycin: X = OH, R =

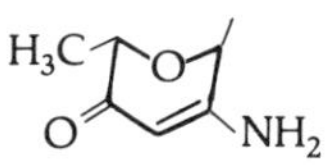

Figure 53 Type-II anthracyclines.

of DNA synthesis. None of the compounds studied effected the conversion of preribosomal RNA into lower weight RNA species (Crooke et al., 1978). The obvious structural differences between Type I and Type II anthracyclines are the presence of a 10-carbomethoxy group and the greater number of sugar residues in the Type II compounds. If the carbomethoxy group is removed, as in 10-decarbomethoxy marcellomycin and 10-decarbomethoxy rudolfomycin, a significant decrease in the potency for inhibition of nucleolar RNA synthesis occurs, with concomitant loss in antitumor activity (Du Vernay et al., 1979). Reduction in the number of sugar residues, for example, in the conversion of marcellomycin into pyrromycin (Equation 46), causes complete loss of selectivity. Thus pyrromycin is a Type I anthracycline. It inhibits DNA synthesis and RNA synthesis in intact cells at approximately the same concentration as doxorubicin and carminomycin but it lacks their antitumor activity in mice.

(46)

The effects of daunorubicin, doxorubicin, and rubidazone on DNA synthesis have been studied in detail. It was found that DNA polymerase α, the putative replicative polymerase, is inhibited in preference to DNA polymerase β, the repair polymerase (Sartiano et al., 1979). Observations that daunorubicin was more lethal to cells in the late S-phase than in G_1 or G_2 prompted the suggestion that it can find access to double helical DNA only during the process of replication (Silvestrini et al., 1970). Correlation between the degrees of inhibition of DNA polymerases and elevations of DNA melting temperatures indicated that intercalation is the mechanism by which daunorubicin and related anthracyclines inhibit these polymerases (Sartiano et al., 1979). Furthermore, the earlier studies on daunorubicin showed that it satisfied the criteria for an intercalating agent, including hypsochromic and bathochromic changes in the visible spectrum, increased melting temperature of DNA, increased viscosity, and decreased buoyant density of DNA (Zunino et al., 1972), uncoiling of supercoiled closed circular DNA (Waring, 1970), and change in the fluorescence emission spectrum (Calendi et al., 1965). Based on the X-ray diffraction patterns of fiber DNA-daunorubicin complexes, an intercalation model was proposed in which the daunosamine moiety lies in the major groove of the double helix. The protonated amine of this sugar is attracted electrostatically to a negatively charged phosphate group of the DNA and the hydroxyl group binds with another phosphate group (Pigram et al., 1971).

The intercalation model described above is consistent with the observation that N-acetyldaunorubicin and related analogs bind less strongly to DNA and are less effective inhibiting the incorporation of adenine into DNA than daunorubicin (Zunino et al., 1972; Di Marco et al., 1971). However, the N-trifluoroacetyl derivative of doxorubicin and its valerate ester (AD-32, Figure 54) have impressive antitumor activity in mice. It is stated also that they are not hydrolyzed to doxorubicin (Israel et al., 1975). This result suggests that intercalation into double helical DNA might not be the only mechanism by which anthracyclines produce their antitumor effect. Other studies

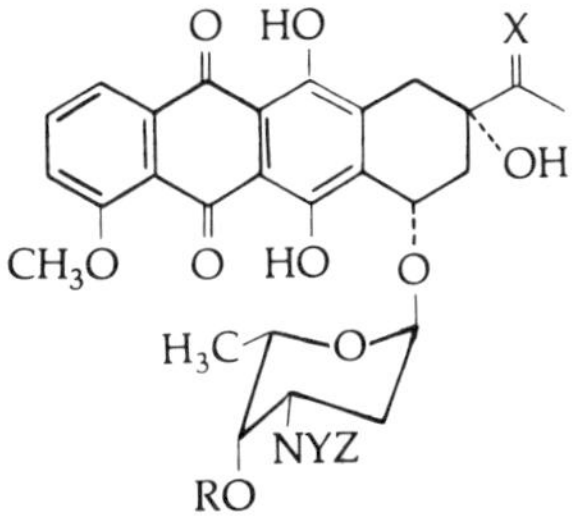

Rubidazone: R = Y = Z = H, X = $NNHCC_6H_5$ (with O double bond)
N,N-Dimethyldaunorubicin: X = O, R = H, Y = Z = CH_3
4'-O-Tetrahydropyranyldaunorubicin: X = O, Y = Z = H, R =

Figure 54 Derivatives of daunorubicin.

support the idea of additional modes of action. One especially intriguing study involved adriamycin covalently attached to agarose beads that were larger than L-1210 leukemia cells. This preparation produced cytotoxicity to the cells under conditions where no adriamycin entered them. It was concluded from this study that adriamycin can exert a cytotoxic effect solely by interaction at the cell surface (Tritton & Yee, 1982). Earlier reports have shown that adriamycin affects a number of membrane activities, including lectin interaction (Murphree et al., 1976), glycoprotein synthesis (Kessel, 1979), phospholipid structure and organization (Tritton et al., 1978), and transport of small molecules (Dasdia et al., 1979).

Daunorubicin, doxorubicin, and certain analogs are unique among clinical antitumor agents in producing serious, total-dose-related cardiotoxicity. Studies at the biochemical level have indicated that it might be caused by the generation of hydroxyl radicals following an NADH-dependent reduction of the drug in cardiac mitochondria (Sato et al., 1977; Lown et al., 1977). The particular susceptibility of cardiac tissues to anthracyclines may be related to the presence of both mitrochondrial and microsomal activating systems and a deficit in catalase activity, which is important to the prevention of free radical toxicity (Donehower et al., 1979).

The clinical importance of doxorubicin, together with its serious side effects, has provided a powerful stimulus to the search for better analogs. Semisynthetic analogs of daunorubicin and doxorubicin with structural variations in both the anthraçyclinone and sugar moieties have been prepared. The antitumor activities of some of these analogs against mouse leukemias are listed in Tables 10 and 11. Because of the considerable variation among these assays, only the maximum effect of each analog and the corresponding maximum effect of daunorubicin or doxorubicin in the same assay is given. Thus the effects of analogs should be compared only in their relationship to the standard compound. Table 10 shows that structural changes at a variety of positions are compatible with the retention or enhancement of activity of daunorubicin analogs. Rubidazone and related benzoylhydrazones show increased activity against P-388 leukemia and so do certain N,N-dialkyl derivatives of daunorubicin. The N-peptidyl derivatives have superior activities against L-1210 leukemia and the 4'-O-tetrahydropyranyl derivative appears to be extremely active in this assay (Figure 54). Particularly surprising from the structure-activity viewpoint is the appreciable activity shown by the 9-deacetyl analog (Scheme 21) and the analog in which daunosamine is replaced by the β-aminopropionyl ester of the 7-hydroxyl group.

The antitumor activities of certain doxorubicin analogs are given in

Table 10 Activities of Daunorubicin Analogs Against Murine Leukemias[a]

Analog	P-388 Maximum Effect (% T/C)	L-1210 Maximum Effect (% T/C)	Reference
13-Dihydro		138(147)	Cassinelli et al., 1979
13-Deoxy	160(167)		Smith et al., 1978
9-Deacetyl	228(171)		Penco et al., 1977
9-Epi-deacetyl	171(171)		Penco et al., 1977
4-O-Demethyl (carminomycin)		162(162)	Cassinelli et al., 1978
13-Dihydro-4-O-demethyl		168(162)	Cassinelli et al., 1978
Benzoylhydrazone (rubidazone)	192(167)		Tong et al., 1978
4-Cl-Benzoylhydrazone	211(167)		Tong et al., 1978
4-CH$_3$O-Benzoylhydrazone	211(167)		Tong et al., 1978
4'-Deoxy		162(162)	Arcamone et al., 1976
4'-O-Methyl	156(169)		Cassinelli et al., 1979
4'-O-Tetrahydropyranyl		>474(191)	Umezawa et al., 1979
N-Lauroyl	240(251)		Arcamone, 1976
N-Glycinyl		194(183)	Baurain et al., 1980
N-Leu		>325(183)	Baurain et al., 1980
N-Leu-Leu		310(183)	Baurain et al., 1980
N-Ala-Leu		293(183)	Baurain et al., 1980
N,N-Dimethyl	214(160)		Tong et al., 1979
N,N-Dibenzyl	259(160)		Tong et al., 1979
7-OCOCH$_2$CH$_2$NH$_2$ (Dedaunosaminyl)	169(160)		Acton et al., 1979

[a] The maximum effect for each compound is given with the maximum effect of daunorubicin control in parentheses beside it. Optimal doses of analog and control are not necessarily the same. T/C = life span treated/life span control × 100.

Table 11 Activities of Doxorubicin Analogs Against Murine Leukemias[a]

Analog	P-388 Maximum Effect (% T/C)	L-1210 Maximum Effect (% T/C)	Reference
4-Demethoxy	261(>300)		Israel et al., 1975
13-Deoxy	164(197)	154(157)	Smith et al., 1978
14-Valerate		125(154)	Israel et al., 1975
14-Octanoate	247(215)		NCI data
14-Nicitinoate	229(215)		NCI data
N-COCF$_3$	189(187)		NCI data
N-COCF$_3$, 14-valerate	221(195)		NCI data
4'-Deoxy	225(>300)	177(155)	Arcamone et al., 1976
4'-O-Methyl	270(254)	312(187)	Cassinelli et al., 1979
4'-Epi	229(221)	150(166)	Arcamone, 1980
4'-Epi-4'-O-methyl		187(187)	Cassinelli et al., 1979
4'-O-Tetrahydropyranyl		>800(>458)	Umezawa et al., 1979
N,N-Dimethyl	164(197)		Tong et al., 1979
N,N-Diethyl	199(197)		Tong et al., 1979

[a] The maximum effect for each compound is given with the maximum effect of doxorubicin control in parentheses beside it. Optimal doses of analog and control are not necessarily the same. % T/C = life span treated/life span control × 100.

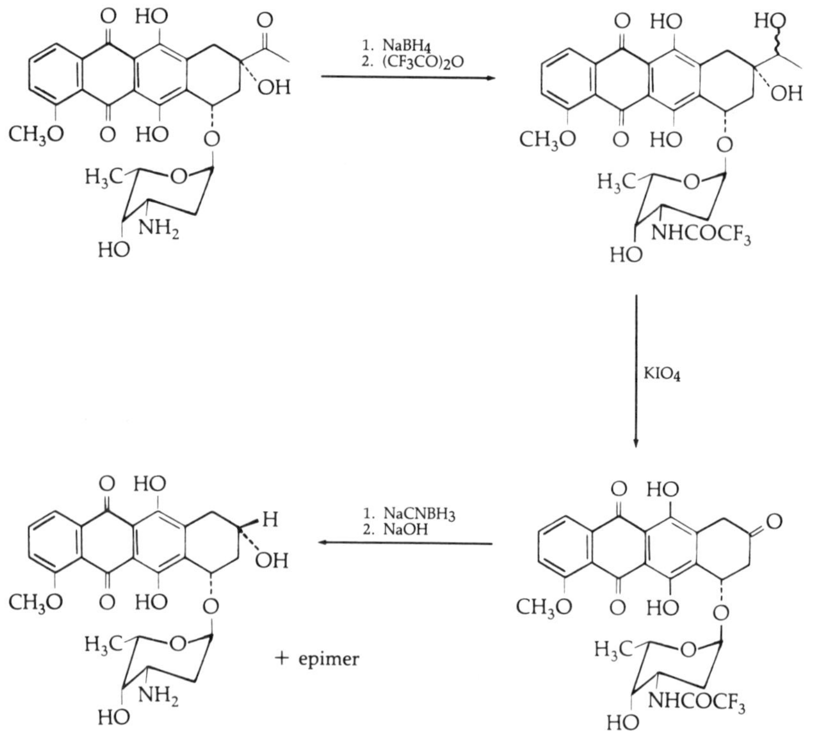

Scheme 21 Preparation of 9-deacetyldaunorubicin.

Table 11. One of the most studied compounds in this table is the N-trifluoroacetyl-14-valerate (AD-32; Figure 55). It has been suggested that the enhanced activity of this compound is caused mainly by the trifluoroacetyl substituent (Israel et al., 1975). This idea is consistent with the poor activity of the 14-valerate and the high activity of the N-trifluoroacetyl derivative. The 4'-0-tetrahydropyranyl (Figure 55) and the 4'-O-methyl derivatives also show much better activity than doxorubicin in preliminary assays. Increased lipophilicity may be an important factor in this respect.

The search for improved analogs has included the preparation of glycosides that involve novel sugar moieties with important aglycones such as daunomycinone (Acton et al., 1974), doxorubicinone (adriamycinone) (Arcamone et al., 1975a), and ε-pyrromycinone (Essery & Doyle, 1980). The reactive primary hydroxyl group of doxorubicinone must be protected by methoxytritylation (Smith et al., 1976) or dioxolane formation (Arcamone et al., 1975a)

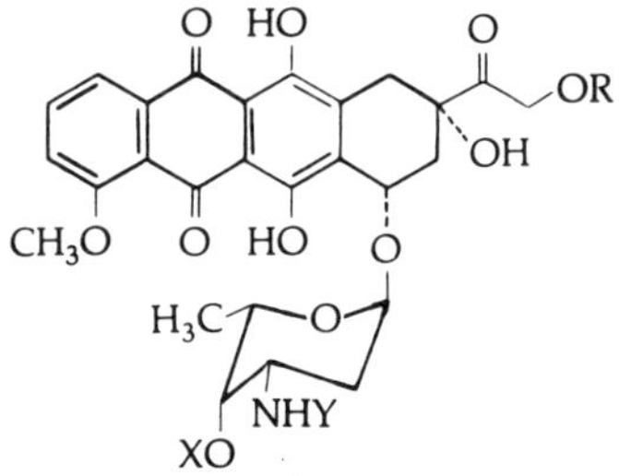

N-Trifluoroacetyldoxorubicin: R = X = H, Y = COCF$_3$
AD-32: R = COC$_4$H$_9$, X = H, Y = COCF$_3$
4'-O-Methyldoxorubicin: R = Y = H, X = CH$_3$

Figure 55 Derivatives of doxorubicin.

before the 7-hydroxyl group can be reacted with an activated sugar to give a glycoside. Usually the 1-bromo sugars, protected on hydroxyl groups with p-nitrobenzoates and on amino groups with trifluoroacetyl groups, are used. Promoters for the Koenigs-Knorr coupling reactions include mercuric compounds and silver triflate. Both α and β anomers are obtained and their proportion depends on the sugar and experimental conditions. The synthesis of 4'-epidoxorubcin (Scheme 22) produced predominately the desired α-anomer. Both anomers had antitumor activity, although, as expected, the α-anomer was more active, (Arcamone et al., 1975b).

Type II anthracyclines have not yet been the subject of exhaustive analog studies, although a beginning has been made. Cytotoxicity was decreased by the demethoxycarbonylation of anthracyclines such as aclacinomycin A, marcellomycin, and rudolfomycin which lack an 11-hydroxyl group (Tanaka et al., 1980; Du Vernay et al., 1979), but 4-O-methylation of aclacinomycin A preserved cytotoxicity (Tanaka et al., 1980). The antitumor activity of some Type II anthracyclines is given in Table 12.

As already noted, the cardiotoxicity of doxorubicin and daunorubicin has limited the total dosage that can be administered to cancer patients. Recent studies in animals have shown that the 4'-O-tetrahydropyranyl derivatives of doxorubicin, aclacinomycin, and AD-32 are less cardiotoxic and caused less alopecia than doxorubicin, daunorubicin, 4'-epidoxorubicin, and rubidazone (Dantchev et al., 1979).

The preparation of novel anthracyclines by controlled biosynthesis appears to be possible now that the biosynthetic pathways are known in some detail. One study showed that 1-hydroxy-13-dihydrodaunomycin and its

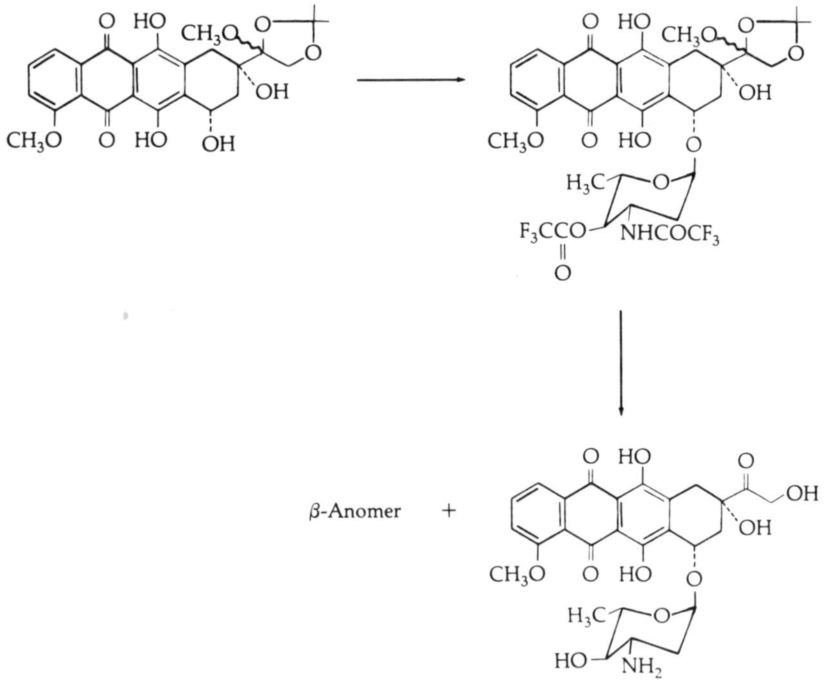

Scheme 22 Synthesis of 4'-epidoxorubicin.

Table 12 Activities of Certain Type II Anthracyclines Against L-1210 Leukemia in Mice[a]

Compound	Optimal Dose (mg kg^{-1} day^{-1})	Maximum Effect (% T/C)
Aclacinomycin A	4[b]	203
MA144-51	4[b]	213
Cinerubin A	1.5[b]	163
Rhodirubin B	2.5[b]	164
Marcellomycin	0.8[c]	150
Rudolfomycin	0.8[c]	143

[a]The tumor inoculum of 10^6 ascites cells was given to BDF$_1$ female mice; % T/C = median lifespan treated/median lifespan controls × 100.
[b]Treated daily on days 0-9 (Oki, 1977).
[c]Treated daily on days 1-5 (Du Vernay et al., 1979).

N-formyl derivative could be obtained from a blocked mutant of *Streptomyces coeruleorubidus* acting on ε-pyrromycinone (Equation 47; Yoshimoto et al., 1980). Another use of biotransformation has been to duplicate the human metabolism of daunorubicin and doxorubicin in cultures of microorganisms (Marshall et al., 1978).

Nogalamycin is an anthracycline whose structure differs from those previously described in having a complex substituent at the 1,2-positions and a novel sugar attached at O-7 (Equation 48; Wiley et al., 1977b). Although its modest antitumor activity and undesirable side effects have prevented it from becoming a clinical agent, nogalamycin has served as the basis for transformations that have yielded compounds comparable or superior to doxorubicin in antitumor effect (but not in potency) in preliminary screens. One of the best of these compounds, 7-O-methylnogarol, was prepared by demethoxycarbonylation followed by methanolysis of the glycoside (Equation 48; Wiley et al., 1977a).

Nogalamycin

Because of the relatively low yields obtained in the production of doxorubicin by fermentation, numerous ingenious total syntheses have been devised and can be found in a recent authoratative review (Kelly, 1979). More general aspects of the anthraclines have also been reviewed (Brockmann, 1963; Di Marco, 1967; Remers, 1979; Arcamone, 1980).

High interest in the total synthesis of anthracyclines led to the preparation of novel anthraquinones that incorporated some, but not all, of the structural features of daunorubicin. Among these compounds, two were significant in that they inhibited DNA synthesis in cultured L-1210 cells at low concentrations (Henry, 1974). Their structures (Figure 56) fit nicely the putative DNA receptor for daunorubicin. Based on this knowledge, two groups investigated the preparation of anthraquinones independently with substituents that could fit this receptor (Zee-Cheng & Cheng, 1978; Murdock et al., 1979). The 1,4-bis[(aminoalkyl)amino]-9,10-anthracenediones (Figure 57) became lead compounds for further refinements when it was shown that they were active against both leukemias and solid tumors in mice. Introduction of hydroxyl groups at positions 5 and 8 enhanced the antitumor activity. As shown in Table 13, the optimum activity was obtained for a compound of this type with 2-[(2-hydroxyethylamino) ethyl] amino substituents at positions 1 and 4 (Figure 57; Murdock et al., 1979). This compound, known as dihydroxyanthracenedione, is in Phase III clinical trial. The corresponding analog with 2-(aminoethyl) amino substituents was also highly active. The synthesis of

Figure 56 Synthetic analogs with partial daunorubicin structures.

Lead compound

Dihydroxyanthracenedione

Figure 57 Antitumor anthroquinones.

Table 13 Activity of Anthraquinones Against P-388 Leukemia and B-16 Melanoma[a]

[structure: anthraquinone with R, NHR' substituents]

		P-388 Leukemia		B-16 Melanoma	
R	R'	Optimal Dose (mg/kg)	ILS (% T/C)	Optimal Dose (mg/kg)	ILS (% T/C)
H	CH_2CH_3	100	0	—	—
H	$CH_2CH_2N(CH_3)_2$	100	68	—	—
H	$CH_2CH_2NH_2$	25	144	12.5	82
H	$CH_2CH_2NHCH_2CH_2OH$	25	154	6.2	129 (1/10)
HO	$CH_2CH_2N(CH_3)_2$	50	122	12.5	70
HO	$CH_2CH_2NH_2$	3.1	173 (4/6)	0.78	>433 (7/10)
HO	$CH_2CH_2NHCH_2CH_2OH$	1.6	>500 (4/5)	0.8	>272 (8/10)
HO	$CH_2CH_2NH(CH_2)_3OH$	12.5	191 (2/6)	6.2	357 (5/10)
HO	$CH_2CH_2NHCH_2\underset{\underset{OH}{\vert}}{C}HCH_2OH$	50	>422 (4/6)	12.5	79

[a] Abstracted from Murdock et al., 1979. Long-term survivors in parentheses.

dihydroxyanthracenedione from 2,3-dihydroxyqinazarine is shown in Equation 49 (Murdock et al., 1979).

Anthracene derivatives without the quinone functionality have antitumor activity. The 9,10-disubstituted anthracenes which are noteworthy in this respect appear to have the ability to intercalate into DNA with the substituent groups rotating to bind in grooves of the double helix. A bis(2-thiopseudourea) derivative of 9,10-dimethylanthracene (Figure 58; Saunders & Saunders, 1971) received clinical study in 1971 but had undesirable phototoxicity (Frei et al., 1971). More recently the bis(2-imidazolinylhydrazone) of anthracene-9,10-dicarboxaldehyde (Figure 58), known as bisanthrene or "orange crush,"

[Equation 49: structure of dihydroxyquinizarine → (H2NR) → intermediate → (Chloranil) → final product, with R = $CH_2CH_2NHCH_2CH_2OH$]

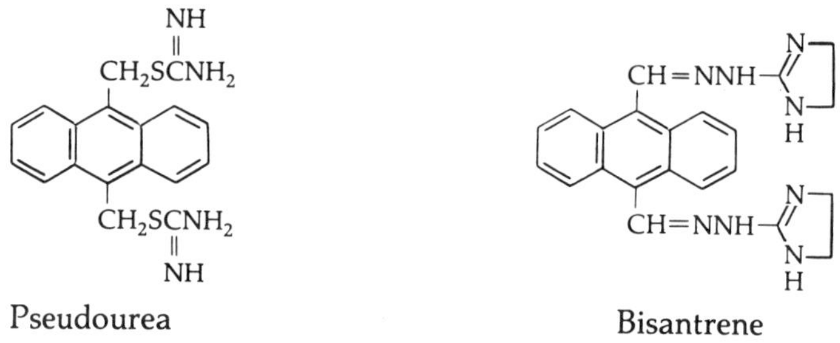

Figure 58 Antitumor anthracenes.

has shown good antitumor activity in animals (Citarella et al., 1982) and in Phase I clinical studies (Alberts et al., 1982). The synthesis of this compound from anthraquinone is shown in Equation 50 (Murdock et al., 1982).

Bleomycin and related glycopeptides have remarkable structural features that form the basis of an unique mode of antitumor action. As shown in Figure 59, bleomycin A_2 has at the right side of its structure a bithiazole functionality linked by an amide bond to an amine that bears a sulfonium group. This part of the molecule is involved in binding to double helical DNA in which the bithiazole is intercalated and the sulfonium is attracted electrostatically to a phosphate anion. Evidence of intercalation of the bithiazole functionality includes fluorescence quenching (Kasai et al., 1978), chemical shift differences in the NMR spectrum (Chien et al., 1977) and changes in the linear dichroism spectrum (Povirk et al., 1979). The left side of bleomycin A_2 contains a variety of functionalities that act in concert to provide strong chelation of metals. Thus in the naturally occurring blue copper (II) complex (Figure 60) two of the six ligands are provided by the L-β-aminoalaninecarboxamide substituent, one by a nitrogen in the pyrimidine ring, two by the L-erythro-β-hydroxyhistidine residue, and one by the carbonyl group of the 3-O-carbamoyl-α-D-mannopyranoside unit (Takita et al., 1978). Treatment of com-

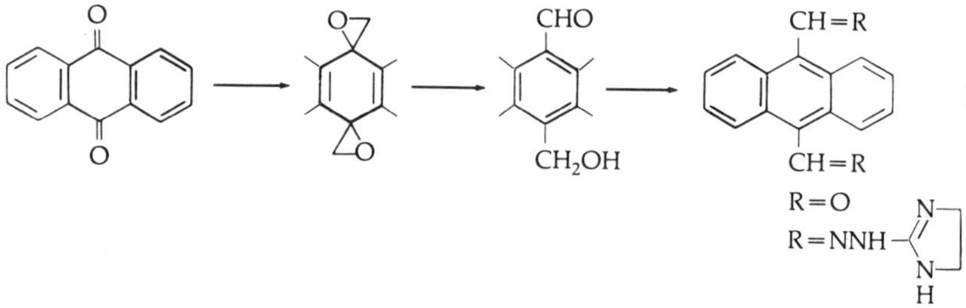

(50)

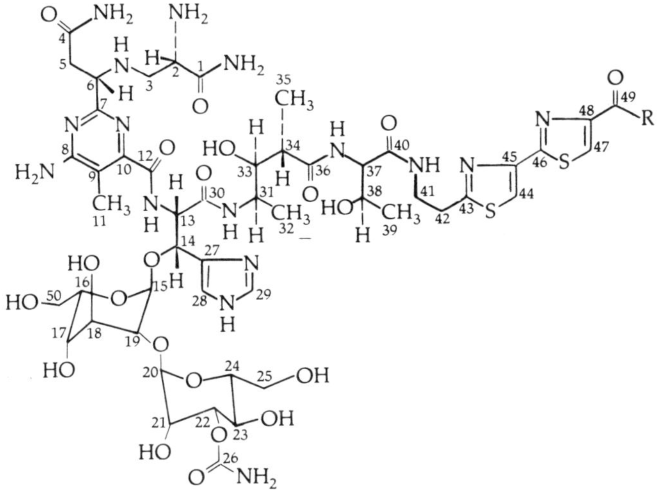

Bleomycinic acid, R = OH

Bleomycin A$_2$, R = NH(CH$_2$)$_3\overset{+}{S}$(CH$_3$)$_2$

Bleomycin B$_2$, R = NH(CH$_2$)$_4$NH$\overset{\overset{NH}{\|}}{C}NH_2$

PEP-Bleomycin, R = NH(CH$_2$)$_3$NHCH$_2$CH$_2$-⟨pyridine⟩

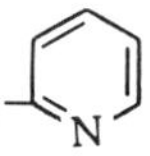

Figure 59 Structures of some bleomycins.

plexes of this type with reducing agents or other complexing agents results in the white copper-free bleomycin.

There is now compelling evidence that the cytotoxic effect of bleomycins is related to their complexes with iron. The Fe(II)-bleomycin complex is air-sensitive and readily oxidized to Fe(III)-bleomycin (Sausville et al., 1976; Sausville et al., 1978). This process produces free radicals that have been detected by ESR spectrometry (Dabrowiak et al., 1979; Lown, 1979). At high concentration of Fe(II)-bleomycin in solution (1.0 m), hydroxyl radicals are formed, whereas at lower concentrations superoxide and hydroperoxy radicals are formed (Suguira & Kikuchi, 1978). These highly reactive species cause strand scission of the DNA, which can lead to the release of purine and pyrimidine bases (Suzuki et al., 1968; Haidle, 1971). The binding of molecular oxygen to Fe(II)-bleomycin is viewed as a process in which the weak interaction between the carbonyl oxygen of the carbamoyloxy group and iron (Figure 60) is replaced by a stronger molecular oxygen-iron interaction (Takita et al., 1978). Formation of free radicals in proximity to the DNA is made possible by the intercalation of the bithiazole part of the bleomycin molecule.

The role of Cu(II)-bleomycin in the degradation of DNA is an intriguing question. It has been observed that copper-free bleomycin and the copper-complex lead to DNA strand scission in cells (Suzuki et al., 1968). Cu(II)-bleomycin, however, does not cause scission of isolated DNA. The recent identification of a protein capable of reducing Cu(II) to Cu(I) in the complex has led to the suggestion that the complex dissociates with a second protein that receives the Cu(I) and the liberated bleomycin binds to cellular Fe(II) to produce the complex responsible for DNA degradation (Takahashi et al., 1977).

Resistance of cells to bleomycin and related compounds is based on their content of bleomycin hydrolase, an aminopeptidase that converts the carboxamide group of the L-aminoalaninecarboxamide substituent into the corresponding carboxylic acid known as deamido bleomycin (Umezawa et al., 1972). This conversion occurs only in the metal-free form of bleomycin and not in the complex. Related studies with the Cu(II) complex formed from deamido bleomycin suggested a pH-dependent ligand change in which the amino group binds to Cu at pH 9.6 but the carboxylate binds at pH 6.9 (Figure 61). At pH 6.9 the spin concentration of free radicals formed from the in-

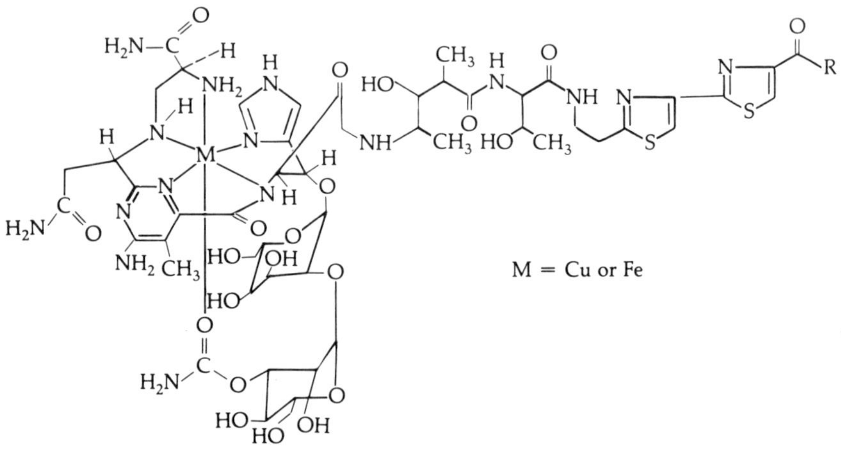

Figure 60 Metal complexes of bleomycin.

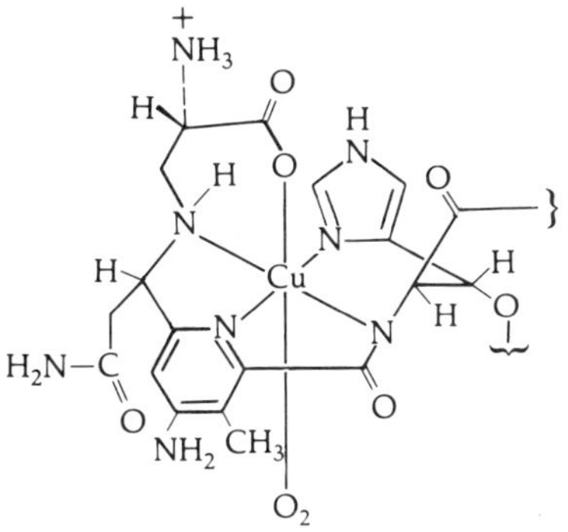

Figure 61 Copper (II) complex of deamido bleomycin and oxygen.

teraction of oxygen with Fe(II)-deamido bleomycin was only about 1% of that formed from the corresponding Fe(II)-bleomycin system. Thus the lower antitumor activity of deamido bleomycin appears to result from the less effective activation of oxygen by its Fe(II)-complex (Sugiura et al., 1979). The sensitivity of various types of tumor cell is related to their content of bleomycin hydrolase; for example, squamous cell carcinomas have low levels of this enzyme and are sensitive to bleomycins. In contrast, squamous cell sarcomas have high levels of this enzyme and are resistant (Umezawa, 1976).

Copper-free bleomycins are converted into isobleomycins in the presence of mild base. This transformation involves migration of the carbamoyl group from the 3-hydroxyl group to the 2-hydroxyl group of the mannose residue (Nakayama et al., 1973). Under the same alkaline conditions Cu(II)-bleomycins undergo a slow, irreversible transformation into epibleomycins. This epimerization takes place at the trisubstituted carbon atom attached to the pyrimidine ring (No. 6 in Figure 59). Epibleomycins retain about 25% of the antitumor activity of the parent bleomycins (Muraoka et al., 1976).

The bleomycin family of compounds contains a number of analogs that differ from bleomycin A_2 only in the terminal amine substituent; for example, bleomycin B_2, which is present in the commercial preparation of bleomycin marketed as Blenoxane, has agmatine as the terminal amine. Other examples include bleomycin A_6 with spermine and bleomycin A_5 with spermidine (Table 14). Phleomycins differ from bleomycins only in the bithiazole portion of the molecule, wherein one of the double bonds is hydrogenated (the 44,45 bond in Figure 59). Thus phleomycin D_1 is the 44,45-dihydro analog of bleomycin B_2. Phleomycins can be distinguished from belomycins by differences in the relative intensities of the ultraviolet absorption maxima in the

Table 14 Terminal Amine Components of Bleomycins[a]

Bleomycin	Terminal Amine
Natural bleomycins	
A_1	$NH(CH_2)_3SOCH_3$
A_2	$NH(CH_2)_3\overset{+}{S}(CH_3)_2$
Demethyl-A_2	$NH(CH_2)_3SCH_3$
A_2'-a	$NH(CH_2)_4NH_2$
A_2'-b	$NH(CH_2)_3NH_2$
A_2'-c	$NH(CH_2)_2\text{-imidazole}$
A_5	$NH(CH_2)_3NH(CH_2)_4NH_2$
A_6	$NH(CH_2)_3NH(CH_2)_4NH(CH_2)_3NH_2$
B_2'	NH_2
B_2	$NH(CH_2)_4NH\overset{\underset{\|}{NH}}{C}NH_2$
Biosynthetic bleomycins	
	$NH(CH_2)_3N(CH_3)_2$
	$NH(CH_2)_3NH(CH_2)_3NH_2$
	$NH(CH_2)_3N(CH_2)_3NH_2$ with CH_3 on N
	$NH(CH_2)_3NH(CH_2)_3NHC_4H_9$
	$NH(CH_2)_3NH(CH_2)_3NHCHC_6H_5$ with CH_3
	$NH(CH_2)_3NH\overset{\underset{\|}{NH}}{C}NH_2$
	$NH(CH_2)_3$–morpholino
	$NH(CH_2)_3\overset{+}{N}(CH_3)_3$
	$NH(CH_2)_3NH(CH_2)_3N(CH_3)_2$
	$NH(CH_2)_3NH(CH_2)_2CH_2OH$
	$NH(CH_2)_3NH(CH_2)_2CH_2OCH_3$
	$NH(CH_2)_3NHCH_2C_6H_5$
	$NH(CH_2)_3NHCHC_6H_5$ with CH_3
	$NHCH_2CHNH_2$ with CH_3
	$NH(CH_2)_3NHCH(CH_2)_2NH_2$
	$NH(CH_2)_3NHCH_3$

Table 14 (Continued).

Bleomycin	Terminal Amine
	$NH(CH_2)_3N\langle\bigcirc\rangle$
	$NH(CH_2)_3N\langle\rangle NH$
	$NH(CH_2)_3NH-\langle\bigcirc\rangle$
	$NH(CH_2)_3NH(CH_2)_4NHCCH_2CH_2CO_2H$
	$NH(CH_2)_3NHCH_2CH_2-\bigcirc_N$

[a] Abstracted from Umezawa, 1976.

290–300 nm region to that at 245 nm. The phleomycins have significant antitumor activity and the original phleomycin mixture was proposed for clinical trial (Bradner & Pindell, 1962). It had to be withdrawn, however, because it showed kidney toxicity in dogs (Ishizuka et al., 1966).

The members of other families of antitumor antibiotics closely resemble the bleomycins and phleomycins. Typically, these compounds conserve the bithiazole functionality and the metal complexing region but show structural variations in the residues that link these key portions of the molecule. Variations in the terminal amine residues tend to be the same as those found in the bleomycins and phleomycins; for example, the cliomycin family contains compounds that are the same as bleomycins except that the L-threonine residue (carbons 37–40) is replaced by (s)-amino-(1-hydroxycyclopropyl)-acetic acid (Umezawa et al., 1980). The tallysomycins are related to the bleomycins with the following changes in structure: 4-amino-3-hydroxypentanoic acid instead of 4-amino-3-hydroxy-2-methylpentanoic acid (no 35-methyl group) and hydroxyl groups at positions 41 and 42 [in the 2'-(2-aminoethyl)2,4'-bithiazole carboxylic acid unit] with 4-amino-4,6-dideoxy-L-talose linked to the 41-hydroxyl group. Tallysomycin B has spermidine as the terminal amine, whereas tallysomycin A has β-lysine-spermidine (Konishi et al., 1977; Takita et al., 1978). Zorbamycin, independently discovered and named YA-65X, is related to the phleomycins by its partially reduced thiazole ring (44,45-dihydro). It differs from phleomycins in having a hydroxymethyl group substituted at C-32, a methyl group substituted at C-39, and no hydroxyl group at C-50 (6-deoxy-L-gulose instead of gulose). The terminal amine is 3-aminopropionamidine (Ihashi, 1973a; Ito et al., 1972; Ihashi et al., 1973b). The zorbonamycins are identical with the zorbamycins except that they have the bithiazole functionality (Ihashi et al., 1973a). Total structures for the platomycins and victomycin, compounds believed to be related to bleomycin, have not been published.

The fact that bleomycins with a variety of terminal amines show good antitumor activity has stimulated the preparation of novel bleomycin analogs (Table 14). Successful approaches have included partial synthesis from bleomycinic acid, the bleomycin fragment lacking the terminal amine, and controlled biosynthesis in which the new terminal amine is added to the fermentation. Bleomycinic acid has been prepared from bleomycin A_2 chloride in a two-step chemical process that involves pyrolysis to the demethyl analog followed by cyanogen bromide degradation of the copper complex of this intermediate (Scheme 23; Takita et al., 1973; Umezawa et al., 1973). An alternative method for bleomycinic acid is the incubation of bleomycin B_2 with species like *Fusarium roseum* that contain the enzyme acylagmatine amidohydrolase (Umezawa et al., 1972). Formation of an amide bond between bleomycinic acid and the new terminal amine is promoted by water-

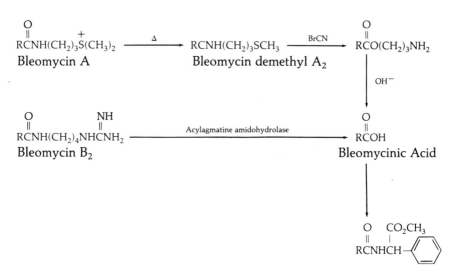

Scheme 23 Preparation of semisynthetic bleomycins.

soluble carbodiimides such as 1-ethyl-3-(dimethylaminopropyl)-carbodiimide. An example of a new bleomycin analog prepared in this manner is copper (II)-α-carbomethoxybenzylaminobleomycin hydrochloride (Scheme 23; Umezawa, 1976). In practice, the preparation of bleomycin analogs by controlled biosynthesis has been much more fruitful than by chemical transformation. Adding certain new amines to *Streptomyces verticillus* fermentations led to their incorporation into the product as the terminal unit, with suppression of the natural bleomycins (Fujii et al., 1974). Some of these amines are listed in Table 14. Certain of the corresponding bleomycins have pronounced activity in experimental tumor systems; one of them, the N-[2-(β-pyridyl)ethyl]-1,3-diaminopropane analog known as PEP-bleomycin (Figure 54; Takita et al., 1978), is showing considerable promise in clinical trials.

The aureolic acid group of antitumor antibiotics includes mithramycin (aureolic acid), the chromomycins, the olivomycins, and variamycin (Figure 62; Berlin et al., 1968). Mithramycin is the only member of the group approved for use in the United States and its former importance in treating testicular tumors has yielded to newer agents. However, it is valuable for controlling cancer-induced hypercalcemia and calciurea that do not respond to other agents. Chromomycin A_3 is used in Japan and olivomycin A, in the Soviet Union (Remers, 1979). Aureolic acids complex with double helical DNA to prevent the action of DNA-directed RNA polymerase (Mueller et al., 1971). The evidence for intercalation is conflicting. Thus complexation with DNA caused changes in the visible spectra of the antibiotics and enhanced their fluorescence (Kerstin, 1968; Hill, 1976). The buoyant density of the DNA decreased and the thermal transition temperature increased (Kajiro & Kamiyama, 1965). The sedimentation curves for binding with supercoiled circular DNA from bacteriophage were distinctly different from those given by intercalative compounds such as ethidium and daunorubicin (Waring, 1970). Aureolic acids have a unique requirement for magnesium or other divalent cations in order to bind to DNA. It has been suggested that these ions are needed to counteract electrostatic repulsion between phosphate groups on the DNA and the anionic site on the antibiotic (Nayak et al., 1973).

The sugar chains of the aureolic acid compounds are essential for DNA binding and antitumor activity (Sedov et al., 1969). Analogs in which some of the sugars have been cleaved bind weakly and are inactive. In the only study of systematic analog formation published to date it was found that a variety of simple structural changes on olivomycin A were consistent with

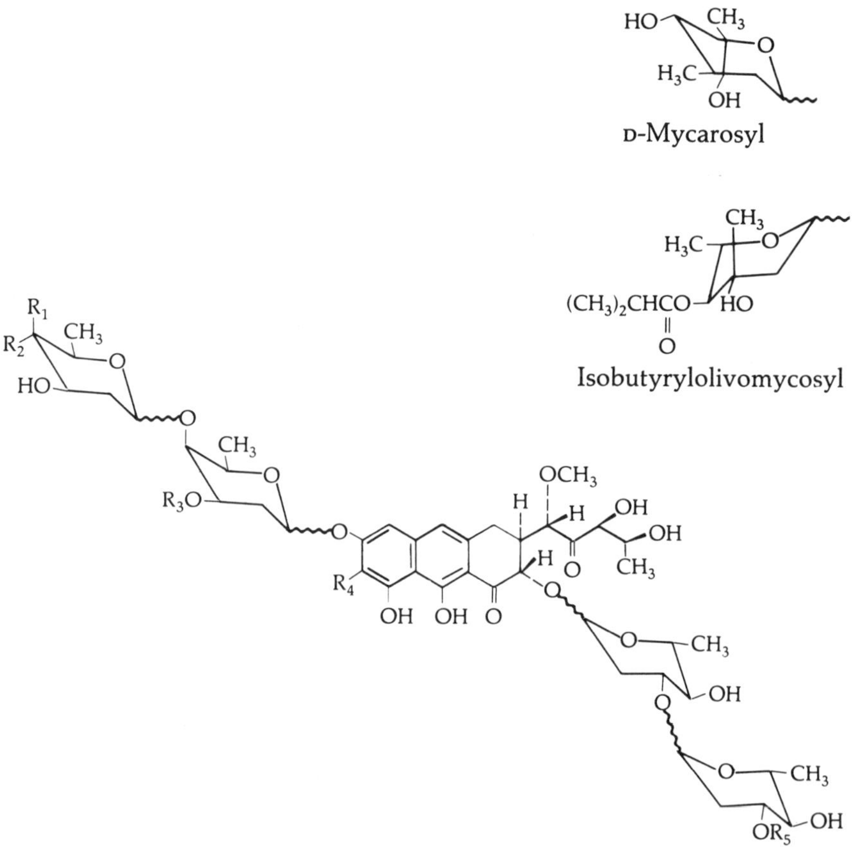

Aureolic acid (mithramycin): R_1H, R_2 = OH, R_3 = H, R_4 = CH_3, R_5 = D-mycarosyl
Olivomycin A: R_1 = CH_3O, R_2 = H, R_3 = CH_3CO, R_4 = H, R_5 = isobutyrylolivomycosyl
Chromomycin A_3: R_1 = CH_3O, R_2 = H, R_3 = CH_3CO, R_4 = CH_3, R_5 = isobutyrylolivomycosyl

Figure 62 Compounds of the aureolic acid group.

retention of antitumor activity in mice. Increased activity against P-388 leukemia was provided by side-chain carbonyl derivatives such as methoxime and methylimine, whereas small decreases in activity were obtained by methylation of the phenolic hydroxyls, ketal formation with the vicinal hydroxyls in the side chain, or reduction of the ketone carbonyls. Electron-withdrawing groups in the benzenoid ring destroyed activity (Kumar et al., 1980).

OTHER COMPOUNDS THAT INTERFERE WITH DNA STRUCTURE AND FUNCTION

In addition to the intercalating agents described in the preceding section, certain other drugs bind to DNA and disrupt its normal function. The exact mechanisms of action of these agents, which include campthothecin, streptonigrin, neocarcinostatin, and the platinum complexes, are unknown, but all can cause extensive damage and degradation to DNA.

Camptothecin was isolated from *Camptotheca acuminata*, a small Chinese tree (Wall et al., 1966), which was so potent that even the crude extracts were active against L-1210 leukemia in mice. Camptothecin is an alkaloid that contains five rings (Scheme 24). It has one basic nitrogen (quinoline) and one neutral nitrogen (α-pyridone). The lactone ring is opened readily in sodium hydroxide solution to give the water-soluble salt of the carboxylic acid. This salt was selected for clinical trial because of its solubility but it proved to be inactive. Treatment with acid promotes recyclization to the insoluble lactone (Scheme 24).

Natural analogs of camptothecin are rare; structural variations are limited to positions 9 and 10 of the benzenoid ring (Wani & Wall, 1969; Govindachari & Viswanathan, 1972). The 10-hydroxy analog (Figure 63) is equi-

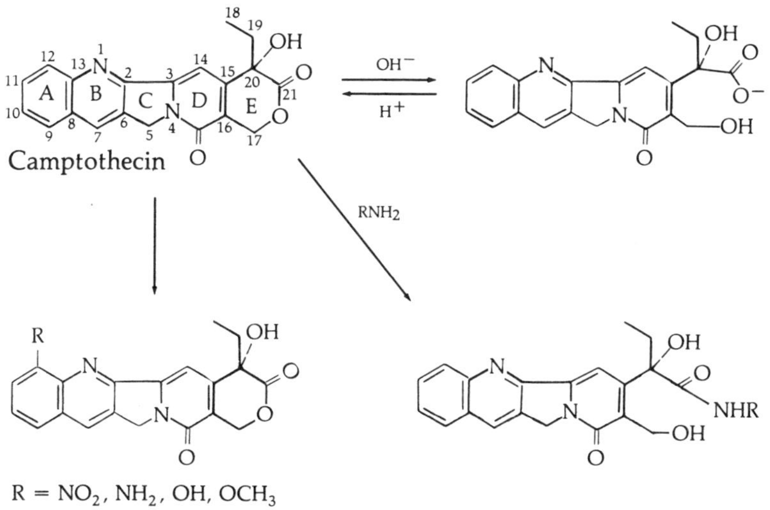

$R = NO_2, NH_2, OH, OCH_3$

Scheme 24 Reactions of camptothecin.

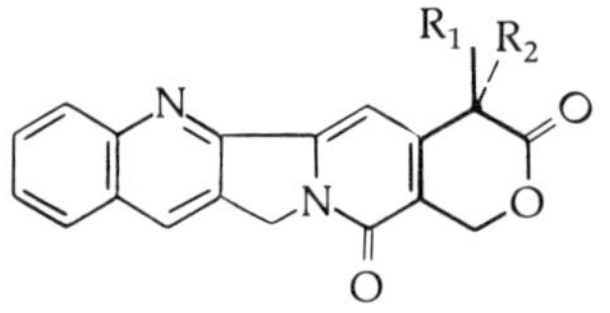

20-Substituted analogs

$R_1 = C_2H_5$; $R_2 = O\overset{\overset{O}{\|}}{C}CH_3$, Cl or H

$R_2 = OH$; $R_1 = CH_2CH=CH_2$, $CH_2C\equiv CH$,

$CH_2C_6H_5$, $CH_2\overset{\overset{O}{\|}}{C}C_6H_5$

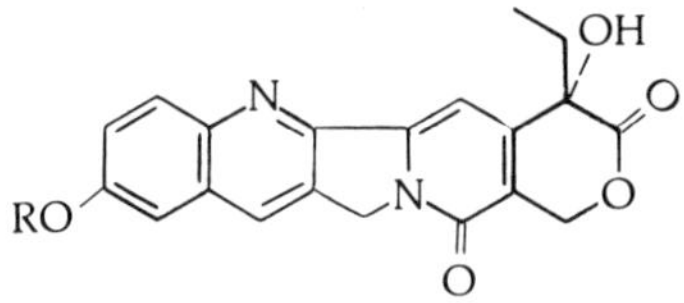

10-Substituted analogs

$R = H$, CH_2CO_2Na, $OCH_2CH_2N(C_2H_5)_2$

Figure 63 Camptothecin analogs.

potent with camptothecin and more effective in prolonging the lives of tumored mice. A variety of semisynthetic analogs has been prepared and tested. Among them the 20-acetate, 20-chloro analog and 20-deoxy analog (Figure 63) were less active than the parent compound. The lactol, obtained by sodium borohydride reduction, was less active also. These results led to the hypothesis that an α-hydroxy lactone is an "absolute" requirement for activity (Wall & Wani, 1980). Treatment of camptothecin with amines produces a lactone-ring opening with an amide formation. The amides generally are less active than the lactone; the best compound, the N-methyl derivative (Scheme 24), shows about three-fifths of the activity of camptothecin (Wall & Wani, 1980). Water-soluble ethers of the 10-hydroxy group of 10-hydroxycamptothecin were prepared for clinical studies (Figure 63) and were active but less potent than the parent compound (Wani et al., 1980). Substituents have been introduced at the 12-position of camptothecin by a route that involves nitration and reduction to the amine, diazotization, and Sandmeyer or related reactions. The hydroxy and methoxy derivatives obtained in this way are reported to show improved activity against the Ehrlich ascites carcinoma (Pan et al., 1975).

A number of totally synthetic analogs of camptothecin have been prepared (Wall et al., 1972; Danishefsky et al., 1973; Plattner et al., 1974). They range in complexity from bicyclic structures that contain the elements of rings D and E to fully elaborated compounds such as dl-campthothecin and its iso- and homoanalogs. In general, only the complete pentacyclic analogs showed good antitumor effects. Isoteric compounds, which included 12-azacamptothecin, were synthesized but were less active than camptothecin (Wani et al., 1980). The one type of analog that showed increased activity against L1210 leukemia was that in which the 20-ethyl substituent was replaced by various allyl, propargyl, benzyl, and phenacyl groups. It was claimed that the allyl analog had the best activity in this series (Sugasawa et al., 1976).

Camptothecin, a potent inhibitor of DNA and RNA synthesis, causes single-strand breaks in DNA that can lead to fragmentation. The source of DNA cleavage may be a direct effect of the antibiotic or an indirect effect in which binding renders the DNA more susceptible to endonucleases (Horwitz, 1975). Although the mode of binding to DNA is uncertain, it has been suggested that the planar aromatic portion intercalates, and this action places the α-hydroxylactone functionality in an orientation favorable to form a covalent bond with a nucleophilic group on the DNA (Wall & Wani, 1980). More evidence is needed to support this hypothesis.

Streptonigrin is a substituted quinolinequinone (Figure 64) isolated from cultures of *Streptomyces flocculus* (Rao & Cullen, 1960). It is active against experimental tumors such as sarcoma 180 and Walker carcinoma 755 and has produced remissions in a variety of human cancers (Harriss et al., 1965). Its serious myelosuppression, however, prevented acceptance into clinical practice (Hackethal et al., 1961). Attempts to devise less toxic analogs led to the preparation and antitumor testing of streptonigrin methyl ester (Humphrey & Dietrick, 1963; Rivers et al., 1966) and the isopropylidene derivative of azastreptonigrin (Figure 64; Kremer & Laszlo, 1967). Neither of these analogs proved to be superior to streptonigrin.

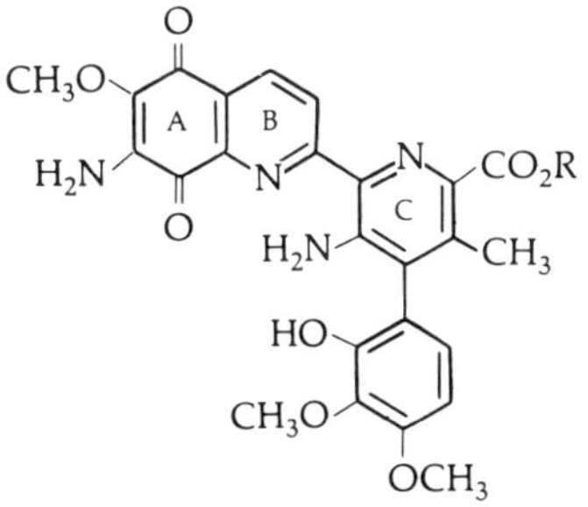

Streptonigrin, R = H
Methyl ester, R = CH_3

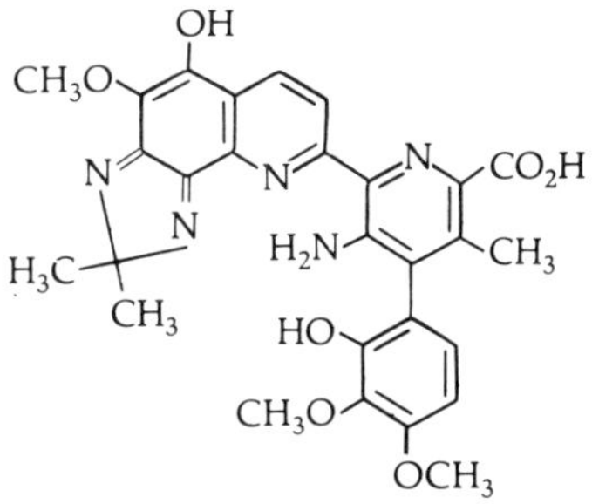

Isopropylidene azastreptonigrin

Figure 64 Streptonigrin and analogs.

In mammalian cells low levels of streptonigrin inhibit mitosis and replication (Young & Hodas, 1965). There is a decrease in high molecular weight DNA and extensive breakage and rearrangement of chromosomes (Radding, 1963; Cohen et al., 1963). The lethal event appears to be DNA degradation rather than inhibition of DNA synthesis because cell death occurs at drug levels that allow synthesis (White & White, 1964). Another highly cytotoxic effect of streptonigrin is interference with oxidative phosphorylation. A marked depletion of cellular ATP occurs, but oxygen is consumed and hydrogen peroxide is formed (Miller et al., 1967).

Model studies with simple 5,8-quinolinequinones have helped to elucidate the DNA cleavage process (Lown & Sim, 1976). These quinones accept hydrogen from reduced pyridines such as NADH to form the corresponding hydroquinones. Metal complexes of these species are reoxidized to the quinones with formation of oxidizing species which include superoxide radical and hydroxyl radial. The latter species is thought to be responsible for DNA cleavage. Protection of the DNA is provided by the enzymes superoxide dismutase and catalase, which prevent the formation of hydroxyl radicals, or by free radical scavengers. The mode of streptonigrin binding to DNA is unclear. Obvious possibilities such as intercalation or alkylation have been eliminated by suitable experiments on DNA-streptonigrin complexes (Cone et al., 1975; Mizuno & Gilboe, 1970). The inability of streptonigrin to intercalate is surprising in view of the coplanarity of its A, B, and C rings (Figure 64; Chiu & Lipscomb, 1975).

Streptonigrin biosynthesis has been elucidated by ingenious studies based partly on the feeding of glucose labeled uniformly with ^{13}C and the analysis of ^{13}C-^{13}C spin-coupling patterns (Gould & Cane, 1982). They suggest that the quinoline quinone portion of streptonigrin is formed from glucose in a complex process that involves the condensation of erythrose-4-phosphate with 4-aminoanthranilic acid. The remainder of the molecule is formed from glucose by way of β-methyltryptophan (Gould & Chang, 1980). A possible intermediate in late stages of streptonigrin biosynthesis is lavendamycin, a structurally related antitumor antibiotic (Doyle et al., 1981). These biosynthetic intermediates are shown in Figure 65.

Chemical synthesis of streptonigrin and its simpler analogs has been pursued intensively in recent years. Two recent total syntheses of streptonigrin promise to provide the basis for new analogs at the level of complexity appropriate for antitumor activity (Kende et al., 1981; Basha et al., 1980).

In recent years a number of polypeptide antibiotics with antitumor activity have been isolated. The most thoroughly studied compound among

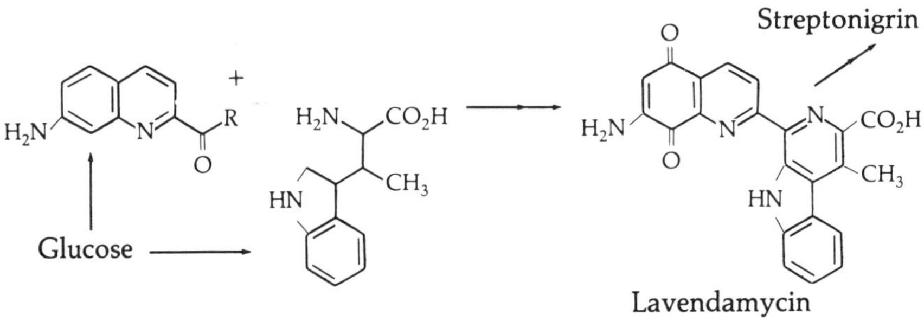

CHEMISTRY OF ANTITUMOR DRUGS

Figure 65 Proposed biosynthetic intermediates for streptonigrin.

them is neocarcinostatin (Ishida et al., 1965). Others include mitomalcin, macromomycin, lymphomycin, and actinocarcin. Neocarcinostatin is an acidic single-chain molecule with 109 amino acid residues. The sequence of this polypeptide chain has been determined (Meienhofer et al., 1972). A nonpeptide component also has been obtained and characterized as 2-hydroxy-5-methoxy-7-methyl-1-naphthalenecarboxylic acid (Edo et al., 1980). The point of attachment of this component to the polypeptide is not known.

Neocarcinostatin inhibits DNA-dependent DNA polymerase and degrades existing DNA by inducing shear-sensitive discontinuities (Sawada et al., 1974). This hardly excisable damage tends to be repaired by an error-prone recombination process (Tatsumi & Nishioka, 1977).

A succinyl derivative of neocarcinostatin enhances antitumor activity, probably because of its increased stability to proteolytic enzymes (Maeda et al., 1978).

Cisplatin and related platinum complexes are potent inhibitors of DNA polymerase. Their antitumor activities and toxicities resemble those of the alkylating agents. Considerable evidence obtained for DNA crosslinking by the platinum complex includes facilitated renaturation, increased sedimentation coefficient, and hyperchromicity of the DNA ultraviolet spectrum (Gale, 1975). Guanine reacts over other bases with the complex, and it is thought that a closed-ring chelate is formed by displacement of the two chloride ligands of the platinum complex by the N-7 and O-6 nucleophilic sites of guanine (Rosenberg, 1980). Recent evidence indicates that only one binding site on closed circular plasmid DNA pSMI is affected by low binding levels. This site appears to be a specific sequence of two or more consecutive dG-dC pairs at a recognition site (Tullius & Lippard, 1981).

The discovery of biologically active platinum complexes resulted from an investigation of the effects of electrical fields on bacteria in which it was observed that *Escherichia coli* cells formed long filaments instead of dividing (Rosenberg et al., 1965). Subsequent experiments revealed that this inhibition of mitosis was caused not by the electric current passing through the suspension of bacteria but by a complex, $Pt(Cl_4)(NH_3)_2$, formed by reaction of a platinum electrode in the presence of ammonium and chloride ions (Rosenberg et al., 1967). This discovery led to the testing of a variety of platinum neutral complexes against biological systems, which included tumor cells, with the eventual result that *cis*-dichlorodiammineplatinum II (cisplatin, DDP, Figure 66) became established as a clinical anticancer agent (Hill et al., 1975; Rozencweig, 1979).

Many other platinum complexes active against experimental tumors generally belong to a structural class of cis-isomers in which one pair of ligands is monodentate or bidentate amines and the other is monodentate anions with moderately good leaving ability, such as chloride, or bidentate anions, such as malonate (Rozencweig, 1979). One of the main problems in chemotherapy with cisplatin is its nephrotoxicity. Thus the search for analogs has made low nephrotoxicity an important goal. An analog reported to have no nephrotoxicity in rats is diamminecyclobutanedicarboxylatoplatinum II (CBDCA), whereas dichlorodihydroxybisisopropylamineplatinum IV (CHIP) and di(aminomethyl)cyclohexaneplatinum(II) sulfate (TNO-6) have degrees of nephrotoxicity intermediate between those of CBDCA and cisplatin (Lelieveld & van Putten, 1981; Figure 66). Cisplatin is prepared by treating K_2PtCl_4 with iodide and ammonia.

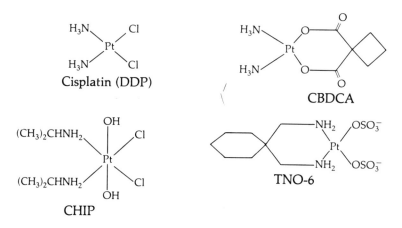

Figure 66 Platinum complexes.

INHIBITORS OF PROTEIN SYNTHESIS

A substantial number of antibacterial agents act by inhibiting protein synthesis at the ribosomal level. Their toxicity is selective because of the substantial differences between the ribosomes of bacterial cells (procaryotes) and mammalian cells (eucaryotes). Certain compounds that exert a selective effect on the ribosomes of eucaryotes have been found. All are highly toxic, but some of them show substantial antitumor activity in animal models and have moderately good therapeutic ratios. The two most important groups with this kind of activity are the tricothecanes and cephalotaxine esters. Each of these groups is discussed below.

Tricothecanes are produced by various species of imperfect fungi. The one known exception to this source is the baccharin family, which is found in higher plants. It is possible, however, that the baccharins actually are formed by fungal infection of the plants from which they were isolated (Doyle & Bradner, 1980). This point needs to be clarified. The parent compound, tricothecin (Figure 67) was isolated from *Trichothecium roseum* in 1948 (Freeman & Morrison, 1948) and the most important member of the

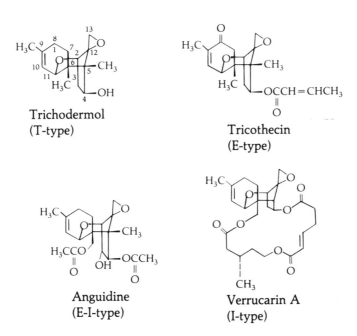

Figure 67   Examples of tricothecanes.

family, anguidine (Figure 67) was isolated from *Fusarium scirpi* in 1961 (Brian et al., 1961). Phase I clinical trials in the United States were completed on anguidine in 1976 (Helman & Slavik, 1976) and it is presently in Phase II.

Compounds in the tricothecane family act at peptidyl transferase centers in the 60 S subunits of 80 S ribosome-m-RNA complexes. Distinctions between individual compounds depend on their relative size and the length of the nascent polypeptide chain present on the t-RNA in the P or A site on the ribosome. Thus small tricothecanes can bind and inhibit the completion of peptide synthesis even when relatively large chains are present, whereas large trichothecanes can bind only in the initiation stage when the nascent chains are small. Based on this reasoning and the numerous experiments that support it (Ueno & Fukushima, 1968; Carrasco et al., 1973; Hansen & Vaughn, 1973; Wei & McLaughlin, 1974; Carter et al., 1976; Carter & Cannon, 1978), the trichothecanes have been divided into three types, according to the stage at which they inhibit polypeptide synthesis (Cundliffe et al., 1974). The smallest trichothecanes, such as trichodermol (Figure 67), which has no substituent on the methyl group at C-15, are able to fit the peptidyl transferase site near the end of translation. Hence they are denoted termination inhibitors (T-type). Structures with slightly larger size, such as tricothecin (Figure 67), are able to inhibit the elongation of smaller polypeptides, although they do not prevent the termination of nearly completed polypeptides. They are called elongation inhibitors (E-type). The largest tricothecin analogs are the macrocyclic types, exemplified by verrucarin A (Figure 67). They can bind only to the transferase centers that have small chains; hence they are initiation inhibitors (I-type). Analogs with sizes between those of tricothecin and verrucarin A tend to inhibit elongation at high concentration but initiation only at low concentration. They are known as mixed I-E-types (Doyle & Bradner, 1980). A representative example is anguidine (see Figure 67).

The detailed mechanism by which trichothecans bind to the ribosome is not known. It has been postulated that they interfere with thiol residues at the active site (Ueno, 1977). Because the 12,13-epoxide function is essential to antitumor activity (Doyle & Bradner, 1980), it is likely that alkylation of thiols occurs. An unusual reaction that involves double-bond participation in the hydrolytic ring opening of the epoxide of triactylscirpinetriol has been observed (Equation 51; Sigg et al., 1965). This process would account for the essential features of the 9,10-double bond and the 12,13-epoxide; there is no evidence that it occurs with thiols on ribosomes.

Structural modifications of semisynthetic tricothecanes of the macrocyclic type have included reduction of the 9,10-double bond, removal of the 12,13-

epoxide, and cleavage of the 3,4-bond. These modifications drastically reduce antitumor activity. Conversion of the 9,10-double bond to an epoxide and the introduction of an 8 β-hydroxy group decrease potency but increase the life span of tumor-bearing rodents. In the side chain the 7'-hydroxyethyl group is more active than the 7'-carbonyl group (Doyle & Bradner, 1980).

Among the nonmacrocyclic tricothecanes, analogs of anguidine have received the most attention. Because anguidine is a diacetate of scirpenetriol, all possible acetate derivatives of scirpenetriol were prepared (Claridge et al., 1979). The most active of these derivatives was the 15-monoacetate, which was 2–4 times as active as anguidine. A natural compound named T-2 toxin, the 8 α-valeroxy analog of anguidine, is slightly more active against experimental tumors (Doyle & Bradner, 1980). Certain totally synthetic tricothecin analogs with simplified structure are marginally active in screening systems. They include an analog with a fully aromatic ring and even an epoxide derived from methylenecyclohexane (Figure 68; Anderson et al., 1977; Fullerton et al., 1976).

The alkaloid cephalotoxine isolated from *Cephalotaxus* species in 1963 (Paudler et al., 1963) is devoid of antitumor activity. Various natural and semisynthetic esters of cephalotaxine (Figure 69) have significant activity (Smith et al., 1980). The two most important esters, harringtonine and homoharringtonine, are active in all the standard prescreens of the National Cancer Institute (Corbett et al., 1977). Clinical studies on these compounds have been hindered in the United States by a scarcity of their plant source. How-

Figure 68 Totally synthetic analogs of tricothecin.

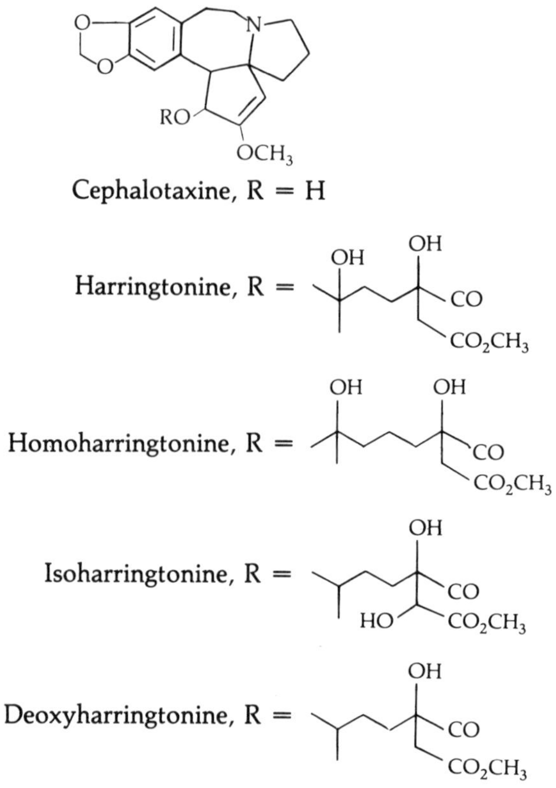

Figure 69 Cephalotaxine esters.

ever, the abundance in China of trees of the *Cephalotaxus* species which contain these esters has provided adequate material for clinical trials. Favorable results have been reported for the treatment of acute leukemia (Smith et al., 1980).

The scarcity of harringtonine and homoharringtonine has stimulated research on their synthesis from cephalotaxine, which is relatively abundant. Their syntheses are difficult because the carboxylic acid moities (Figure 69) contain two asymmetric centers and because their bulky shape hinders acylation of the hydroxyl group on cephalotaxine. A satisfactory synthesis of harringtonine was provided by an indirect route that featured a series of cyclic ketal and hemiketal intermediates, along with acylation of cephalotaxine, before complete elaboration of the acid moiety (Mikolajczak & Smith, 1978; Scheme 25). In this route isobutyraldehyde and malonic acid were condensed

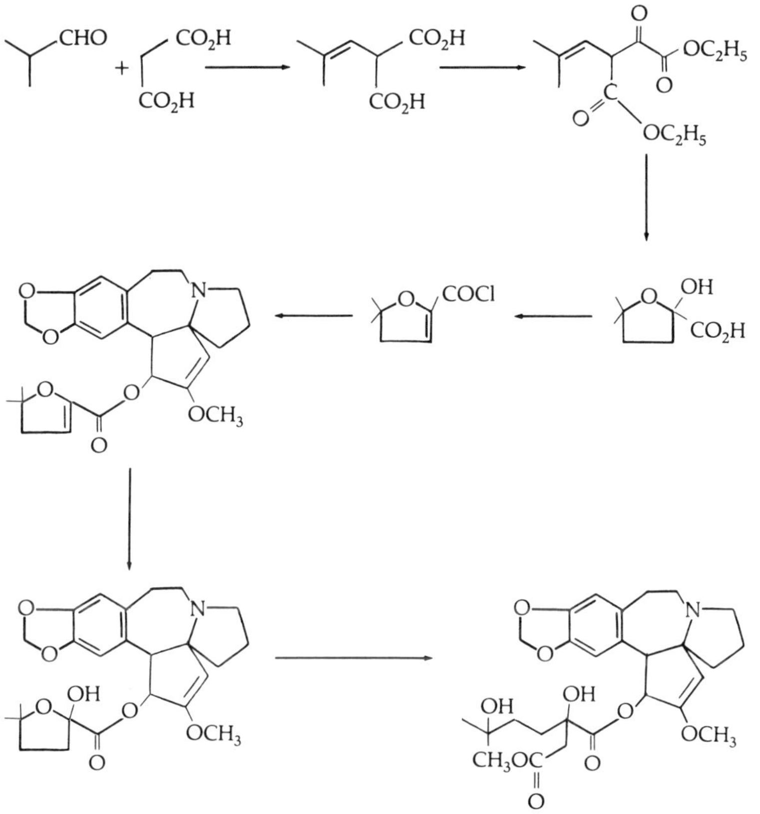

Scheme 25 Synthesis of harringtonine.

to yield a product that was decarboxylated, esterified, and condensed with diethyl oxalate to produce an α-keto ester. Acid hydrolysis of this compound gave a cyclic hemiketal, which was converted into the α,β-unsaturated acid chloride. This derivative was used to acylate cephalotaxine and the resulting ester was hydrated and converted into harringtonine by a Reformatsky reaction. Two elegant total syntheses of cephalotaxine have been reported (Weinreb & Auerbach, 1975; Semmelhack et al., 1975).

Among the known esters of cephalotaxine, harringtonine and homoharringtone show the best antitumor activity (Powell et al., 1972). Isoharringtonine is moderately effective in prolonging the lives of leukemic mice but less potent than harringtonine. Dehydroharringtonine has about one-half the potency of harringtonine. Simple synthetic esters such as the acetate are inac-

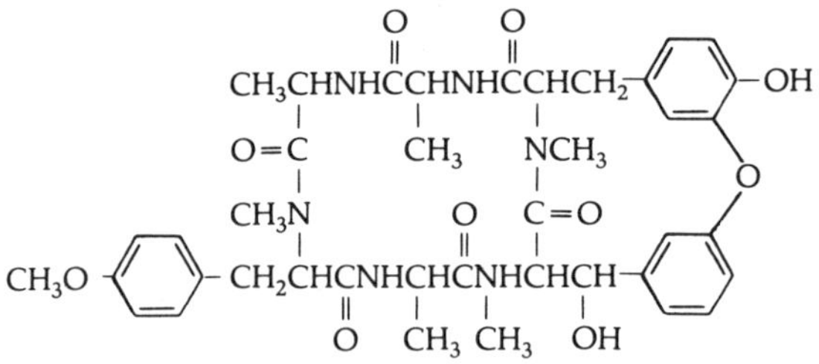

$$\text{CH}_3\text{CHNHCCHNHCCHCH}_2\underset{\underset{\text{O}}{|}}{\overset{\overset{\text{O}}{||}\overset{\text{O}}{||}}{}}\text{—}\langle\text{ring}\rangle\text{—OH}$$

Figure 70 Bouvardin.

tive but the more complex, such as methyl itaconate, show significant activity (Smith et al., 1980).

Cephalotaxine esters inhibit protein synthesis in He La cells (Huang, 1975) mainly in the G_1 and G_2 phases of the cell cycle (Baaske & Heinstein, 1977). The initiation, but not the elongation, of polypeptide chains is prevented. Polyribosomes disappear and are replaced by monosomes and subunits (Tscherne & Pestka, 1975).

Another plant product that inhibits protein synthesis at the initiation stage is bouvardin (Johnson & Chitnis, 1978). This compound has an unusual bicyclic hexapeptide structure in which two N-methyltyrosine residues are coupled to form a rather strained 14-membered ring (Figure 70; Jolad et al., 1977). The mechanism by which this structure influences protein synthesis is unknown.

MITOTIC INHIBITORS

Some highly important antitumor agents act by producing mitotic arrest at metaphase in dividing cells. These agents are typically isolated from higher plants, although certain maytansinoids have been found recently in *Nocardia*. Because they interact with cellular proteins rather than DNA, the mitotic inhibitors do not show mutagenicity. They do produce other serious toxicities like myelosuppression and nerve damage.

Mitotic inhibitors such as dimeric catharanthus alkaloids (vincristine), maytansinoids, podophyllotoxins, and colchicine produce antitumor effects by interacting with tubulin, a protein that is capable of reversible polymerization to microtubules. These microtubules are hollow, tubelike filaments found in mitotic spindles and are essential for cell division (Snyder & McIntosh, 1976). The classical example of a spindle poison is colchicine (Figure 71), a complex tropolone derivative that disrupts the assembly of microtubules

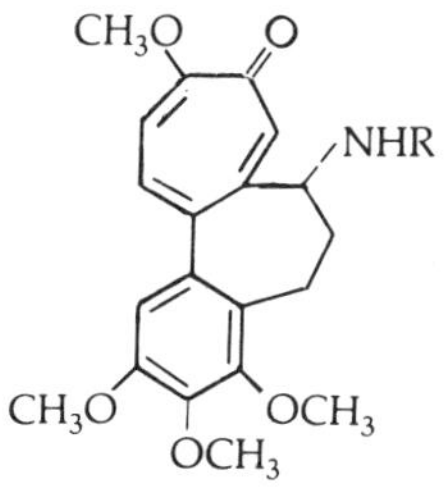

Colchicine, R = COCH$_3$
Demecolcine, R = CH$_3$ **Figure 71** Colchicine.

in dividing cells at a concentration of $10^{-8}M$ (Dustin, 1963) by binding noncovalently with a high affinity site on tubulin. This binding does not affect intact microtubules but it prevents their formation by polymerization of tubulin. The binding site for cochicine also serves as a binding site for podophyllotoxin (Figure 72), a lignan that produces inhibitory effects virtually indistinguishable from those of colchicine (Cornman & Cornman, 1951; Kelly & Hartwell, 1954). Evidence of a common binding site is found in the ability of podophyllotoxin to prevent the binding of colchicine to tubulin. It has been suggested, however, that these compounds occupy overlapping rather than identical sites because tropolone inhibits the binding of colchicine but not of podophyllotoxin (Bhattacharyya & Wolff, 1974). The recent discovery of colchicinelike peptides that function in microtubule regulation by binding to the colchicine site of tubulin promises to shed further light on the process of

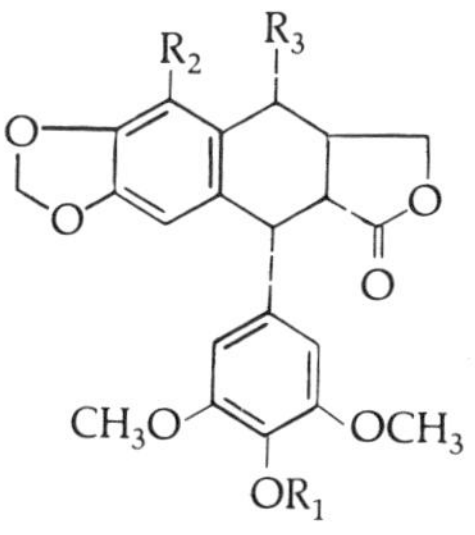

Podophyllotoxin: R$_1$ = CH$_3$, R$_2$ = H, R$_3$ = OH
α-Peltatin: R$_1$ = CH$_3$, R$_2$ = OH, R$_3$ = H
4'-Demethylpodophyllotoxin: R$_1$ = R$_2$ = H, R$_3$ = OH
Podophyllotoxin Glucoside: R$_1$ = CH$_3$, R$_2$ = H,
      R$_3$ = O-D-glucosyl

Figure 72 Some naturally occurring podophyllotoxins.

mitotic inhibition (Lockwood, 1979). Perhaps it will lead to new antitumor drugs that use this mode of action.

Dimeric catharanthus alkaloids such as vincristine (Table 15) and vinblastine also cause mitotic arrest by binding to tubulin and preventing its assembly into microtubules. Apparently their binding site is different from the one that responds to colchicine and podophyllotoxin. The catharanthus alkaloids can cause the dissolution of microtubules in cells. As a consequence, microtubule crystals which contain the alkaloids can be found in the cytoplasm (Bensch & Malawista, 1969). Maytansinoids are thought to have the same tubulin binding site as catharanthus alkaloids. Thus maytansine is a competitive inhibitor of vincristine binding (Adamson et al., 1976). The carbinolamine functionality of maytansine (Figure 74) gives it the ability to alkylate the binding site, which might partly account for its very high potency (Kupchan et al., 1978).

The North American plant *Podophyllum peltatum* (may apple) and related species were long known to have medicinal properties (Kelly & Hartwell, 1954). Extracts such as podophyllum and podophyllin found use as cathartics and gall bladder stimulants in the last century and the lignan podophyllotoxin (Figure 72) was isolated in crystalline form from podophyllin (Podwyssotzki, 1880). Interest in podophyllotoxin and related compounds as potential antitumor agents began in 1942 when their activity against a venereal wart, *Condyloma acuminatum*, was demonstrated (Kaplan, 1942). A comprehensive study of *Podophyllum* species resulted in the isolation and structure elucidation of a variety of lignans and their glycosides (Hartwell & Schrecker, 1958). Early clinical investigations were undertaken for compounds like α-peltatin (Figure 72) but no significant results were obtained (Greenspan et al., 1950). In the meantime the structure and sterochemistry were defined, and it was recognized that the unique configuration at C-2, C-3, and C-4 of these compounds, which resulted in a highly strained *trans*-γ-lactone, contributes to their antimitotic activity (Hartwell & Schrecker, 1958). Picropodophyllin, the C-3 epimeric compound, is almost entirely inactive.

More recent investigations of podophyllotoxin and related compounds demonstrated that glycosides (e.g., podophyllotoxin glucoside, Figure 72) are much more potent than the aglycones for inhibiting mitosis in vitro but less active against animal tumors (Emenegger et al., 1961). Condensation of the sugar moieties of these glycosides with aldehydes such as acetaldehyde or benzaldehyde produced derivatives that are less toxic than the unsubstituted glycosides. A significant breakthrough finally resulted from studies of the total synthesis of podophyllotoxins (Gensler et al., 1954), wherein 1-epi-

Table 15 Structures and Antitumor Activity of Catharanthus Alkaloids

			Substituents			Antitumor Activity[a]	
						Gardner Lymphosarcoma Dose Range for Optimal Activity (mg/kg[b])	B16 Melanoma % T/C at Optimal Dose[c] (survivors)
	R_1	R_2	R_3	R_4	R_5		
Vincristine	CHO	OH	C_2H_5	OCH_3	$COCH_3$	+++, 0.15–0.2	175 (0/10)[d]
Vinblastine	CH_3	OH	C_2H_5	OCH_3	$COCH_3$	++, 0.3	202 (1/10)[d]
Desacetyl vinblastine	CH_3	OH	C_2H_5	OCH_3	H	+++, 0.3–0.5	
Vindesine	CH_3	OH	C_2H_5	NH_2	H	+++, 0.2–0.4	242 (2/10)[d]
N-Methyl vindesine	CH_3	OH	C_2H_5	$NHCH_3$	H	+++, 0.2–0.3	
N-(2-Hydroxyethyl)-vindesine	CH_3	OH	C_2H_5	$NHCH_2CH_2OH$	H	+++, 0.05–0.4	193 (0)[e]
N-[2-(4-Hydroxyphenyl)-ethyl]vindesine	CH_3	OH	C_2H_5	$NH(CH_2)_2C_6H_4$	H	Inactive	290 (8/10)[e]
Bis(N-ethylidine vindesine)-disulfide	CH_3	OH	C_2H_5	$(NHCH_2CH_2S)_2$	H		
Leurosine	CH_3	C_2H_5	3',4'-oxide	OCH_3	$COCH_3$		
Leurosidine	CH_3	C_2H_5	OH	OCH_3	$COCH_3$		
Vinglycinate	CH_3	C_2H_5	C_2H_5	OCH_3	$COCH_2N(CH_3)_2$		

[a] Abstracted from Gerzon, 1980.
[b] C_3H mice treated ip for nine days: +++ = 75–100% inhibition of tumor growth compared with controls; ++ = 50–75% inhibition.
[c] % T/C = median survival treated/median survival controls × 100: survivors on day 56 or 60.
[d] Lilly Research Laboratories. C57BL/6 mice. Average death day of control mice, 19.1 days.
[e] NCI Data. B6D2F₁ male mice. Median death day of control mice, 20 days.

podophyllotoxins were obtained (Kuhn & von Wartburg, 1968). These compounds yielded glycosides that were not significantly different in antitumor activity from the podophyllotoxins (Keller-Juslen et al., 1971). However, the cyclic acetals made from their condensation with aldehydes were much more active against L-1210 leukemia and other animal tumors (Stahelin, 1973). The ethylidene and 2-thenylidene derivatives of the α-D-glucoside of 4'-demethylepipodophyllotoxin (Scheme 26), known as VP 16-213 (etiposide) and VM 26 (teniposide), respectively, are undergoing clinical trials and appear to be promising drugs. Especially noteworthy is their reported activity against small cell lung carcinoma (Carter & Slavik, 1976).

The synthesis of VP 16-213 and VM 26 is illustrated in Scheme 26 (Kuhn & von Wartburg, 1968). In this scheme the benzylcarbonate of 4'-demethylepipodophyllotoxin is condensed with glucose tetraacetate in the presence of boron trifluoride etherate to give the tetraacetyl glucoside. The same product can be obtained from the benzylcarbonate of 4'-demethylpodophyllotoxin, which undergoes epimerization at C-1 in the presence of the Lewis acid catalyst. 4'-Demethylation of podophyllotoxin had been established (Sandoz, 1968). The tetraacetylglucoside is deacetylated by methanolysis in the presence of zinc chloride and the cyclic acetals are formed by acid-catalyzed reaction with the appropriate aldehyde or derived acetal. For the preparation of a simple aliphatic aldehyde derivative such as that of acetaldehyde (VP 16-213) the transacetalization reaction is preferred, whereas for thiopene-2-carboxaldehyde (VM-26) the direct reaction is satisfactory (Keller-Juslen et al., 1971).

The epipodophyllotoxin derivatives VP 16-213 and VM-26 are surprising not only in their enhanced antitumor activity when compared with the corresponding podophyllotoxins but also in their mode of action. In contrast to podophyllotoxins they do not arrest cell mitosis at metaphase. Instead, they arrest cells in the late S or G_2 phase and prevent them from entering mitosis (Stahelin, 1973; Greider et al., 1974). They appear to have a unique mode of action that is not well understood.

Although colchicine is a potent inhibitor of cell mitosis, it is not sufficiently active against tumors to be of clinical significance. Its main use is in terminating acute attacks of gout. Among a large number of colchicine derivatives and analogs that have been screened, demecolcine, the analog in which the N-acetyl group is replaced by methyl (Figure 71), shows the best antitumor activity in experimental systems.

Dimeric catharanthus alkaloids have achieved a prominent place in the chemotherapy of cancer. Vincristine is a drug of choice for seven different

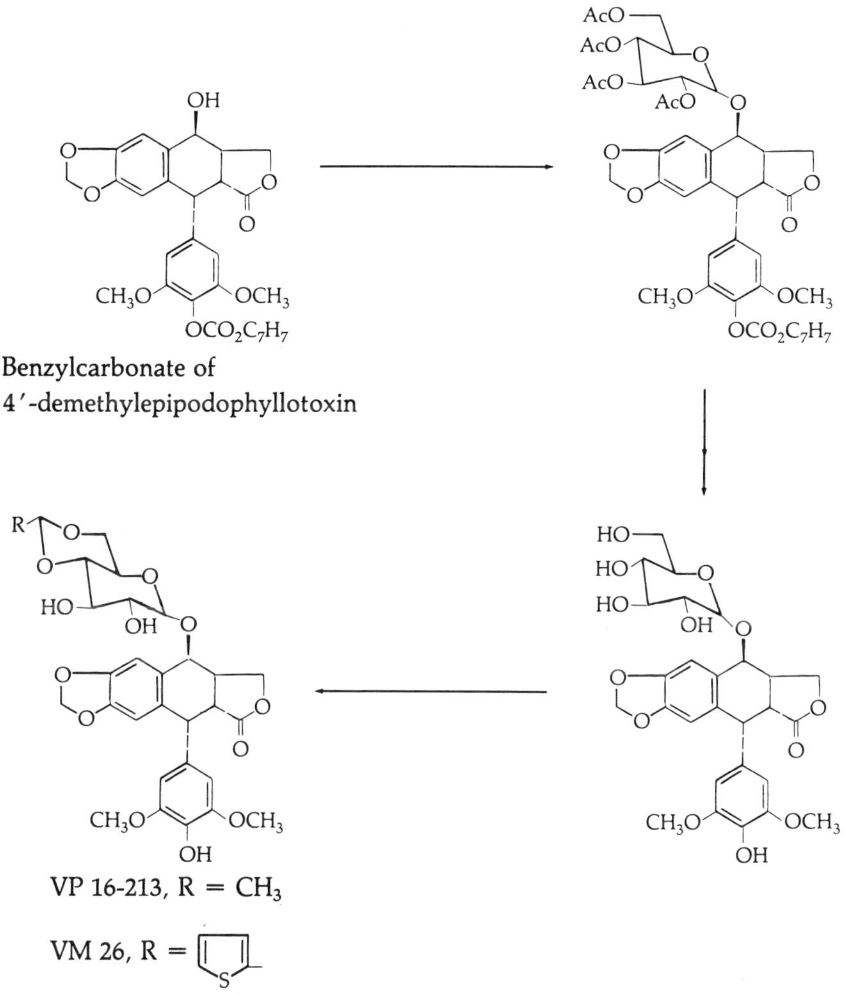

Benzylcarbonate of
4'-demethylepipodophyllotoxin

VP 16-213, R = CH$_3$

VM 26, R = [thienyl]

Scheme 26 Synthesis of VP 16-213 and VM 26.

tumors and vinblastine is an approved agent (Zubrod, 1974). Furthermore, the new semisynthetic compound vindesine is now in clinical trial.

Vinblastine was discovered in 1959 as an antiproliferative factor in leaves of the periwinkle plant, *Catharanthus roseus* (Johnson et al., 1959). It was detected because of its activity in the P-1534 mouse leukemia screen, a system that has since become resistant to vinblastine (Gerzon, 1980). Vincristine, leurosine, and leurosidin were isolated during the next four years (Svoboda, 1961; Svoboda et al., 1962). Vinblastine and vincristine soon became established drugs, but leurosidine and preparations of leurosine, including the sulfate and methiodide, failed in the clinic because of toxicity and poor activity (Gailani et al., 1966; Hodes et al., 1963). The structures of these alkaloids and some semisynthetic analogs are listed in Table 15. These complex structures are composed of an indole moiety known as catharanthine and an indoline moiety known as vindoline (from vinca indoline). The individual alkaloids differ according to substituents on these nuclei. Vincristine and vinblastine differ only in the substituent (R_1) on the indoline nitrogen, which is a formyl group for the former and a methyl group for the latter. This relatively small structural change makes a difference in the potencies of the two compounds (Table 15) and their clinical applications and toxicities. The toxicity patterns are particularly noteworthy because neurotoxicity is dose-limiting for vincristine but minimal for vinblastine (Grobe & Palm, 1972). The latter agent is limited mainly by myelosuppression. Leurosine and leurosidine differ in structure from vinblastine by being epimeric at the 4′-position in the catharanthine moiety. Leurosine differs further by its 3′,4′-epoxide functionality.

The toxic effects of vincristine and vinblastine have stimulated a search for semisynthetic analogs with improved therapeutic properties; the goal is to prepare an analog as active as vincristine but less neurotoxic. Most of the analogs with significant activity prepared so far have involved the vindoline moiety of vinblastine, which is readily transformed in its two ester functionalities. Thus partial hydrolysis gives desacetyl vinblastine (Table 15), an analog that retains the antitumor activity. Reesterification of the liberated 4-hydroxyl group with N,N-dimethylglycine produced vinglycinate (Table 15; Hargrove, 1964). This compound was placed in a clinical trial but withdrawn because of problems with chemical instability. Greater success was obtained with a series of desacetyl vinblastine amides. The most important analog in this group is vindesine, which was first obtained from the ammonolysis of vinblastine (Scheme 27). An alternative route to vindesine was based on hydrazinolysis followed by hydrogenolysis with Raney nickel. Good yields of vindesine and other amides are now obtained by a route that involves conversion of the hydrazide into the corresponding azide, followed by treatment with the appropriate amines (Scheme 27; Barnett et al., 1978).

The activities of vindesine and other selected amides against B-16 melonoma and the Gardner lymphosarcoma are shown in Table 15. Vindesine was chosen for clinical trial because of its high activity against B-16 and its broad spectrum of activity against experimental tumors (Sweeney et al.,

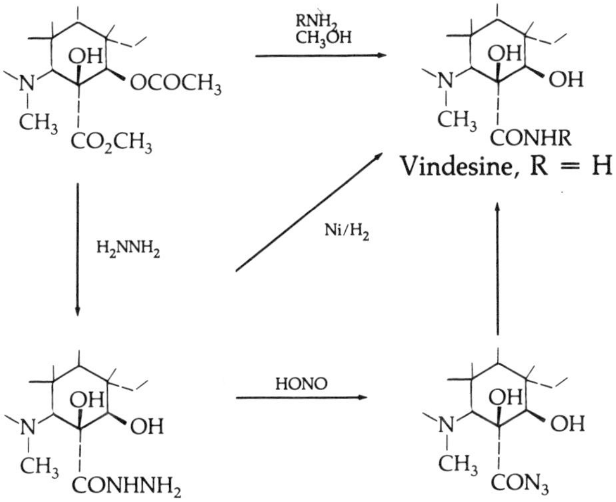

Scheme 27 Preparation of Vindesine and other amides from vinblastine.

1978). The clinical trial appears to be promising; the neurotoxicity of vindesine is less than that of vincristine but greater than that of vinblastine (Dyke & Nelson, 1977). The 2-hydroxyethyl derivative was less active than vindesine against B-16 melanoma. Although the 4-hydroxyphenylethyl derivative showed outstanding activity against B-16 melanoma (Conrad, et al., 1979), it was not chosen for clinical trial because its spectrum of activity was limited (Gerzon, 1980); for example, it showed no inhibition of the Gardner lymphosarcoma (Table 15). One further amide of special interest is bis(N-ethylidene-vindesine)-disulfide (Figure 73). This compound was prepared in the expectation that its disulfide linkage, or the sulfhydryl groups derivable from it, would interfere with thiol functionalities on tubulin (Conrad et al.,

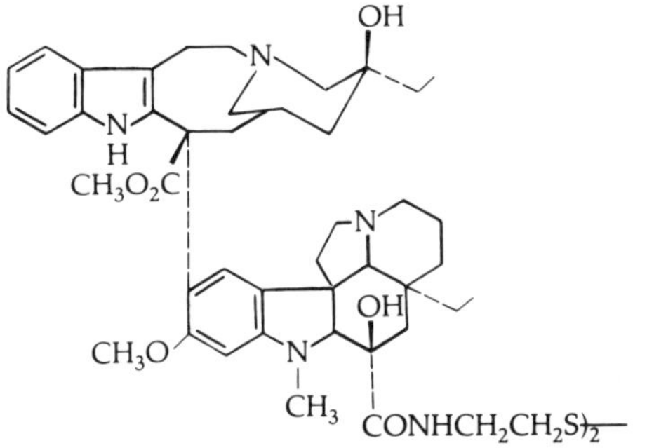

Figure 73 Bridged bis-vindesine disulfide.

1979). It has a profile of antitumor activity similar to that of vindesine, except that it is highly active against a strain of P-388 leukemia resistant to vincristine and maytansine (Wolpert-DeFilippes et al., 1975).

In contrast to the vinblastine analogs, vincristine analogs have not shown improved therapeutic properties. A series of desacetyl amides of vincristine was prepared by low-temperature CrO_3 oxidation of the corresponding N-substituted vindesines. This procedure oxidizes the indoline N-methyl selectively to the formyl group (Barnett, 1978). Unfortunately the resulting vincristine analogs were uniformly less active that the corresponding vindesine derivatives. The earlier CrO_3 oxidation of leurosine provided its indoline-N-formyl derivative (Richter, 1973). This analog, a hybrid of leurosine and vincristine, had only one-tenth of the potency of vinblastine, but it was active against clinical tumors and not neurotoxic (Gerzon, 1980).

Maytansine has provided a great challenge to the natural-products chemists because it occurs in low concentrations in plant species. The original discovery in *Maytenus serrate* probably would not have occurred except for its high potency against tumors in cell culture and mice (Kupchan et al., 1972). After the initial isolation and antitumor testing of maytansine a search was made for other plants that might produce it in greater yield. The species *Maytenus buchanaii* was selected for large-scale extraction for clinical trial material. Meanwhile, efforts at total synthesis resulted in two elegant but lengthy routes (Meyers, et al., 1980; Corey et al., 1980). The most promising solution to the problem of maytansine production appears to be in a discovery that compounds closely related in structure to maytansine, known as ansamitocins (Figure 74), are produced by a microorganism of the *Nocardia* species (Higashide et al., 1977). This discovery opens the way to large-scale production by fermentation.

The name maytansine is derived from its origin in *Maytenus* and its structure, which resembles those of the ansa antibiotics (ansa = basket with a handle). According to X-ray crystallography (Bryan et al., 1973), the maytansine nucleus is a roughly rectangular, 19-membered ring. The two longer sides are approximately parallel and separated by about 5.4 A; a hole is left in the center of the ring (Figure 74). The ester group is above the molecule and oriented to provide steric hindrance to the upper face, which is more hydrophilic than the lower face. An ester group based on an amino acid or a fatty acid is essential for significant antitumor activity. However, considerable variation can be tolerated in the structure of this substituent (Kupchan et al., 1978). A carbinolamine function is needed at C-9. Etherification of the 9-hydroxyl group greatly reduces activity and the absence of the carbinolamine abolishes it. At the present time it is thought that the ester group plays a key role in the

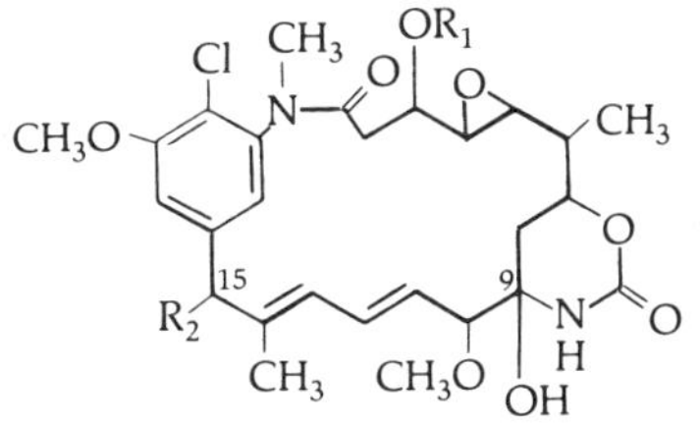

Figure 74 Structures of representative maytansinoids.

Maytansine: $R_1 = \text{COCHNCOCH}_3$ (with CH$_3$ on N and CH$_3$ branch), $R_2 = \text{H}$

Colubrinol: $R_1 = \text{COCHNCOCH(CH}_3)_2$ (with CH$_3$ groups), $R_2 = \text{OH}$

Maytansinol: $R_1 = R_2 = \text{H}$

Ansamitocin P-3: $R_2 = \text{COCH(CH}_3)_2$, $R_2 = \text{H}$

initial binding of maytansine to tubulin, by positioning the maytansine so that its carbinolamine functionality alkylates the tubulin. The C-4,5 epoxide enhances the binding to tubulin (Kupchan et al., 1978).

Some examples of naturally occurring analogs of maytansine are illustrated in Figure 74. These analogs are divided into groups among which are C-15 unsubstituted maytansides with ester groups (maytansine), C-15 substituted maytansides with ester groups (colubrinol), maytansides without ester groups (maytansinol), and the maytansinoids of microbial origin (ansamitocins) such as ansamitocin P3. The ansamitocins correspond to the C-15 unsubstituted maytansides, except that their esters are derived from simple acids like butyrate, isoburate, and isovalerate. They show correspondingly potent antitumor activity (Komoda & Kishi, 1980).

MISCELLANEOUS NATURAL AND SYNTHETIC AGENTS

The extensive natural-product screening programs supported by the National Cancer Institute and certain pharmaceutical companies have uncovered many novel structures that show antitumor activity in one or more animal assays. Some of these structures and their analogs have demonstrated important clinical activity; for example, the catharanthus alkaloids and the epipodophyllotoxins. Others are presently under investigation in preclinical or Phase I protocols. Still others are not candidates for clinical studies but they are important because their unique structural features may be used in synthetic analogs.

Among natural-product classes, the terpenes have yielded a particularly large number of compounds with antitumor activity. A complete survey of the antitumor terpenes is beyond the scope of this chapter; however, an authoritative review was published recently (Cassady & Suffness, 1980). Many of the terpenes have functional groups such as epoxide and α,β-unsaturated carbonyl that can alkylate biological nucleophiles. Helenalin, a sesquiterpene of the pseudoguaianolide type, was given as an example (Figure 1; Lee et al., 1978). Among the diterpenes, tripdiolide and taxol are of interest for their antitumor activities (Figure 75; Kupchan et al., 1972; Wani et al., 1971).

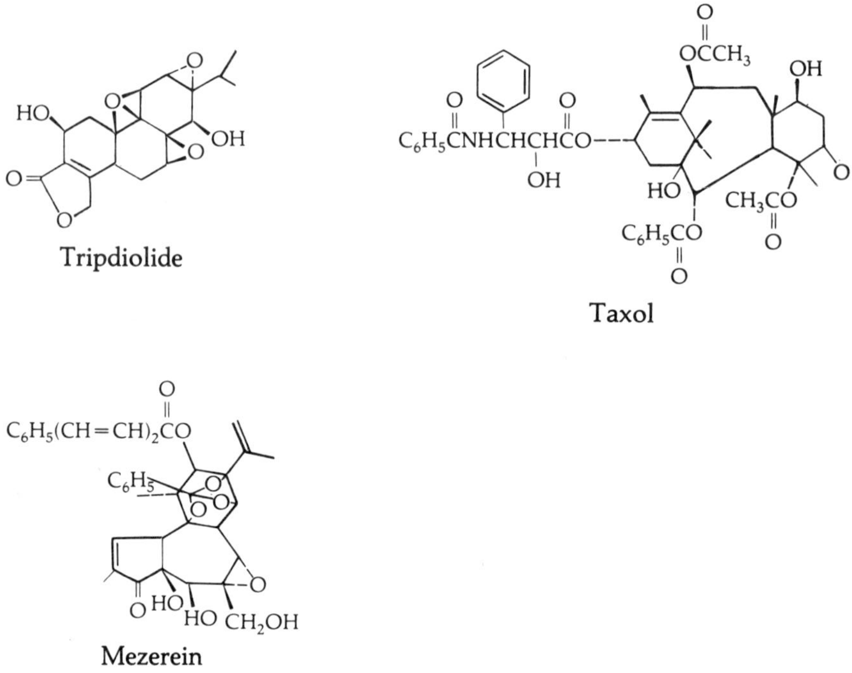

Figure 75 Antitumor diterpenes.

Diterpenes of the phorbol and dephentoxin types also are active. Mezerein (Figure 75) is one of the best of them.

Steroids of the cardenolide, bufadienolide, and withanolide types are cytotoxic. The most active compound among the steroids, 4-β-hydroxywithanolide E (Figure 76) has been selected for further development by the NCI (Cassady & Suffness, 1980). The quassinoids are also biogenetically derived from triterpenoids. Only the Type I quassinoids, which have the picrasane skeleton (Figure 76), show antitumor activity. Within Type I are the Group A quassinoids, such as glaucarubinone, which have an 11,20-oxide bridge, and the Group B quassinoids, such as bruceantin (Figure 76), which have a 13,20-oxide bridge (Cassady & Suffness, 1980). Bruceantin, the quassinoid of broadest activity in rodent tumors (Liao et al., 1976), has been advanced to Phase II clinical trials.

Among the nonterpene natural products not already discussed in this chapter indicine-N-oxide is of particular interest. It is presently in Phase II clinical trials. Indicine-N-oxide (Figure 77) is a member of the pyrrolizidine alkaloids, a group of compounds well known for their ability to cause liver cirrhosis and tumors when ingested by humans and livestock (McLean, 1970).

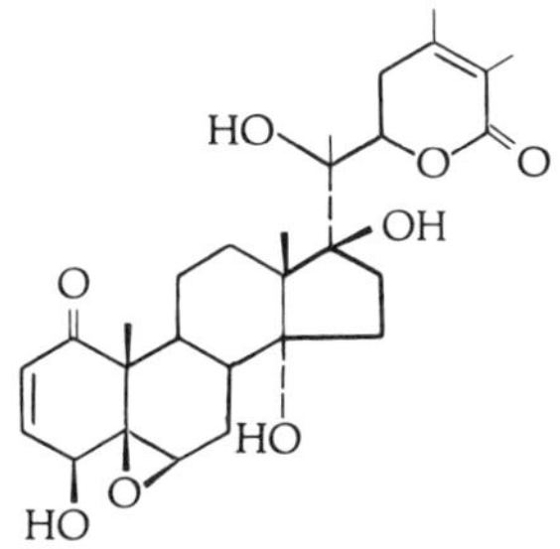

4β-Hydroxywithanolide E

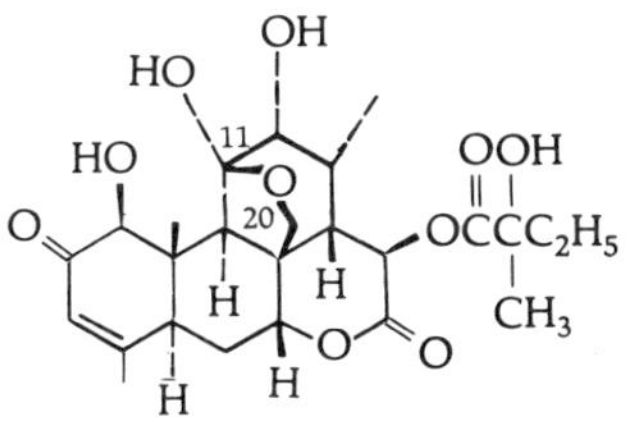

Glaucarubinone

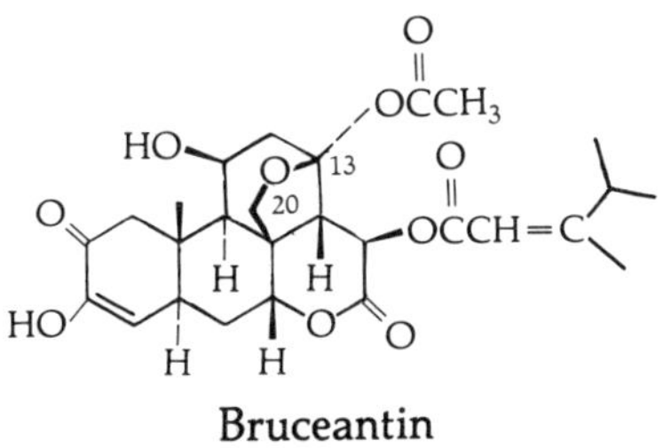

Bruceantin

Figure 76 Antitumor compounds derived from triterpenoids.

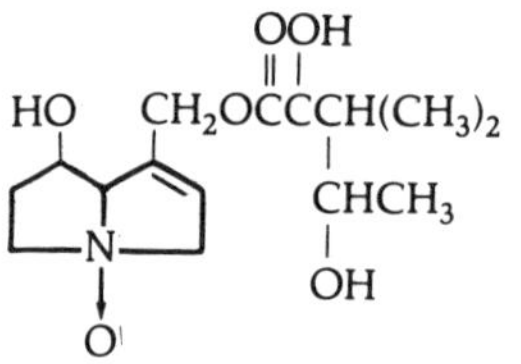

Figure 77 Indicine-*N*-oxide.

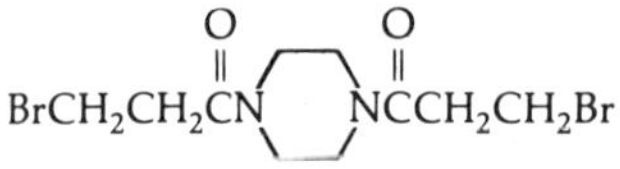

Pipobroman

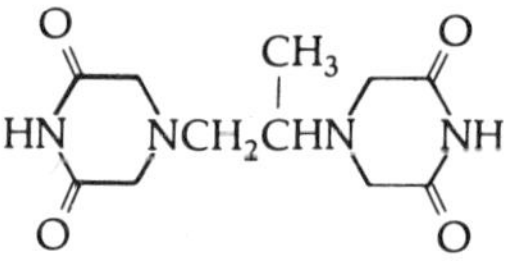

IRCF-159

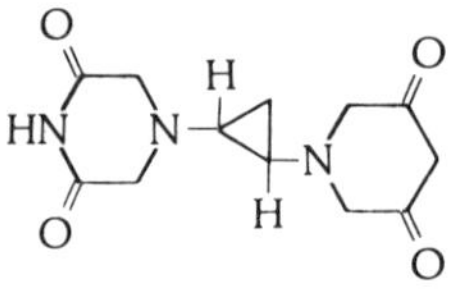

trans-Cyclopropyl analog

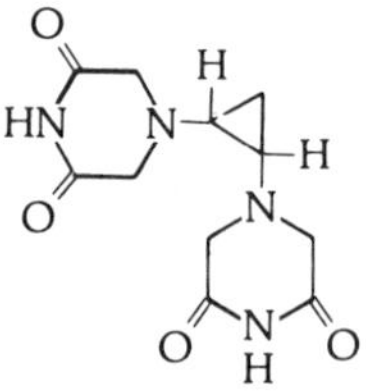

cis-Cyclopropyl analog

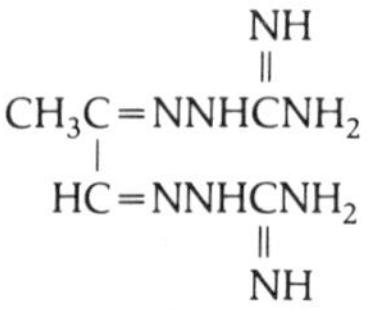

Methylglyoxal bis(guanylhydrazone)

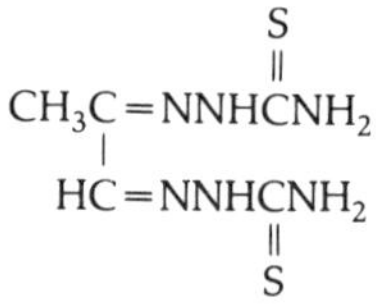

Methylglyoxal bis(thiosemicarbazone)

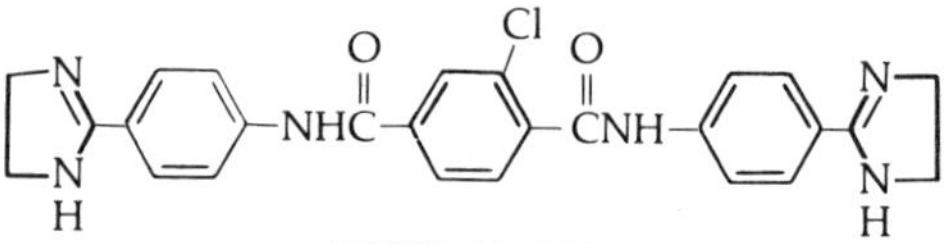

NSC 60,339

Figure 78 Miscellaneous synthetic antitumor compounds.

Many of these compounds have shown antitumor activity, but the first to have both antitumor activity and low hepatotoxicity is indicine-N-oxide (Kugelman et al., 1976). The high doses required of this compound suggest that the active form is probably a metabolite, but studies directed toward this problem have not been completed.

Pipobroman (Figure 78) has sometimes been classified as an alkylating agent, although it does not show alkylating properties in vitro under physiological conditions. It is used against hematological cancers (Bond, 1962). The compound known as ICRF-159 shows an unusual antitumor property by inhibiting the metastasis of Lewis lung tumor in mice without affecting the primary tumor. It is a potent inhibitor of DNA synthesis (Creighton & Birnie, 1970) and is thought to affect the development of blood vessels at invading margins of the primary tumor (Creighton et al., 1969). A recent study of the cyclopropyl analogs of ICRF-159 (Figure 78; Witiak et al., 1978) revealed that in one tumor model the cis isomer had antitumor activity, whereas the trans isomer potentiated tumor growth.

Methylglyoxal bis(guanylhydrazone) (Figure 78) shows antitumor activity in man. Many of its actions are related to spermidine, which it resembles in structure. It inhibits spermidine biosynthesis and transport (Mihich, 1963). Other actions are interference with nuclear and mitochondrial metabolism (Pressman, 1963). Phthalanilides such as NSC 60, 339 (Figure 78) also interfere with oxidative phosphorylation in mitochondira. They showed antitumor activity in clinical trials but were highly toxic (Yesair & Kensler, 1975). The bis(thiosemicarbazones) of α-ketoaldehydes and α-diketones have been extensively studied because of their antitumor activity. They form strong chelates with copper and zinc but the significance of this property is unknown (Booth & Sartorelli, 1967). A representative example is the bis(thiosemicarbazone) of methylglyoxal (Figure 78; French & Friedlander, 1958).

HORMONES

Effects of Hormones on Tumors

Steroid hormones act on target tissues at the level of transcription and stimulate cellular processes by derepression of genetic template operation. The target cells contain specific protein receptors in their cytoplasm with high affinities for the hormones. Binding of the hormones causes a change in the receptor structures, which leads to migration of the resulting complexes into the nucleus. Interaction of the complexes with acceptor sites produces the derepression (Gorski et al., 1965).

The hormone dependence of breast cancer has been known since 1896, when Beatson reported the successful use of oophorectomy in producing remissions in premenopausal women. Huggins' report in 1952 of the efficacy of bilateral adrenalectomy in postmenopausal women opened the modern era of endocrine surgery. More recently, hypophysectomy by surgery or stereotactic placement of yttrium 90 has been claimed to be superior to adrenalectomy (Patterson, 1974; Benabid et al., 1978). These ablative treatments produce their effects by denying estrogens and other hormones to cancer cells that depend on them for stimulation. Only about 30% of breast cancers respond to endocrine surgery or hormonal manipulation with positive responses defined as a 50% reduction in the size of the lesion. Probably many of the other cancers give lesser responses (Stoll, 1978). Recent studies show that the positive response can be correlated with the relative number of estrogen receptors present in the cancer cells. When the presence of an adequate number of estrogen receptors can be demonstrated the response rate to endocrine surgery and tamoxifen (an antiestrogen) is increased to 60% (Saez et al., 1978). The failure of all patients with hormone receptors to respond to this type of treatment has concerned oncologists. It is now thought that breast cancers contain mixed clonogenic populations of hormone-dependent and -independent cells in which only the dependent population is controlled by endocrine ablation (Sluyser, 1978). Fortunately cytotoxic agents such as cyclophosphamide affect both populations equally well. Thus chemotherapy is the most effective treatment modality in a random population (Carter, 1976). Another property of chemotherapy is a marked suppression of ovarian activity, which equals the effect of oophorectomy in premenopausal women (Dao, 1978).

Estrogens given in high levels to postmenopausal women with metastic breast cancer have produced remissions in about 30% of the cases. Ethinyl estradiol and estradiol propionate (Figure 79) have been used for this purpose. The use of estrogens appears paradoxical, but the high levels apparently interfere with the peripheral activity of prolactin, a pituitary hormone that also stimulates breast tissue (Pearson, 1972). Androgens also have been used against metastatic breast cancer in postmenopausal women. The rationale is the inhibition of the release of pituitary gonadotrophins. Their mode of action, however, must be more complex than this because they are active in hypophysectomized patients (Beckett & Brennan, 1959). Androgens are useful in advanced cancer because they stimulate the hematopoietic system

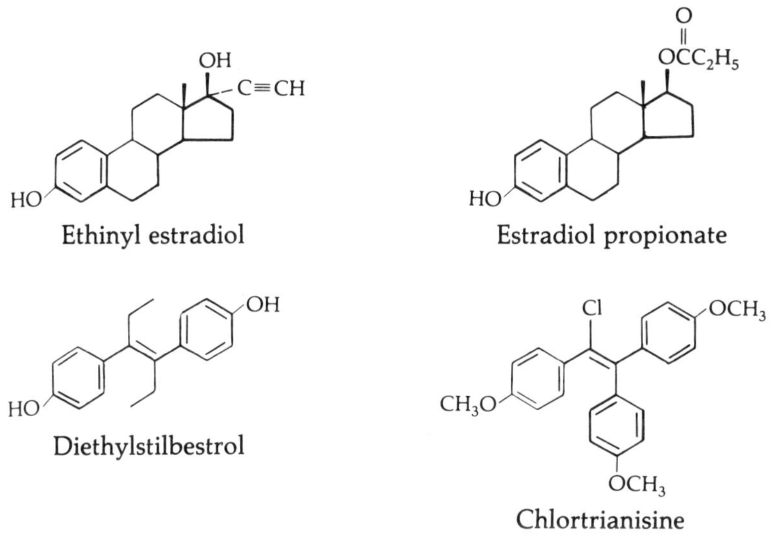

CHEMISTRY OF ANTITUMOR DRUGS

Figure 79 Steroidal and nonsteroidal estrogens.

and reverse bone demineralization. On the bad side are their serious masculinizing effects. Testosterone propionate, 2 α-methyldihydrotestosterone, 17 α-methyltestosterone, fluoxymestrone, and 19-nor-17α-methyltestosterone and testolactone (Figure 80) have been used in androgen therapy.

In the last few years specific antiestrogens that lack androgenic properties have found favor in the treatment of hormone-dependent breast cancer. Tamoxifen (Equation 52) is the leading agent of this type and Nafoxidine (Equation 53) has been used clinically. They probably act on estrogen receptors in the cancer cells, but other factors such as prolactin inhibition may be involved (Pearson, et al., 1978). These two compounds have virtually identi-

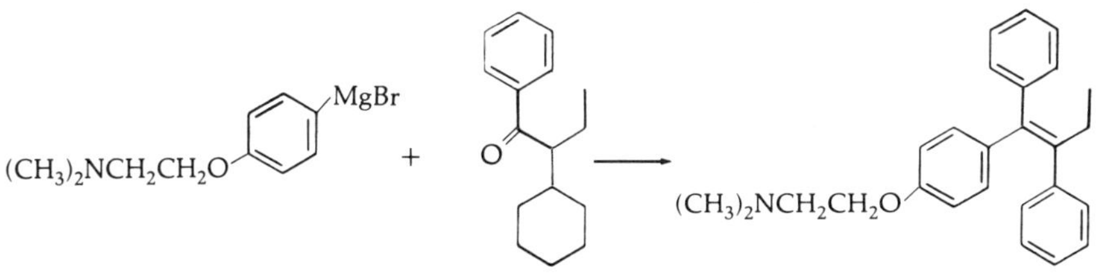

(52)

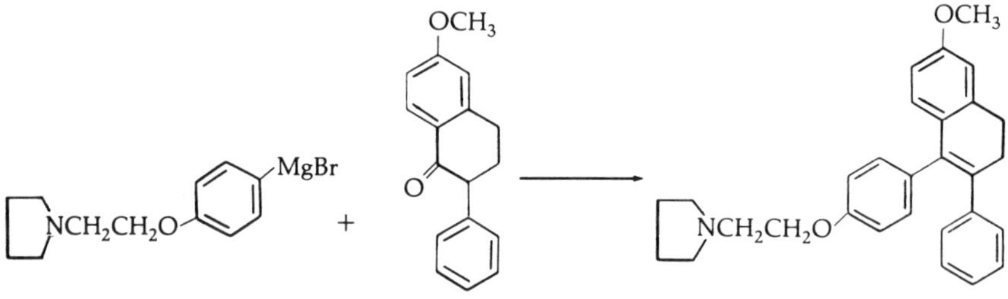

Nafoxidine

(53)

cal conformations in their triarylethylene portions and similar torsional angles in their side chains, according to X-ray diffraction analysis (Camerman et al., 1980). Tamoxifen has two geometrical isomers with somewhat different antiestrogenic activities. They are separated and only the Z-isomer is used.

Estrogens have induced remissions of metastatic prostate cancer. The nonsteroidal compounds diethylstilbestrol and chlortrianisine (Figure 79) are the usual choices. It is not known whether their effects are due to direct competition with peripheral androgens, inhibition of pituitary gonadotrophic hormone, or both (Dao, 1975).

Certain neoplasms stimulated by estrogens can be inhibited by progesterone and its analogs. Thus endometrial carcinoma in about 30% of the cases responds to progesterone, 17α-hydroxyprogesterone caproate, or medroxyprogesterone acetate (Figure 81). Renal-cell carcinoma responds to medroxyprogesterone in a small percentage of cases (Bloom, 1971).

Glucocorticoids produce pronounced acute changes in lymphoid tissues; for example, lymphocytes in the thymus and lymph nodes are dissolved and lymphopenia is produced in the peripheral blood (Kelley & Baker, 1960). This effect is the basis for steroid treatment of leukemias and Hodgkins' disease. It usually results in profound temporary regressions. Prednisone (Figure 81) is generally used, although other glucocorticoids and ACTH are effective (Dougherty & White, 1965). Glucocorticoids, which are given to metastatic prostate cancer patients who have relapsed after castration inhibit the release of ACTH from the pituitary and result in adrenal atrophy and decreased biosynthesis of androgens (Heilman & Kendall, 1944). Prednisone is used in the treatment of breast cancer. It does not have an antineoplastic effect in this condition but helps to alleviate complications such as hypercalcemia and anemia (Dao, 1957).

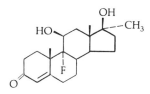

Figure 80 Androgens.

The role of prolactin in stimulating breast cancer has been discussed. In addition to interfering with peripheral prolactin activity by supplying high levels of estrogens, oncologists are investigating the effects of ergoline derivatives as prolactin inhibitors. Compounds such as lergotrile and 8-cyano-8-methylergoline (Figure 82) prevent the release of prolactin from the pituitary (Floss et al., 1973).

Mitotane is the o,p-isomer formed in the chemical synthesis of 1,1-di(chlorophenyl)-2,2-dichloroethane, an insecticide related to DDT (Equation 56). This compound is unique in its highly selective effect on adrenal cortical cells. It causes extensive damage to their mitochondria, which leads to cell death and atrophy of the gland (Hart & Straw, 1971). Clinical use of mitotane is limited to adrenocortical carcinoma (Bergenstal et al., 1960).

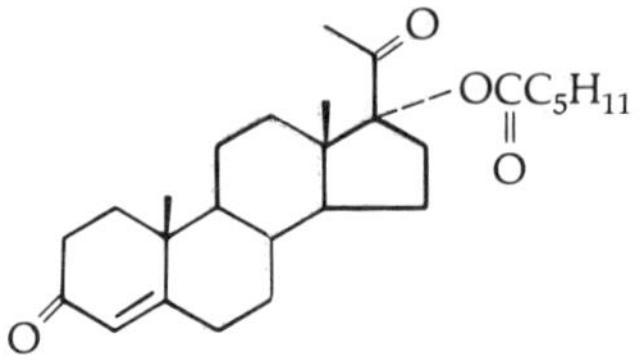

17α-Hydroxyprogesterone caproate

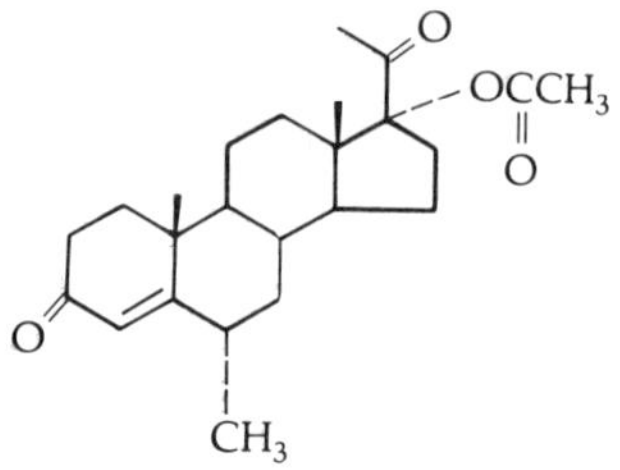

Medroxyprogesterone acetate

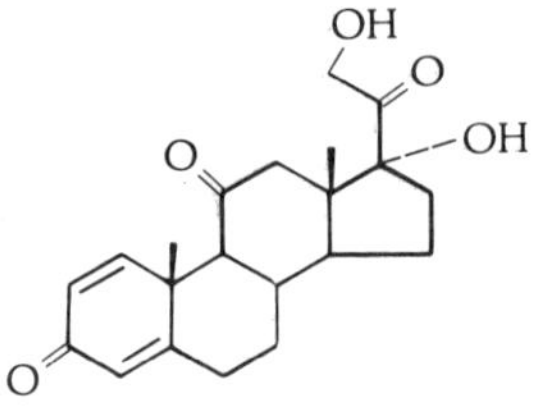

Prednisone

Figure 81 Progestins and glucocorticoid.

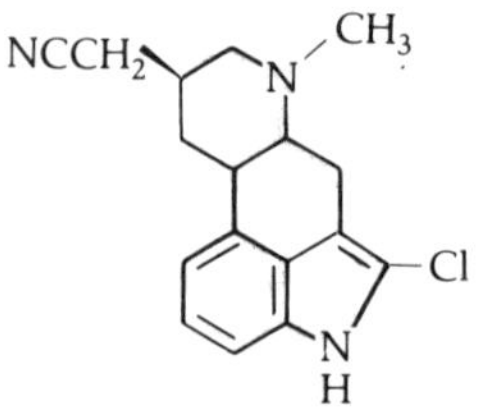

Lergotrile

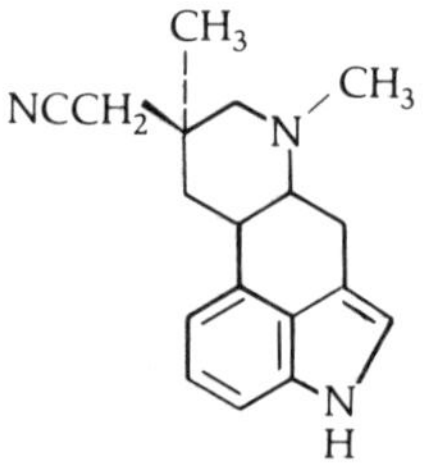

8-Cyanomethyl-8-methylergoline

Figure 82 Prolactin inhibitors.

Synthesis of Hormones

The chemical and microbiological synthesis of steroidal and nonsteroidal sex hormones is a complex and fascinating subject on which important reviews are available (Djerassi, 1963; Fieser & Fieser, 1967; Fried & Edwards, 1972; Blickenstaff et al., 1964). Complete discussion of this topic is beyond the scope of this chapter, but the preparations of specific compounds which have been described in it are given below.

Tamoxifen is prepared by a Grignard reaction between p-(dimethylaminoethoxy)-phenylmagnesium bromide and α-ethylbenzyl phenyl ketone, followed by dehydration in refluxing hydrochloric acid (Imperial Chemical Industries, 1964; Equation 52). The geometrical isomers are separated by recrystallization from petroleum ether and the Z-isomer is formulated as tamoxifen (Bedford & Richardson, 1966). In a related synthesis nafoxidine is formed from 2-phenyl-6-methoxy-tetralone and the appropriate Grignard reagent (Lednicer et al., 1963; Equation 53). Chlortrianisine is prepared by a related process that involves the reaction of anisoin with anisylmagnesium bromide, followed by chlorination of the resulting trianisylethylene (Shelten & Van Campen, 1947; Scheme 28).

The synthesis of diethylstilbestrol also begins with anisoin, which is ethylated with ethyl iodide and strong base. Treatment with ethylmagnesium bromide and acid gives the stilbene intermediate mainly in the trans form. Ether cleavage then completes the synthesis of diethylstilbestrol (Dodds et al., 1938; Scheme 28).

The orally active steroidal estrogen, ethinyl estradiol, is prepared by treating estrone with lithium acetylide (Inhofen et al., 1938; Equation 54). A variety of synthetic routes has been developed for the androgenic agents used in cancer chemotherapy. Thus methyl testosterone is made from dehydroepiandrosterone by a route that involves treatment with excess methylmagnesium bromide and Oppenauer oxidation of the resulting 17 α-methyl compound (Ruzicka et al., 1935; Scheme 29). The corresponding 19-nor androgen, normethandrolone, is prepared by treating estrone methyl ether with methylmagnesium bromide, metal-in-ammonia (Birch) reduction, and acid hydrolysis of the intermediate enol ether (Djerassi et al., 1954; Scheme 30). Dihydrotestosterone propionate is converted into its 2 α-methyl derivative, dromostanolone propionate, by treatment with ethyl formate and base to give the 2-hydroxymethylene intermediate. Catalytic reduction to the 2 α-methyl derivative and base-catalyzed isomerization (Ringold et al., 1959; Scheme 31) follows.

Fluoxymestrone requires an extended synthesis from androstenedione. This process begins with microbiologic 11 α-hydroxylation which is followed by

(54)

Ethinyl estradiol

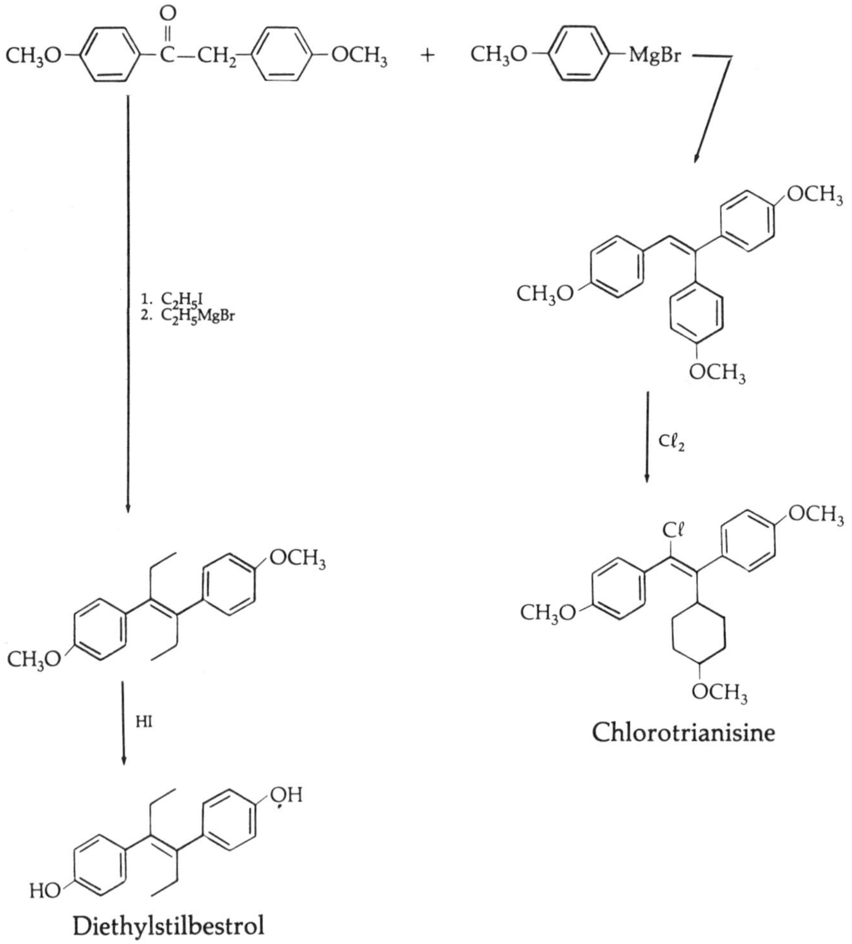

Scheme 28 Synthesis of nonsteroidal estrogens.

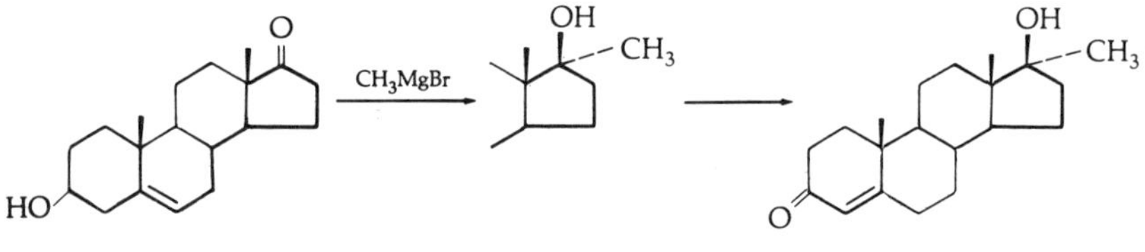

Scheme 29 Synthesis of methyltestosterone.

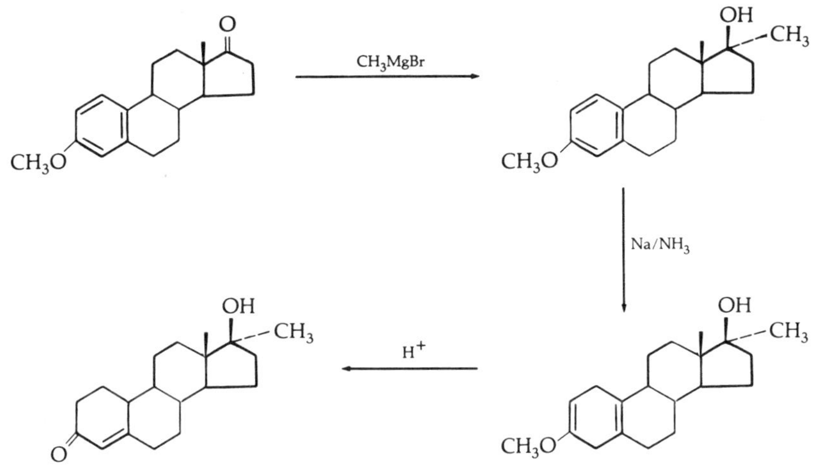

Scheme 30 Synthesis of 19-nor-17α-methyltestosterone.

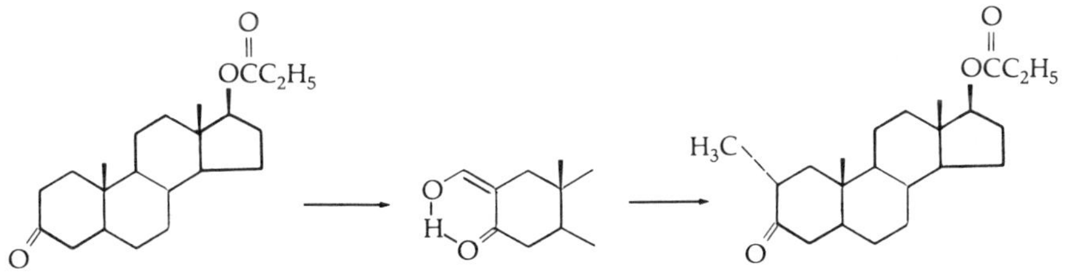

Scheme 31 Synthesis of 2-methyldihydrotestosterone propionate.

oxidation to the trione, protection of the 3-carbonyl group by enamine formation, and selective methylation of the 17-one. The hindered 11-carbonyl group does not react under experimental conditions. Lithium aluminum hydride reduction of the 11-carbonyl group, deprotection, tosylation, and elimination then produce the 9 (11) olefin. Finally, the 9,11-fluorohydrin functionality is introduced by the addition of hydrobromous acid to the double bond, followed by epoxide formation in base, and opening of this epoxide with hydrogen fluoride (Herr et al., 1956; Scheme 32). In contrast to this lengthy synthesis, testolactone is prepared directly from progesterone by microbiological oxidation (Fried et al., 1953; Equation 55).

The progestational agents used against tumors are synthesized from 17α-hydroxypregnenolone. Thus treatment with acetic anhydride under mild con-

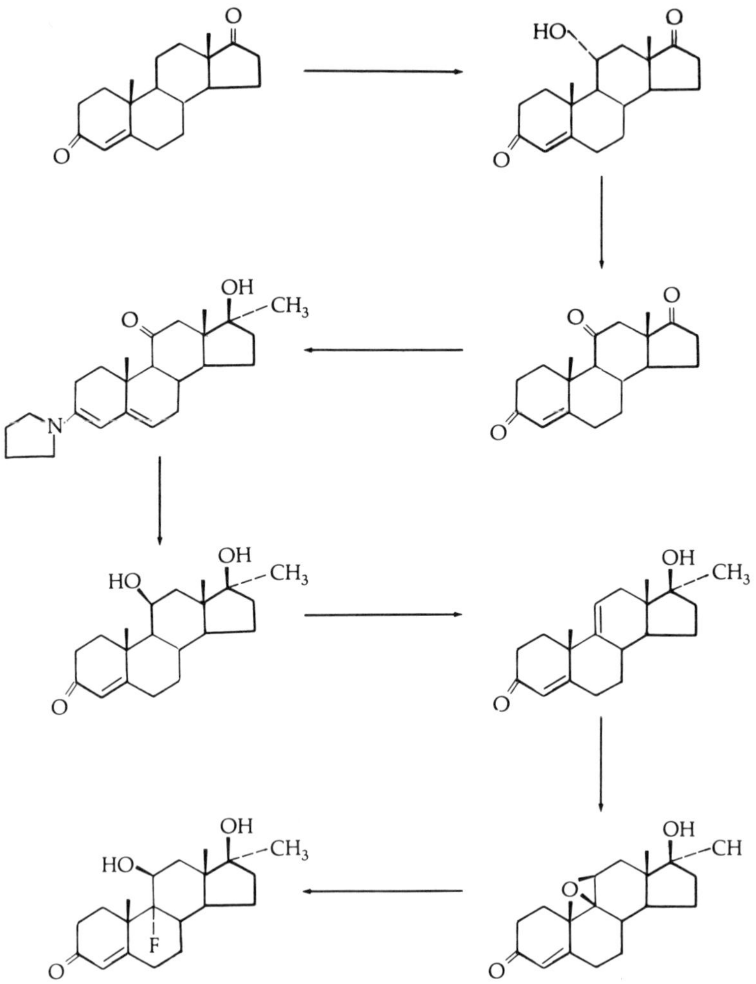

Scheme 32 Synthesis of fluoxymestrone.

ditions and then with caprioc anhydride under forcing conditions yields the 3-acetate, 17-caproate. Ester interchange with methanol removes the 3-acetate and Oppenauer oxidation furnishes hydroxyprogesterone caproate (Babcock et al., 1958; Scheme 33). For the synthesis of medroxyprogesterone acetate hydroxypregnenolone is first subjected to Oppenauer oxidation. The carbonyl groups are then protected as ketals. The 6-methyl group is introduced by epoxidation, followed by treatment of the α-epoxide with methylmagnesium bromide. Removal of the ketals and treatment with base result in the 6 β-methyl-4-en-3-one system, which epimerizes to the more stable 6 α-configuration after treatment with acid. Acetylation under forcing conditions completes the synthesis (Babcock et al., 1958; Scheme 33).

Prednisone, the principal glucocorticoid used against cancer, is prepared by incubating cortisone with *Corynebacterium simplex* (Nobile et al., 1955; Equation 56).

The synthesis of 1,1-di(chlorophenyl)-2,2-dichloroethane is accomplished by treating 2,2-dichloro-1-(o-chlorophenyl) ethanol with chlorobenzene in the presence of sulfuric acid (Haller et al., 1945; Equation 57). Separation of the isomers by chromatography furnishes pure mitotane (Cueto & Brown, 1958).

IMMUNOSTIMULANTS

Cells with neoplastic potential are produced continually in the human body but are detected and destroyed by a properly functioning immune surveillance

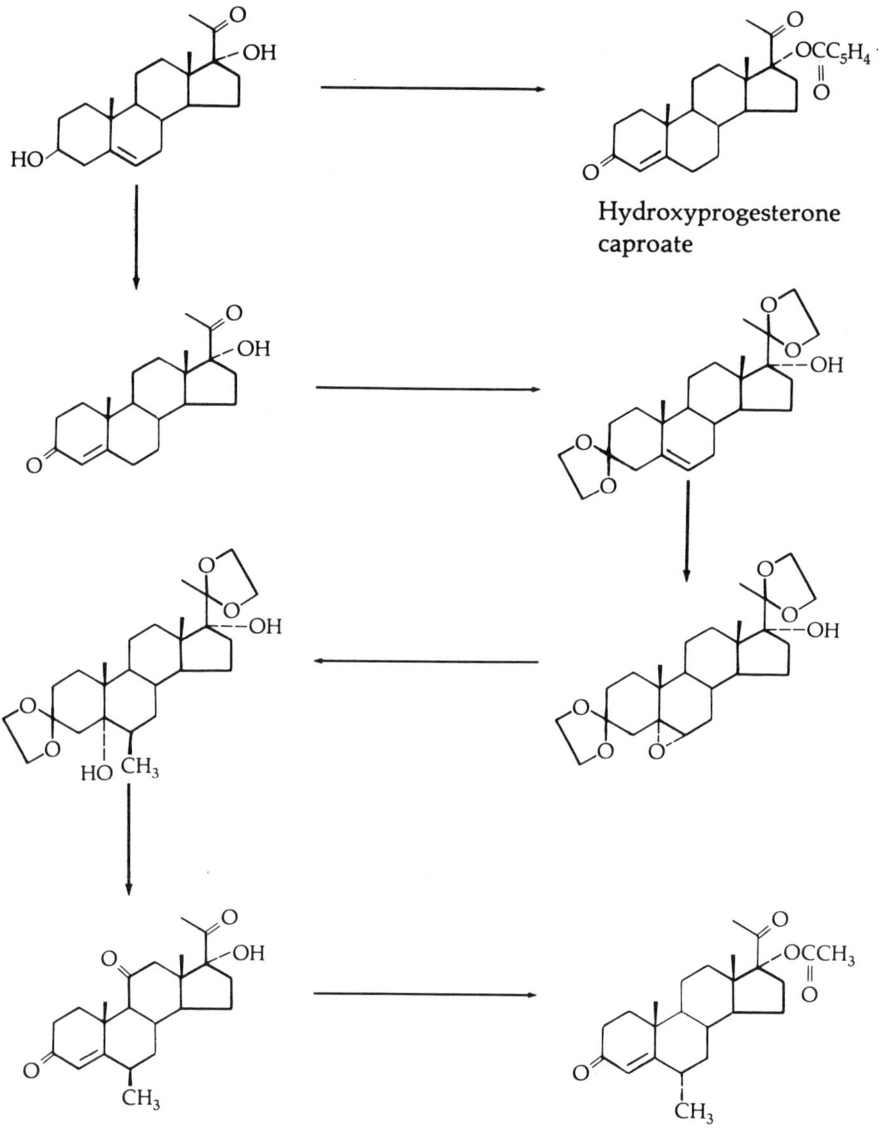

Scheme 33 Synthesis of progestogens.

system. The development of tumors results from a malfunction of this system. Thus a high rate of cancer incidence has been found in organ transplant patients whose immune systems are suppressed by drugs such as azathioprene. Furthermore, a high correlation exists between cancer and immuno-defiency diseases such as bacterial and viral infections (Morton, 1974). Given this correlation between the onset of cancer and lack of immunocompetence, it follows that stimulation of the deficient immune system would reverse and possibly eradicate neoplastic disease. This prospect has fostered a major commitment to research on immunostimulants. Some promising leads have emerged, although it now appears that no major breakthrough is imminent.

The original attempts at immunotherapy were made by Coley in the 1890s who injected bacterial toxins into cancer patients and reported some dramatic results. His rather extravagant claims were not widely accepted; however, the concept of immunostimulation by bacterial substances is valid and his techniques have been revived recently. In order for an immunostimulant to be useful the immune system must have a minimal activity. Patients who show sensitivity to dinitrochlorobenzene are considered possible candidates for immunotherapy (O'Brien, 1972). The most widely used immunostimulant is *Bacillus Calmette-Guerin* (BCG), a live tuberculosis vaccine (Morton, 1974) which has induced remissions in certain patients with melanoma, breast cancer, and leukemia. Unfortunately, it causes a number of undesirable side effects. A purified preparation, the methanol-extracted resin of BCG (MER), has been developed to help reduce them. It now appears that muramylpeptides on the bacterial cell walls are the immune-stimulating components. Other immunostimulants currently in use include *Corynebacterium parvum*, *Bordatella pertussis* vaccine, and synthetic polynucleotides (Goodnight & Morton, 1978).

Considerable emphasis has been given to the use of human interferon in the treatment of cancer. In particular, a large multicenter clinical trial has been designed to evaluate the value of interferon against a variety of tumors. Interferon is a glycoprotein produced by normal cells in response to attack by viruses. The mechanism by which interferon acts against cancer cells is not well understood, but it appears to activate natural killer cells whose function is to destroy cancer and virus-transformed cells.

Simpler chemical structures have pronounced immunostimulant activity. Two compounds are presently under investigation as potential clinical agents. Levamisole (Figure 83) is an anthelmintic agent that was identified as an immunostimulant by Renoux in 1972. It is thought to influence many functions in cell-mediated immunity. Tilorone (Figure 83) is one member of a large

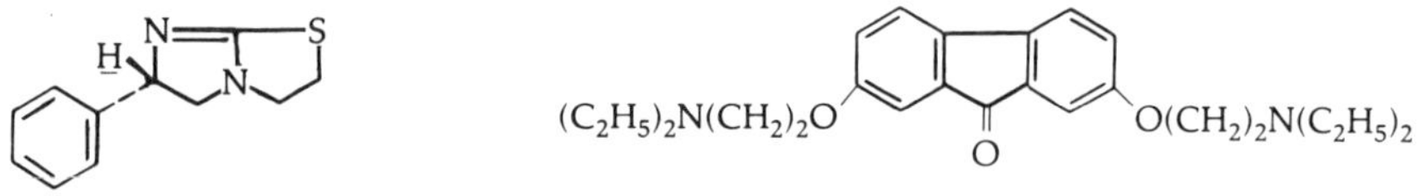

Levamisole

Tilorone

Figure 83 Synthetic immunostimulants.

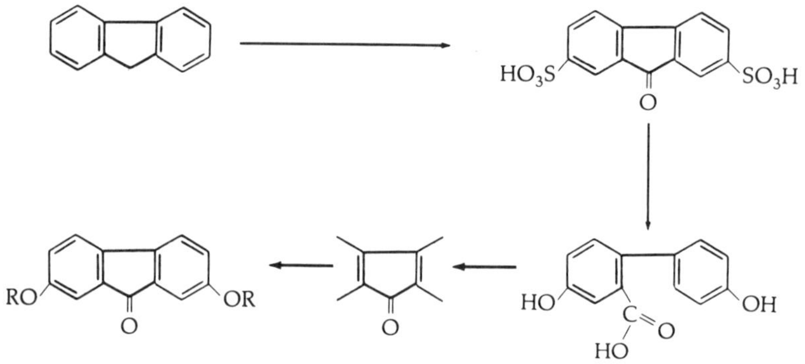

R = (CH₂)₂N(C₂H₅)₂

Scheme 34 Synthesis of tilorone.

family of synthetic compounds that affects the immune system. It stimulates the production of interferon and enhances antibody formation. It differs from levamisole by inhibiting cell-mediated immunity, whereas levamisole stimulates it (Sanders, 1974).

Tilorone was prepared from fluorene by a sequence that involved disulfonic acid formation and oxidation with permanganate to the corresponding fluorenone. Sodium hydroxide fusion then gave the dihydroxybiphenyl-2-carboxylic acid, which was recyclized with zinc chloride and alkylated on the phenolic hydroxyls with diethylaminoethyl chloride (Andrews et al., 1974; Scheme 34).

RADIOSENSITIZING AND RADIOPROTECTING COMPOUNDS

The destruction by radiation of cells in certain solid tumors is limited by resistance of hypoxic cells in the tumor center. One way to overcome this lim-

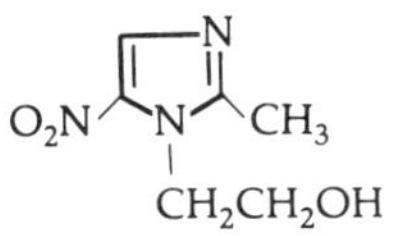

Metronidazole: O₂N-[imidazole]-CH₃, N-CH₂CH₂OH

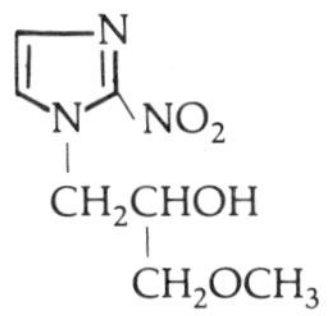

Misonidazole: [imidazole]-NO₂, N-CH₂CHOH-CH₂OCH₃

WR-2721: H₂N(CH₂)₃NH(CH₂)₂SPO₃H

Figure 84 Radiosensitizing and radioprotecting agents.

itation has been the use of hyperbaric oxygen techniques in which the radiosensitizing effects of oxygen are increased (Gray et al., 1953). More recently, organic compounds that can mimic the effects of oxygen have been used as adjuvants to radiotherapy. The main type of compound used for radiosensitization is the nitroimidazole. One of this type, metronidazole (Figure 84), received a Phase II clinical trial in 1974 (Urtasun et al., 1976). It showed a positive effect against glioblastomas. Subsequently, more effective radiosensitizers based on misonidazole (Figure 82), were advocated for clinical study (Urtasun et al., 1977).

Radioprotective agents are of potential value in selectively protecting normal cells in the presence of irradiated solid tumors. One group of compounds characterized as having superior radioprotective effectiveness is the phosphorothioate (Akerfeldt, 1963). In this group WR-2721 (Figure 82) has shown desirable radioprotective activity in animals (Yuhas, 1973).

REFERENCES

Abrams, J. T., Barker, R. L., Jones, W. E., Vallender, H. W. & Woodard, F. N. (1949), *J. Soc. Chem. Ind. London*, 68, 280-284.

Acton, E. M., Fujiwara, A. N. & Henry, D. W. (1974), *J. Med. Chem.* 17, 659-660.

Acton, E. M., Tong, G. L., Mosher, C. W., Smith, T. H. & Henry, D. W. (1979), *J. Med. Chem.*, 22, 922-925.

Adamson, R. H., Sieber, S. M., Wang-Peng, J. & Wood, H. B. (1976), *Proc. Am. Assoc. Cancer Res.*, 17, 42.

Akerfeldt, A. (1963), *Acta Radiol. Ther. Phys. Biol.*, 1, 465-470.

Alberto, P., Rozencweig, M. Gangji, D., Brugarolas, A., Cavalli, F., Siegenthaler, P., Hansen, H. H. & Sylvester, R. (1978), *Eur. J. Cancer*, 14, 195-201.

Alberts, D. S., Mackel, C., Pocelinko, R. & Salmon, S. E. (1982), *Cancer Res.*, 42, 1170-1175.

Ammann, C. A. & Safferman, R. S. (1958), *Antibiot. Chemother.*, 8, 1-7.

Anderson, W. K. & Corey, P. F. (1977a), *J. Med. Chem.*, 20, 812-818.

Anderson, W. K. & Corey, P. R. (1977b), *J. Med. Chem.*, **20**, 1691-1694.
Anderson, W. K., La Voie, E. J. & Lee, G. E. (1977), *J. Org. Chem.*, **42**, 1045-1050.
Andrews, E. R., Fleming, R. W., Grisar, J. M., Kihm, J. C., Wenstrup, D. L. & Mayer, G. D. (1974), *J. Med. Chem.*, **17**, 882-886.
Angier, R. B., Boothe, J. H. & Curran, W. V. (1959), *J. Am. Chem. Soc.*, **81**, 2814-2818.
Angier, R. B., Boothe, J. H., Hutchings, B. L., Mowat, J. H., Semb, J., Stokstad, E. L. R., Subbarow, Y., Waller, C. W., Cosulich, D. B., Fahrenbach, M. J., Hultquist, M. E., Kuh, E., Northey, E. H., Seeger, D. R., Stickels, J. P. & Smith, J. M. (1946), *Science*, **103**, 667-669.
Anzai, K., Nakamura, G. & Suzuki, S. (1957), *J. Antibiot. Tokyo*, **10**, 201-204.
Arcamone, F. (1980), in *Anticancer Agents Based on Natural Product Models* (Cassady, J. M. & Douros, J. D., Eds.) pp. 1-41, Academic, New York.
Arcamone, F., Penco, S., Redaelli, S. & Hanessian, S. (1976), *J. Med. Chem.*, **19**, 1424-1425.
Arcamone, F., Penco, S. & Vigevani, A. (1975a), *Cancer Chemother. Rep.*, **6**, 123-129.
Arcamone, F., Penco, S., Vigevani, A., Redaelli, S., Franchi, G., Di Marco, A., Casazza, A. M., Dasdia, T., Formelli, F., Necco, A. & Soranzo, C. (1975b), *J. Med. Chem.*, **18**, 703-707.
Arnold, H. & Bourseaux, F. (1958), *Angew. Chem.*, **70**, 539-544.
Arnold, H., Bourseaux, F. & Brock, N. (1958), *Nature London*, **181**, 931.
Atkinson, M. R., Eckermann, G. & Stephenson, J. (1965), *Biochem. Biophys. Acta*, **108**, 320-322.
Atwell, G. J., Cain, B. F. & Seelye, R. N. (1972), *J. Med. Chem.*, **15**, 611-615.
Baaske, D. M. & Heinstein, P. (1977), *Antimicrob. Agents Chemother.*, **12**, 298-300.
Babcock, J. C., Gutsell, E. S., Herr, M. E., Hogg, J. A., Stucki, J. C., Barnes, L. E. & Dulin, W. E. (1958), *J. Am. Chem. Soc.*, **80**, 2904-2905.
Baker, B. R. (1967), *Design of Active-Site Directed Irreversible Enzyme Inhibitors*, Wiley, New York.
Baker, B. R. (1968), *J. Med. Chem.*, **11**, 483-486.
Baker, B. R. (1969), *Acc. Chem. Res.*, **2**, 129-136.
Baker, B. R., Schaub, R. E., Joseph, J. P. & Williams, J. H. (1955), *J. Am. Chem. Soc.*, **77**, 12-15.
Bardos, T. J., Chmielewicz, Z. F. & Hebborn, P. (1969), *Ann. N.Y. Acad. Sci*, **163**, 1006-1025.
Barnett, C. J., Cullinan, G. J., Gerzon, K., Hoying, R. C., Jones, W. E., Newlon, M. W., Poore, G. A., Robison, R. L., Sweeney, M. J., Todd, G. C., Dyke, R. W. & Nelson, R. L. (1978), *J. Med. Chem.*, **21**, 88-96.
Basha, F. Z., Hibino, S., Kim, D., Pye, W. E., Wee, T. T. & Weinreb, S. M. (1980), *J. Am. Chem. Soc.*, **102**, 3962-3964.
Baugh, C. M. & Krumdieck, C. L. (1971), *Ann. N.Y. Acad. Sci.*, **186**, 7-28.
Baurain, R., Masquelier, M., Deprez-De Campeneere, D. & Trouet, A. (1980), *J. Med. Chem.*, **23**, 1171-1174.
Beaman, A. G. & Robins, R. K. (1961), *J. Am. Chem. Soc.*, **83**, 4038-4044.
Beatson, G. T. (1896), *Lancet*, **2**, 104-107.

Beckett, V. L. & Brennen, M. J. (1959), *Surg. Gynecol. Obstet.*, **109**, 235–239.
Bedford, G. R. & Richardson, D. N. (1966), *Nature London*, **212**, 733–734.
Behr, W. & Hartmann, G. (1965), *Biochem. Z.*, **343**, 519–527.
Belikova, A. M., Zarytova, V. F. & Grineva, N. I. (1967), *Tetrahedron Lett.*, 3557–3562.
Benabid, A. L., Schaerer, J., de Rougemont, J., Pallo, D., Barge, M., Chirossel, J. P., Clamadieu, M. M. & des Sablons, C. H. U. (1978), *Proc. Int. Symp. Endocrine-Related Cancer, Lyon, France, June, 1977* (Mayer, M., Saez, S. & Stoll, B. A., Eds.), pp. 111–117, Imperial Chemical Industry, London.
Benu, M. H. (1958), *J. Chem. Soc.* (London), 2800–2806.
Bennett, L. L., Jr. (1975), in *Handbook of Experimental Pharmacology* (Sartorelli, A. C. & Johns, D. G., Eds.), Vol. 38, Part 2, pp. 511, Springer-Verlag, Berlin.
Bennett, L. L., Jr., Schabel, F. M., Jr. & Skipper, H. E. (1956), *Arch. Biochem. Biophys.*, **64**, 423–436.
Bensch, K. G. & Malawista, S. E. (1969), *J. Cell. Biol.*, **40**, 95–107.
Benvenuto, J. A., Lee, K., Hall, S. W., Benjamin, R. S. & Loo, T. L. (1978), *Cancer Res.*, **38**, 3867–3870.
Bergel, F. & Stock, J. A. (1954), *J. Chem. Soc.*, 2409–2417.
Bergel, F. & Stock, J. A. (1953), *Ann. Rep. Br. Emp. Cancer Campaign*, **31**, 6.
Bergel, F., Stock, J. A., Wade, R., Johnson, J. M., Hopewood, W. & Black, M. H. (1959), *Ann. Rep. Br. Emp. Cancer Campaign*, **37**, 26.
Bergenstal, D. M., Herz, R., Lipsett, M. B. R. & Moy, R. H. (1960), *Ann. Int. Med.*, **53**, 672–682.
Bergman, W. & Feeney, R. J. (1951), *J. Org. Chem.*, **16**, 981–987.
Bergman, W. & Stempien, M. F. (1957), *J. Org. Chem.*, **22**, 1575–1577.
Berlin, Yu. A., Kiseleva, O. A., Kolosov, M. N., Shemyakin, M. M., Soifer, V. S., Vasina, I. V., Yartseva, I. V. & Kuznetsov, V. D. (1968), *Nature London*, **218**, 193–194.
Bertino, J. R., Booth, B. A., Cashmore, A. R., Bieber, A. L. & Sartorelli, A. C. (1964), *J. Biol. Chem.*, **239**, 479–485.
Bhattacharyya, B. & Wolff, J. (1974), *Proc. Nat. Acad. Sci. U.S.A.*, **71**, 2627–2631.
Blickenstaff, R. T., Ghosh, A. C. & Wolff, G. C. (1964), *Total Synthesis of Steroids*, Academic, New York.
Bloom, H. J. G. (1971), *Br. J. Cancer*, **25**, 250–265.
Bodey, G. P., Freireich, E. J., Monto, R. W. & Hewlett, J. S. (1969), *Cancer Chemother. Rep.*, **53**, 59–66.
Bond, J. V. (1962), *Proc. Am. Assoc. Cancer Res.*, **3**, 306.
Booth, B. A. & Sartorelli, A. C. (1967), *Mol. Pharmacol.*, **3**, 290–302.
Boothe, J. H., Mowat, J. H., Waller, C. W., Angier, R. B., Semb, J. & Gazzola, A. L. (1952), *J. Am. Chem. Soc.*, **74**, 5407–5409.
Boothe, J. H., Semb, J., Waller, C. W., Angier, R. B., Mowat, J. H., Hutchings, B. L., Stokstad, E. L. R. & Subbarow, Y. (1949), *J. Am. Chem. Soc.*, **71**, 2304–2308.
Bradner, W. T. & Pindell, M. H. (1962), *Nature London*, **196**, 682–683.
Brian, P. W., Dawkins, A. W., Grove, J. E., Hemming, H. G., Lowe, D. & Norris, G. L. F. (1961), *J. Exp. Bot.*, **12**, 1–12.

Brink, J. J. & LePage, G. A. (1964), *Cancer Res.*, **24**, 1042-1049.
Brockman, R. W. (1963), *Adv. Cancer Res.*, **7**, 129-234.
Brockman, R. W., Schabel, F. M., Jr. & Montgomery, J. A. (1977), *Biochem. Pharmacol.*, **26**, 2193-2196.
Brockman, R. W., Shaddix, S., Lacter, R. W., Jr. & Schabel, F. M., Jr. (1970a), *Cancer Res.*, **30**, 2358-2368.
Brockman, R. W., Shaddix, S. C., Williams, M. & Struck, R. F. (1976), *Cancer Treat. Rep.*, **60**, 1317-1324.
Brockman, R. W., Sidwell, R. W., Arnett, G. & Shaddix, S. (1970b), *Proc. Soc. Exp. Biol.*, **133**, 609-614.
Brockmann, H. (1960), *Fortschr. Chem. Org. Naturst.*, **18**, 1-54.
Brockmann, H. (1963), *Fortschr. Chem. Org. Naturst.*, **21**, 121-182.
Brockmann, H., Ammann, J. & Muller, W. (1966), *Tetrahedron Lett.*, 3595-3597.
Brockmann, H. & Franck, B. (1954), *Chem. Ber.*, **87**, 1767-1779.
Brockmann, H. & Franck, B. (1956), *Angew. Chem.*, **68**, 68-69.
Brockmann, H., Grone, H. & Pampus, G. (1958), *Chem. Ber.*, **91**, 1916-1920.
Brockmann, H. & Grubhofer, N. (1950), *Naturwissenschaften*, **37**, 494-496.
Brockmann, H., Hocks, P. & Muller, W. (1967), *Chem. Ber.*, **100**, 1051-1062.
Brockmann, H. & Lackner, H. (1964), *Tetrahedron Lett.*, 3523-3525.
Brockmann, H. & Lackner, H. (1968), *Chem. Ber.*, **101**, 2231-2243.
Brockmann, H. & Manegold, J. H. (1960), *Chem. Ber.*, **93**, 2971-2982.
Brockmann, H. & Manegold, J. H. (1962), *Chem. Ber.*, **95**, 1081-1093.
Brockmann, H. & Manegold, J. H. (1964), *Naturwissenschaften*, **51**, 383-384.
Brockmann, H. & Seela, F. (1971), *Chem. Ber.* **104**, 2751-2771.
Brockmann, H. & Sunderkotter, W. (1960), *Naturwissenschaften*, **47**, 229-230.
Brockmann, H. & Waehneldt, Th. (1961), *Naturwissenschaften*, **48**, 717.
Brooks, P. & Lawley, P.D. (1961), *Biochem. J.*, **80**, 496-503.
Brown, B. G. & Weliky, V. S. (1953), *J. Org. Chem.*, **23**, 125-126.
Brown, S. S. & Timmis, G. M. (1958), *Ann. Rep. Br. Emp. Cancer Campaign*, **37**, 48.
Brown, S. S. & Timmis, G. M. (1959), *Ann. Rep. Br. Emp. Cancer Campaign*, **37**, 29.
Bryan, R. F., Gilmore, C. J. & Haltiwanger, R. D. (1973), *J. Chem. Soc. Perkin II*, 897-902.
Bukhari, M. A., Everett, J. L. & Ross, W. C. J. (1971), *Biochem. Pharmacol.*, **21**, 963-967.
Bukva, N. F. & Gass, G. H. (1967), *Cancer Chemother. Rep.*, **51**, 431-433.
Burchenal, J. H., Lester, R. A., Riley, J. B. & Rhoads, C. P. (1948), *Cancer*, **1**, 399-411.
Burchenal, J. H., Myers, W. P. L., Craver, L. F. & Karnovsky, D. A. (1949), *Cancer*, **2**, 1-17.
Cain, B. F., Atwell, G. J. & Denny, W. A. (1975), *J. Med. Chem.*, **18**, 110-117.
Caldwell, I. C. (1969), *J. Chromat.*, **44**, 331-341.
Calendi, E., Di Marco, A., Reggiani, M., Scarpinato, B. M. & Valentini, L. (1965), *Biochem. Biophys. Acta*, **103**, 25-49.
Camerman, N., Chan, L. Y. Y., & Camerman, A. (1980), *J. Med. Chem.*, **23**, 941-945.

Camiener, G. W. (1967), *Biochem. Pharmacol.*, **16**, 1691–1702.
Canellakis, E. S., Shaw, Y. H., Hanners, W. E. & Schwartz, R. A. (1976), *Biochem. Biophys. Acta*, **418**, 277–289.
Carrasco, L., Barbacid, M. & Vasquez, D. (1973), *Biochem. Biophys. Acta*, **312**, 368–376.
Carrol, F. I., Philip, A., Blackwell, J. T., Taylor, D. J. & Wall, M. E. (1972), *J. Med. Chem.*, **15**, 1158–1161.
Carter, C. J. & Cannon, M. (1978), *Eur. J. Biochem.*, **84**, 103–111,
Carter, C. J., Cannon, M. & Smith, K. E. (1976), *Biochem. J.*, **154**, 171–178.
Carter, S. K. (1976), in *Breast Cancer: Trends in Research & Treatment* (Henson, J. C., Mattheieu, W. H. & Rozencweig, M., Eds.), pp. 193–215, Raven, New York.
Carter, S. K. & Slavik M. (1976), *Cancer Treat. Rev.*, **3**, 43.
Cassady, J. M. & Suffness, M. (1980), in *Anticancer Agents Based on Natural Product Models* (Cassady, J. M. & Douros, J. D., Eds.), pp. 201–269, Academic, New York.
Cassinelli, G., Grein, A., Masi, P., Suarato, A., Bernardi, L., Arcamone, F., DiMarco, A., Casazza, A. M., Pratesi, G. & Soranzo, C. (1978), *J. Antibiot. Tokyo*, **31**, 178–184.
Cassinelli, G., Ruggieri, D. & Arcamone, F. (1979), *J. Med. Chem.*, **23**, 121–123.
Cater, D. B. & Philipps, F. S. (1954), *Nature London*, **174**, 121–123.
Cerny, A., Semonsky, M., Jelinek, V., Francova, V., Raz, K. & Franc, Z. (1967), *Coll. Czech. Chem. Comm.*, **32**, 1035–1044.
Chabner, B. A., De Vita, V. T., Considine, N. & Oliverio, V. T. (1969), *Proc. Soc. Exp. Biol.*, **132**, 1119–1122.
Champe, S. P. & Benzer, S. (1962), *Proc. Nat. Acad. Sci. U.S.A.*, **48**, 532–546.
Chang, P. K. (1965), *J. Med. Chem.*, **8**, 884.
Chaykovsky, M., Modest, E. J. & Sengupta, S. K. (1977), *J. Heterocycl. Chem.*, **14**, 661–664.
Chien, M., Grollman, A. P. & Horwitz, S. B. (1977), *Biochemistry*, **16**, 3641–3647.
Chiu, Y. Y. H. & Lipscomb, W. N. (1975), *J. Am. Chem. Soc.*, **97**, 2525–2530.
Chu, M. Y. & Fischer, G. A. (1962), *Biochem. Pharmacol.*, **11**, 423–430.
Chou, F., Khan, A. H. & Driscoll, J. S. (1976), *J. Med. Chem.*, **19**, 1302–1308.
Cihak, A., Skoda, J. & Sorm, F. (1964), *Coll. Czech. Chem. Commun.*, **29**, 300–308.
Citarella, R. V., Wallace, R. E., Murdock, K. C., Angier, R. B. & Durr, F. E. (1982), *Cancer Res.*, **42**, 440–444.
Claridge, C. A., Bradner, W. T. & Schmitz, H. (1979), *Cancer Chemother. Pharmacol.*, **2**, 181–182.
Cohen, M. M., Shaw, M. W. & Craig, A. P. (1963), *Proc. Natl. Acad. Sci. (Wash.)*, **50**, 16–24.
Cohen, R. M. & Wolfenden, R. (1971), *J. Biol. Chem.*, **246**, 7561–7565.
Cohen, S. S., Flaks, J. G., Barner, H. D., Loeb, M. R. & Lichtenstein, J. (1958), *Proc. Nat. Acad. Sci. U.S.A.*, **44**, 1004–1012.
Cone, R., Masan, S. K., Lown, J. W. & Morgan, A. R. (1975), *Can. J. Biochem.*, **54**, 219–223.

Connors, T. A. (1969), *Cancer Res.*, **29**, 2443–2447.
Connors, T. A. (1975), in *Handbook of Experimental Pharmacology* (Sartorelli, A. C. & Johns, D. J., Eds.), Vol. 38, Part 2, p. 19, Springer-Verlag, Berlin.
Connors, T. A., Hickman, J. A., Jarman, J., Melzack, D. H. & Ross, W. C. J. (1975), *Biochem. Pharmacol.*, **24**, 1665–1670.
Conrad, R. A., Cullinan, G. J., Gerzon, K. & Poore, G. A. (1979), *J. Med. Chem.*, **22**, 391–400.
Conrad, R. A., Gerzon, K. & Poore, G. A. (1978), *Am. Chem. Soc.*, *10th Cent. 12th Great Lakes Reg. Meet.*, Indianapolis IN, Abstr. MEDI 18.
Cooney, D. A., Jayaram, H. N., Ryan, J. A. & Bono, V. H. (1974), *Cancer Chemother. Rep. Part 1*, **58**, 793–802.
Corbett, T. H., Griswold, D. P., Roberts, B. J., Peckham, J. C. & Schabel, F. M. (1977), *Cancer*, **40**, 2660–2680.
Corey, E. J., Weigel, L. O., Chamberlin, A. R., Cho, H. & Hua, D. H. (1980), *J. Am. Chem. Soc.*, **102**, 6613–6615.
Corey, J. & Parker, S. (1979), *Biochem. Pharmacol.*, **28**, 867–871.
Cornman, I. & Cornman, M. E. (1951), *Ann. N.Y. Acad. Sci.*, **51**, 1443.
Cosulich, D. B., Seeger, D. R., Fahrenbach, M. J., Collins, K. R., Roth, B., Hultquist, M. & Smith, J. M. (1953), *J. Am. Chem. Soc.*, **75**, 4675–4680.
Cosulich, D. B. & Smith, J. M. (1948), *J. Am. Chem. Soc.*, **70**, 1922–1926.
Crawford, L. V. & Waring, M. J. (1967), *J. Mol. Biol.*, **25**, 23–30.
Creasey, W. A. (1975), in *Handbook of Experimental Pharmacology* (Sartorelli, A. C. & Johns, D. G., Eds.), Vol. 38, Part 2, pp. 232–256, Springer-Verlag, Berlin.
Creighton, A. M. & Birnie, G. D. (1970), *Int. J. Cancer*, **5**, 47–54.
Creighton, A. M., Hellmann, K. & Whitecross, S. (1969), *Nature London*, **222**, 384–385.
Crooke, S. T., Du Vernay, V. H., Galvan, L. & Prestayko, A. W. (1978), *Mol. Pharmacol.*, **14**, 290–298.
Cueto, C. & Brown, J. H. V. (1958), *Endocrinology*, **62**, 326–333.
Cundliffe, E., Cannon, M. & Davies, J. (1974), *Proc. Nat. Acad. Sci. U.S.A.*, **71**, 30–34.
Cunningham, K. G., Hutchinson, S. A., Manson, W. & Spring, F. S. (1951), *J. Chem. Soc.*, 2299–2300.
Dabrowiak, J. C., Greenaway, F. T., Santillo, F. S. & Crooke, S. T. (1979), *Biochem. Biophys. Res. Commun.*, **91**, 721–729.
Danenberg, P. V. & Heidelberger, C. (1976), *Biochemistry*, **15**, 1331–1337.
Danielli, J. F. (1954), *Ciba Foundation Symposium on Leukaemia Research*, p. 263, London.
Danielli, J. F. (1959), *Ann. Rep. Br. Emp. Cancer Campaign*, **37**, 575.
Danishefsky, S., Quick, J. & Horwitz, S. B. (1973), *Tetrahedron Lett.*, 2525–2528.
Dantchev, D., Paintrand, M., Hayat, M., Bourut, C. & Mathe, G. (1979), *J. Antibiot. Tokyo*, **32**, 1085–1086.
Dao, T. L. (1957), *Third National Cancer Conference Proceedings*, pp. 292–296, Lippincott, Philadelphia.
Dao, T. L. (1975), in *Handbook of Experimental Pharmacology* (Sartorelli, A. C. & Johns, D. G., Eds.), Vol. 38, Part 2, pp. 170–192, Springer-Verlag, Berlin.

Dao, T. L. (1978), in *Proc. Int. Symp. Endocrine-Related Cancer*, Lyon, France, June 1977 (Mayer, M., Saez, S. & Stoll, B. A., Eds.), pp. 37–50, Imperial Chemical Industries, London.
Dasdia, T., Di Marco, A., Goffredi, M., Minghetti, A. & Necco, A. (1979), *Pharmacol. Res. Commun.*, 11, 19–29.
Davis, W. & Ross, W. C. J. (1965), *J. Med. Chem.*, 8, 757–759.
Davoll, J. (1960), *J. Chem. Soc. London*, 131–138.
Davoll, J. & Johnson, A. M. (1970), *J. Chem. Soc.*, 997–1002.
Dawid, I. B., French, T. C. & Buchanan, J. M. (1963), *J. Biol. Chem.*, 238, 2178–2185.
De Graw, J., Marsh, J. P., Acton, E. M., Crews, O. P., Mosher, C. W., Fujiwara, A. N. & Goodman, L. (1965), *J. Org. Chem.*, 30, 3404–3409.
De Voe, S. E., Rigler, N. E., Shay, A. J., Martin, J. H., Boyd, T. C., Backus, E. J., Mowat, J. H. & Bohonus, N. (1957), *Antibiot Ann.*, 1956–1957, 730–735.
Di Marco, A. (1967), *Antibiotics. I. Mechanism of Action* (Gottlieb, D. & Shaw, P. D., Eds.), pp. 191–210, Springer-Verlag, Berlin.
Di Marco, A., Silvestrini, R., Di Marco, S. & Dasdia, T. (1965), *J. Cell. Biol.*, 27, 545–550.
Di Marco, A., Soldati, M., Fioretti, A. & Dasdia, T. (1964), *Cancer Chemother. Rep.*, 38, 39–47.
Di Marco, A., Zunino, F., Silvestrini, R., Gambarucci, C. & Gambetta, R. A. (1971), *Biochem. Pharmacol.*, 20, 1323–1328.
Djerassi, C. (1963), *Steroid Reactions*, Holden-Day, San Francisco.
Djerassi, C., Miramontes, L., Rosenkranz, G. & Sondheimer, F. (1954), *J. Am. Chem. Soc.*, 76, 4092–4094.
Dodds, E. C., Goldberg, L., Lawson, W. & Robinson, R. (1938), *Nature London*, 142, 34.
Donehower, R. C., Myers, C. E. & Chabner, B. A. (1979), *Life Sciences* 25, 1–14.
Doskocil, J., Paces, V. & Sorm, F. (1967), *Biochem. Biophys. Acta.*, 145, 771–779.
Dougherty, T. F. & White, A. (1965), *Am. J. Anat.*, 77, 81–116.
Doyle, T. W., Balitz, D. M., Grulich, R. E., Nettleton, D. E., Gould, S. J., Tann, C. H. & Moews, A. E., (1981), *Tetrahedron Lett.*, 4595–4598.
Doyle, T. W. & Bradner, W. T. (1980), in *Anticancer Agents Based on Natural Product Models* (Cassady, J. M. & Douros, J. D., Eds.), pp. 43–72, Academic, New York.
Druckery, H., Preussmann, R., Ivankovic, S., Schmahl, D., Afkham, J., Blum, G., Mennel, H. D., Muller, M., Petropoulas, R. & Schneider, H. (1967), *Z. Krebsforsch.*, 69, 103–201.
Dunn, W. J., III, Greenberg, M. J. & Callejas, S. S. (1976), *J. Med. Chem.*, 19, 1299–1301.
Duschinsky, R., Gabriel, T., Tautz, W., Nussbaum, A., Hoffer, M., Greenberg, E., Burchenal, J. H. & Fox, J. J. (1967), *J. Med. Chem.*, 10, 47–58.
Duschinsky, R., Pleven, E. & Heidelberger, C. (1957), *J. Am. Chem. Soc.*, 79, 4559–4568.
Dustin, P., Jr. (1963), *Pharmacol. Rev.*, 15, 449–480.
Du Vernay, V. H., Essery, J. M., Doyle, T. W., Bradner, W. T. & Crooke, S. T. (1979), *Mol. Pharmacol.*, 15, 341–356.

Dyke, R. W. & Nelson, R. L. (1977), *Cancer Treat. Rev.*, **4**, 135.
Earl, R. A. & Townsend, L. B. (1972), *J. Heterocycl. Chem.*, **9**, 1141–1143.
Edo, K., Katamine, S., Kitane, F., Ishida, N., Koide, Y., Kusano, G. & Nozoe, S. (1980), *J. Antibiot. Tokyo*, **33**, 347–351.
Eidinoff, M. L., Knoll, J. E., Marano, B. & Cheong, L. (1958), *Cancer Res.*, **18**, 105–109.
Eidinoff, M. L., Knoll, J. E., Marano, B. J. & Klein, D. (1959), *Cancer Res.*, **19**, 738–745.
Elion, G. B., Burgi, E. & Hitchings, G. H. (1952), *J. Am. Chem. Soc.*, **74**, 411–414.
Elion, G. B. & Hitchings, G. H. (1955), *J. Am. Chem. Soc.*, **77**, 1676.
Elion, G. B. & Hitchings, G. H. (1965), in *Advances in Chemotherapy* (Goldin, A., Hawking, R. & Schnitzer, R. J., Eds.), Vol. 2, pp. 91–177, Academic, London.
Elion, G. B. & Hitchings, G. H. (1967), *Adv. Chemother.*, **2**, 91–177.
Eliott, R. D., Temple, C., Frye, J. L. & Montgomery, J. A. (1971), *J. Org. Chem.*, **36**, 2818–2823.
Emmenegger, H., Stahelin, H., Rutschmann, J., Renz, J. & von Wartburg, A. (1961), *Arzneim. Forsch.*, **11**, 327–333; 459–469.
Engle, T. W., Zon, G. & Egan, W. (1979), *J. Med. Chem.*, **22**, 897–899.
Essery, J. M. & Doyle, T. W. (1980), *Can. J. Chem.*, **58**, 1869–1874.
Ettlinger, L., Gaumann, E., Hutter, R., Keller-Schierlein, W., Kradolfer, F., Neipp, L., Prelog, V., Reusser, P. & Zahner, H. (1959), *Chem. Ber.*, **92**, 1867–1879.
Everett, J. L. & Kon, G. A. R. (1950), *J. Chem. Soc.*, 3131–3135.
Everett, J. L., Roberts, J. J. & Ross, W. C. J. (1953), *J. Chem. Soc.*, 2386–2392.
Everett, J. L. & Ross, W. C. J. (1949), *J. Chem. Soc.*, 1972–1983.
Falco, E. A. & Hitchings, G. H. (1956), *J. Am. Chem. Soc.*, **78**, 3143–3145.
Farber, S. (1958), *Ciba Foundation Symposium on Amino Acids and Peptides with Antimetabolic Activity*, London, p. 138.
Farber, S., Diamond, L. K., Mercer, R. D., Sylvester, R. F. & Wolff, J. A. (1948), *New Engl. J. Med.*, **238**, 787–793.
Farber, S., Toch, R., Craig, A. W. & Jackson, H. (1956), *Adv. Cancer Res.*, **4**, 1–71.
Farmer, P. B. & Cox, P. J. (1975), *J. Med. Chem.*, **18**, 1106–1110.
Fieser, L. F. & Fieser, M. (1967), *Steroids*, Reinhold, New York.
Fishbein, W. N., Carbone, P. P., Owens, A. H., Jr., Kelly, M. G., Rall, D. P. & Tarr, N. (1964), *Cancer Chemother. Rep.*, **42**, 19–24.
Floss, H. G., Cassady, J. M. & Robbers, J. E. (1973), *J. Pharm. Sci.*, **62**, 699–715.
Formica, J. V. & Apple, M. A. (1976), *Antimicrob. Agents Chemother.*, **9**, 214–221.
Fox, P. A., Panasci, L. C. & Schein, P. S. (1977), *Cancer Res.*, **37**, 783–787.
Freeman, G. G. & Morrison, R. I. (1948), *Nature London*, **162**, 30.
Frei, E., III, Luce, J. K. & Loo, T. L. (1971), *Cancer Chemother. Rep.*, **55**, 91–97.
French, F. A. & Friedlander, B. L. (1958), *Cancer Res.*, **18**, 1290–1300.
Fried, J. & Edwards, J. A. (1972), *Organic Reactions in Steroid Chemistry*, Vols. 1 & 2, Van Nostrand-Reinhold, New York.
Fried, J., Thomas, R. W. & Klingsberg, A. (1953), *J. Am. Chem. Soc.*, **75**, 5764.
Friedman, O. M., Klacc, D. L. & Seligman, A. M. (1954), *J. Am. Chem. Soc.*, **76**, 916–917.
Fahrenbach, M. J., Collings, K. H., Hultquist, M. E. & Smith, J. M. (1954), *J. Am. Chem. Soc.*, **76**, 4006–4010.

Fuji, A., Takita, T., Shimada, N. & Umezawa, H. (1974), *J. Antibiot. Tokyo*, **27**, 73–77.
Fuller, W. & Waring, M. J. (1964), *Ber. Bunsenges. Phys. Chem.*, **68**, 805–808.
Fullerton, D. S., Chen, C. M. & Hall, I. H. (1976), *J. Med. Chem.*, **19**, 1391–1395.
Furth, J. J. & Cohen, S. S. (1968), *Cancer Res.*, **28**, 2061–2067.
Futterman, B., Derr, J., Beisler, J. A., Abbasi, M. M. & Voytek, P. (1978), *Biochem. Pharmacol.*, **27**, 907–909.
Gailani, S. D., Armstrong, J. G., Carbone, P. P., Tan, C. & Holland, J. F. (1966), *Cancer Chemother. Rep. Part 3*, **3**, 9.
Gale, G. R. (1975), in *Handbook of Experimental Pharmacology* (Sartorelli, A. C. & Johns, D. G., Eds.), Vol. 38, Part 2, pp. 829–893, Springer-Verlag, Berlin.
Gale, G. R., Ostrander, W. E. & Atkins, L. M. (1968), *Biochem. Pharmacol.*, **17**, 1823–1832.
Gasson, E. J., McCombie, H., Williams, A. H. & Woodward, F. N. (1948), *J. Chem. Soc.*, 44–46.
Gause, G. F., Sveshnikova, M. A., Vkholina, R. S., Gavrilina, G. A., Filicheva, V. A. & Gladkikh, E. G. (1973), *Antibiotiki Moscow*, **18**, 675–678.
Gensler, W. J., Samour, C. M. & Wang, S. Y. (1954), *J. Am. Chem. Soc.*, **76**, 315–316.
Gerzon, K. (1980), in *Anticancer Agents Based on Natural Product Models* (Cassady, J. M. & Douros, J. D., Eds.), pp. 271–317, Academic, New York.
Gerzon, K., Cochran, J. E., White, A. L., Monahan, R., Krumkalns, E. V., Scroggs, R. E. & Mills, J. (1959), *J. Med. Chem.*, **1**, 223–243.
Gilman, A. & Philips, F. S. (1946), *Science*, **103**, 409–415.
Gitterman, C. O., Rickes, E. L., Wolf, D. E., Madas, J., Zimmerman, S. B., Stoudt, T. H. & Demny, T. C. (1970), *J. Antibiot. Tokyo*, **23**, 305–310.
Goldin, A., Humphreys, S. R., Venditti, J. M. & Mantel, N. (1959), *J. Nat. Cancer Inst.*, **22**, 811–823.
Goldin, A., Serpick, A. A. & Mantel, N. (1966), *Cancer Chemother. Rep.*, **50**, 173–218.
Golden, A. & Wood, H. B., Jr. (1969), *Ann. N.Y. Acad. Sci.*, **163**, 954–1005.
Goldin, A., Wood, H. B., Jr. & Engle, R. R. (1968), *Cancer Chemother. Rep. Part 2*, **1**, 1–269.
Goodman, L. (1974), in *Basic Principles in Nucleic Acid Chemistry* (Ts'O, P.O.P., Ed.), Vol. 1, Chapter 2, Academic, London.
Goodman, L., De Graw, J., Kislink, R. L., Friedkin, M., Pastore, E. J., Crawford, E. J., Plante, L. T., Al-Nahas, A., Morningstar, J. F., Kwok, G., Wilson, L., Donovan, F. & Ratzan, J. (1964), *J. Am. Chem. Soc.*, **86**, 308–309.
Goodman, L. E., Kramer, S. P., Gaby, S. D., Bahal, D., Solomon, R. D., Williamson, C. E., Miller, J. I., Sass, S., Whitten, B. & Seligman, A. (1962), *Cancer*, **15**, 1056–1061.
Goodnight, J. E., Jr. & Morton, D. L. (1978), *Ann. Rev. Med.*, **29**, 231.
Gorski, J., Noteboom, W. D. & Nicolette, J. A. (1965), *J. Cell. Comp. Physiol.*, **66**, 91–109.
Gould, S. J. & Cane, D. E. (1982), *J. Am. Chem. Soc.*, **104**, 343–346.
Gould, S. J. & Chang, C. C. (1980), *J. Am. Chem. Soc.*, **102**, 1702–1706.
Govindachari, T. R. & Viswanathan, N. (1972), *Indian J. Chem.*, **10**, 453–454.

Graham, F. L. & Whitmore, G. F. (1970), *Cancer Res.*, **30**, 2636–2644.
Gray, L. H., Conger, A. D., Ebert, M., Hornsey, S. & Scott, O. C. A. (1953), *J. Radiol.*, **26**, 634–648.
Greenspan, E. M., Leiter, J. & Shear, M. J. (1950), *J. Nat. Cancer Inst.*, **10**, 1295–1317.
Greider, A., Maurer, R. & Stahelin, H. (1974), *Cancer Res.*, **34**, 1788–1793.
Grobe, H. & Palm, D. (1972), *Monatsschr. Kinderheilk.*, **120**, 23–27.
Habermann, V. & Sorm, F. (1958), *Coll. Czech. Chem. Commun.*, **23**, 2201–2206.
Hackethal, C. A., Golbey, R. B., Tan, C. T. C., Karnofsky, D. A. & Burchenal, J. H. (1961), *Antibiot. Chemother.*, **11**, 178–183.
Haddow, A., Kon, G. A. R. & Ross, W. C. J. (1948), *Nature London*, **162**, 824.
Haidle, C. W. (1971), *Mol. Pharmacol.*, **7**, 645–652.
Hakala, M. T., Law, L. W. & Welch, A. D. (1956), *Proc. Am. Assoc. Cancer Res.*, **2**, 113.
Haller, H. L., Bartlett, P. D., Drake, N. L., Newman, M. S., Cristol, S. J., Eaker, C. M., Hayes, R. A., Kilmer, G. W., Magerlein, B., Mueller, G. P., Schneider, A. & Wheatley, W. (1945), *J. Am. Chem. Soc.*, **67**, 1591.
Handschumacher, R. E., Calabresi, P., Welch, A. D., Bono. V. H., Fallon, H. J. & Frei, E. (1962), *Cancer Chemother. Rep.*, **21**, 1–18.
Handschumacher, R. E. & Pasternak, C. A. (1958), *Biochem. Biophys. Acta*, **30**, 451–452.
Handschumacher, R. E., Skoda, J. & Sorm, F. (1963), *Coll. Czech. Chem. Commun.*, **28**, 2983–2990.
Haneiski, T., Okazaki, T., Hata, T., Tamura, C., Nomura, M. Naito, A., Seki, I. & Arai, M. (1972), *J. Antibiot. Tokyo*, **24A**, 797–799.
Hanka, L. J., Evans, J. S., Mason, D. J. & Dietz, A. (1966), *Antimicrob. Agents Chemother.*, 619–624.'
Hansch, C., Hatheway, G. J., Quinn, F. R. & Greenberg, N. (1978), *J. Med. Chem.*, **21**, 574–577.
Hansen, B. S. & Vaughan, M. H., Jr. (1973), *Fed. Proc. Fed. Am. Soc. Exp. Biol.*, **32**, 494.
Hanze, A. R. (1967), *J. Am. Chem. Soc.*, **89**, 6720–6725.
Harbers, E., Chaudhuri, N. K. & Heidelberger, C. (1959), *J. Biol. Chem.*, **234**, 1255–1262.
Hargrove, W. W. (1964), *Lloydia*, **27**, 340–345.
Harriss, M. N., Medrek, T. J., Golomb, F. M., Gumport, S. L., Postel, A. H. & Wright, J. C. (1965), *Cancer*, **18**, 49–57.
Hart, M. M. & Straw, J. A. (1971), *Steroids* **17**, 559–574.
Hartlieb, J. (1974), *Z. Krebsforsch*, **81**, 1–6.
Hartman, S. C., *J. Biol. Chem.* (1963), **238**, 3036–3047.
Hartmann, G. & Coy, V. (1962), *Angew. Chem.*, **74**, 501.
Hartwell, J. M. & Schrecker, A. W. (1958), *Fortschr. Chem. Org. Naturst.*, **15**, 83–166.
Hatheway, G. J., Hansch, C., Kim, K. H., Milstein, S. R., Schmidt, C. L., Smith, R. N. & Quinn, F. R. (1978), *J. Med. Chem.*, **21**, 563–574.
Heidelberger, C. (1975), in *Handbook of Experimental Pharmacology* (Sartorelli,

A. C. & Johns, D. G., Eds.), Vol. 38, Part 2, pp. 193-231, Springer-Verlag, Berlin.
Heidelberger, C., Parsons, D. G. & Remy, D. C. (1964), *J. Med. Chem.*, **7**, 1-5.
Heilman, F. R. & Kendall, E. C. (1944), *Endocrinology*, **34**, 416-420.
Helman, L. J. & Slavik, M. (1976), *Anguidine Clinical Brochure*, Investigative Drug Branch, National Cancer Institute, Bethesda, Maryland.
Hendry, J. A., Homer, R. F., Rose, F. L. & Walpole, A. (1951), *Br. J. Pharmacol.*, **6**, 235-255.
Henry, D. W. (1974), *Cancer Chemother. Rep.*, **4**, 5-9.
Herr, M. E., Hogg, J. A. & Levin, R. H. (1956), *J. Am. Chem. Soc.*, **78**, 500-501.
Herr, R. R., Jahnke, H. K. & Argoudelis, A. D. (1967), *J. Am. Chem. Soc.*, **89**, 4808-4809.
Hessler, E. J. & Jahnke, H. K. (1970), *J. Org. Chem.*, **35**, 245-246.
Higashide, E., Asai, M., Ootsu, K., Tanida, S., Kozai, Y., Hasegawa, T., Kishi, T., Sugino, Y. & Yoneda, M. (1977), *Nature London*, **270**, 721-722.
Hill, B. T. (1976), *Anal. Biochem.*, **70**, 635-638.
Hill, J. M., Loeb, E. & Mac Lellan, A. (1975), *Cancer Chemother. Res.*, **59**, 647-659.
Hiller, S. A., Zhuk, R. A., & Lidak, M. Yu. (1967), *Dokl. Acad. Nauk, U.S.S.R.*, **176**, 332-335.
Himmelweit, F. (Ed.) (1956), *The Collected Papers of Paul Ehrlich*, Vol. 1, pp. 596-618, Pergamon, London.
Hirschberg, E., Gellhorn, A. & Gump, W. S. (1958), *Ann. N.Y. Acad. Sci.*, **68**, 888.
Hisamatsu, T. & Uchida, S. (1977), *Gann*, **68**, 819-824.
Hitchings, G. H. & Burchall, J. J. (1965), *Adv. Enzymol.*, **27**, 417-468.
Hitchings, G. H. & Elion, G. B. (1962), U.S. Patent 3,056,785.
Hodes, M. E., Rohn, R. J., Bond, W. H. & Yardley, J. (1963), *Cancer Chemother. Rep.*, **28**, 53-55.
Hocksema, H., Slomp, G. & Van Tamelen, E. E. (1964), *Tetrahedron Lett.*, 1787-1795.
Hoffer, M., Duschinsky, R., Fox, J. J. & Young, N. (1959), *J. Am. Chem. Soc.*, **81**, 4112-4113.
Hoffman-La Roche & Co. (1962), Belg. Pat 618,638.
Hohorst, H. J., Peter, G. & Struck, R. F. (1976), *Cancer Res.*, **36**, 2278-2281.
Hollstein, U. (1974), *Chem. Rev.*, **74**, 625-652.
Hori, M., Ito, E., Takita, T., Koyama, G., Takeuchi, T. & Umezawa, H. (1974), *J. Antibiot. Tokyo*, **17A**, 96-99.
Horwitz, S. B. (1971), *Prog. Subcell. Biol.*, **2**, 40-47.
Horwitz, S. B. (1975), in *Antibiotics. III. Mechanism of Action of Antimicrobial and Antitumor Agents* (Corcoran, J. W. & Hahn, F. E., Eds.), pp. 48-57, Springer-Verlag, Berlin.
Horwitz, S. B., Chang, S. C., Grollman, A. P. & Borkovec, A. B. (1971), *Science*, **174**, 159.
Howie, G. A., Stamos, I. K. & Cassady, J. M. (1976), *J. Med. Chem.*, **19**, 309-313.
Hrushesky, W. J. (1976), *Med. Ped. Oncol.*, **2**, 441.
Huang, M. T. (1975), *Mol. Pharmacol.*, **11**, 511-519.
Huggins, C. & Bergenstal, D. M. (1952), *Cancer Res.*, **12**, 134-141.

Humphrey, E. W. & Dietrick, F. S. (1963), *Cancer Chemother. Rep.*, **33**, 21–26.
Hunter, J. H. (1963), U.S. Patent 3,116,282.
Hurley, L. H. (1977), *J. Antibiot. Tokyo*, **30**, 349–370.
Hurley, L. H., Gailola, C. & Zmyenski, M. (1977), *Biochem. Biophys. Acta*, **475**, 521–535.
Hutchings, B. L., Neowat, J. H., Oleson, J. J., Stokstad, E. L. R., Boothe, J. H., Waller, C. W., Angier, R. B., Semb, J. & Subbarow, Y. (1947), *J. Biol. Chem.*, **170**, 323–328.
Hyman, R. W. & Davidson, N. (1970), *J. Mol. Biol.*, **50**, 421–438.
Ihashi, Y., Abe, H. & Ito, Y. (1973a), *Agric. Biol. Chem.*, **37**, 2283–2287.
Ihashi, Y., Abe, H. & Ito, Y. (1973b), *Agric. Biol. Chem.*, **37**, 2277–2280.
Ikehara, M., Tesugi, S. & Kaneko, M. (1967), *Chem. Commun.*, 17–18.
Imperial Chemical Industries (1964), Belgian Patent 637,389.
Inagaki, A., Nakamura, T. & Wakisaka, G. (1969), *Cancer Res.*, **29**, 2169–2176.
Inhofen, H. H., Logeman, W., Holway, W. & Serini, A. (1938), *Chem. Ber.*, **71**, 1024–1032.
Ishida, N., Miyazaki, K., Kumagai, K. & Rikimaru, M. (1965), *J. Antibiot. Tokyo*, Ser. A, **18**, 68–76.
Ishidate, M., Kobayashi, K., Sakurai, Y., Sato, H. & Yoshida, T. (1951), *Proc. Jpn. Acad.*, **27**, 493.
Ishizuka, M., Takayama, H., Takeuchi, T. & Umezawa, H. (1966), *J. Antibiot. Tokyo*, Ser. A., **19**, 260–271.
Israel, M., Modest, E. J. & Frei, E., III. (1975), *Cancer Res.*, **35**, 1365–1368.
Ito, Y., Ihashi, Y., Kawabe, S., Abe, H. & Okuda (1972), *J. Antibiot. Tokyo*, **25**, 360–361.
Iyengar, B. S., Lin, H. J., Cheng, L., Remers, W. A. & Bradner, W. T. (1981), *J. Med. Chem.*, **24**, 975–981.
Jarman, M. & Ross. W. C. J. (1967), *Chem. Ind. London*, 1789–1790.
Jarjaram, H. N., Cooney, D. A., Ryan, J. A., Neil, G., Dion, R. J. & Bono, V. H. (1975), *Cancer Chemother. Rep. Part 1*, **59**, 481–491.
Johns, D. G., Sartorelli, A. C., Bertino, J. R., Iannotti, A. T., Booth, B. A. & Welch, A. D. (1966), *Biochem. Pharmacol.*, **15**, 400–403.
Johns, D. G. & Valerino, A. M. (1971), *Ann. N.Y. Acad. Sci.*, **186**, 378–386.
Johnson, I. S., Wright, H. F. & Svoboda, G. H. (1959), *J. Lab. Clin. Med.*, **54**, 830.
Johnson, R. K. & Chitnis, M. P. (1978), *Proc. Am. Assoc. Cancer Res.*, **19**, 218.
Johnston, T. P., McCaleb, G. S., Clayton, S. D., Frye, J. L., Krauth, C. A. & Montgomery, J. A. (1977), *J. Med. Chem.*, **20**, 279–290.
Johnston, T. P., McCaleb, G. S. & Montgomery, J. A. (1963), *J. Med. Chem.*, **6**, 669–681.
Johnston, T. P., McCaleb, G. S. & Montgomery, J. A. (1975), *J. Med. Chem.*, **18**, 104–110.
Johnston, T. P., McCaleb, G. S., Oplinger, P. S. & Montgomery, J. A. (1963), *J. Med. Chem.*, **9**, 892–911.
Johnston, T. P., McCaleb, G. S., Oplinger, P. S. & Montgomery, J. A. (1971), *J. Med. Chem.*, **14**, 600–614.

Jolad, S. D., Hoffmann, J. J., Torrance, S. J., Wiedhopf, R. M., Cole, J. R., Arora, S. K., Bates, R. B., Gargulio, R. L. & Kriek, G. R. (1977), *J. Am. Chem. Soc.*, **99**, 8040–8044.
Jones, R., Jr., Jonsson, V., Browning, M., Lessner, H., Price, C. C. & Sen, A. K. (1968), *Ann. N.Y. Acad. Sci.*, **68**, 1133–1150.
Kajiro, Y. & Kamiyama, M. (1965), *Biochem. Biophys. Res. Commun.*, **19**, 433.
Kalman, T. I. & Bardos, T. J. (1970), *Mol. Pharmacol.*, **6**, 621–630.
Kanai, T., Kojima, T., Maruyama, O. & Ichino, M. (1970), *Chem. Pharm. Bull.*, **18**, 2569–1571.
Kandaswamy, T. S. & Henderson, J. F. (1962), *Nature London*, **195**, 85.
Kaplan, I. W. (1942), *New Orleans Med. Surg. J.*, **94**, 388.
Kasai, H., Naganawa, H., Takita, T. & Umezawa, H. (1978), *J. Antibiot.*, **31**, 1316–1320.
Katz, E. (1967), in *Antibiotics* (Gottlieb, D. & Shaw, P. D., Eds.), Vol. 2, pp. 267–341, Springer-Verlag, Berlin.
Katz, E. (1974), *Cancer Chemother. Rep. Part 1*, **58**, 83–91.
Katz, E. & Goss, W. A. (1959), *Biochem. J.*, **73**, 458–465.
Katz, E., Williams, W. K., Mason, K. T. & Mauger, A. G. (1977), *Antimicrob. Agents Chemother.*, **11**, 1056–1063.
Kaufman, H. E. (1965), *Prog. Med. Virol.*, **7**, 116–159.
Keller-Juslen, C., Kuhn, M., von Wartberg, A. & Stahelin, H. (1971), *J. Med. Chem.*, **14**, 936–940.
Kelley, M. G. & Hartwell, J. L. (1954), *J. Nat. Cancer Inst.*, **14**, 967–1010.
Kelley, R. & Baker, W. (1960), in *Biological Activities of Steroids in Relation to Cancer* (Pincus, G. & Vollmer, E. P., Eds.), pp. 427–443, Academic, New York.
Kelly, T. R. (1979), *Ann. Rep. Med. Chem.*, **14**, 288–298.
Kende, A. S., Lorah, D. P. & Boarman, R. J. (1981), *J. Am. Chem. Soc.*, **103**, 1271–1273.
Kennedy, B. J. (1970), *Cancer*, **26**, 755–766.
Kent, R. J. & Heidelberger, C. (1972), *Molec. Pharmacol.*, **8**, 465–467.
Kersten, H. (1974), in *Handbook of Experimental Pharmacology*, Vol. 38, Part 2, pp. 47–64, Springer-Verlag, Berlin.
Kersten, W. (1968), *Abh. Deut. Akad. Wiss., Berlin Kl. Med.*, 593–598.
Kersten, W. & Kersten, H. (1962), *Z. Physiol. Chem.*, **327**, 234–242.
Kessel, D. (1967), *Proc. Am. Assoc. Cancer Res.*, **8**, 36.
Kessel, D. (1967), *Proc. Am. Assoc. Cancer Res.*, **29**, 687–696.
Kessel, D. (1979), *Mol. Pharmacol.*, **11**, 306–312.
Kessel, D. & Wodinsky, I. (1970), *Mol. Pharmacol.*, **6**, 251–254.
Khan, A. H. & Driscoll, J. S. (1976), *J. Med. Chem.*, **19**, 313–317.
Kidd, J. G. (1953), *J. Exp. Med.*, **98**, 565.
Kidder, G. W., Dewey, V. C., Parks, R. E., Jr. & Woodside, G. L. (1949), *Science*, **109**, 511–514.
Kimball, A. P. & Wilson, M. J. (1968), *Proc. Soc. Exp. Biol.*, **127**, 429–432.
Kisliuk, R. L. (1960), *Nature London*, **188**, 584–585.
Kohn, K. W. (1966), *J. Mol. Biol.*, **19**, 266–287.

Kohn, K. W. (1977), *Cancer Res.*, **37**, 1450-1454.
Kohn, K. W., Waring, M. J., Glaubiger, D. & Friedman, O. M. (1966), *Cancer Res.*, **35**, 71-76.
Komoda, Y. & Kishi, T. (1980), in *Anticancer Agents Based on Natural Product Models* (Cassady, J. M. & Douros, J. D., Eds.) pp. 353-389, Academic, New York.
Kon, G. A. R. & Roberts, J. J. (1950), *J. Chem. Soc.*, 978-982.
Konishi, M., Saito, K., Numata, K., Tsuno, T., Asama, K., Tsukiura, H., Naito, T. & Kawaguchi, H. (1977), *J. Antibiot. Tokyo*, **30**, 789-805.
Korman, S. & Tendler, M. D. (1965), *J. New Drugs*, **5**, 275-285.
Knott, R. & Taunton-Rigby, A. (1971), *126th ACS Natl. Meeting, Washington, D.C., Abstr.* MEDI 27.
Krakoff, I. H., Brown, N. C. & Reichard, P. (1968), *Cancer Res.*, **28**, 1559-1565.
Kremer, W. B. & Laszlo, J. (1967), *Cancer Chemother. Rep.*, **51**, 19-23.
Kugelman, M., Liu, W. C., Axelrod, B., McBride, T. J. & Rao, K. V. (1976), *Lloydia*, **37**, 125-128.
Kuh, E. & Seeger, D. R. (1954), U.S. Patent 2,670,347.
Kuhn, M. & von Wartburg, A. (1968), *Helv. Chim. Acta.*, **51**, 163-168.
Kumar, V., Remers, W. A. & Bradner, W. T. (1980), *J. Med. Chem.*, **23**, 376-379.
Kunimoto, S., Masuda, T., Kanbayashi, N., Hamada, M., Naganawa, H., Miyamoto, M., Takeuchi, T. & Umezawa, H. (1980), *J. Antibiot. Tokyo*, **33**, 665-667.
Kupchan, S. M., Court, W. A., Dailey, R. F., Jr., Gilmore, C. J. & Bryan, R. F. (1972), *J. Am. Chem. Soc.*, **94**, 7194-7195.
Kupchan, S. M., Eakin, M. A. & Thomas, A. M. (1971), *J. Med. Chem.*, **14**, 1147-1152.
Kupchan, S. M., Komoda, Y., Court, W. A., Thomas, G. J., Smith, R. M., Karim, A., Gilmore, C., Haltiwanger, R. C. & Bryan, R. F. (1972), *J. Am. Chem. Soc.*, **95**, 1354-1356.
Kupchan, S. M., Sneden, A. T., Braufman, A. R., Howie, G. A., Rebhun, L. I., McIvor, W. E., Wang, R. W. & Schnaitman, T. C. (1978), *J. Med. Chem.*, **21**, 31-37.
Kusakabe, Y., Nogateu, J., Shibuya, M., Kawaguchi, O., Hirose, C. & Shirato, M. (1972), *J. Antibiot. Tokyo*, **25A**, 44-47.
Langen, P., Kowollik, G., Schutt, M. & Etzold, G. (1969), *Acta Biol. Med. Ger.*, **23**, K19-22.
Larionov, L. F. (1960), *Cancer Progress*, p. 211, Butterworths, London.
Larionov, L. F., Khoklov, A. S., Shkowinskaja, E. N., Vasina, O. S., Troosheikina, V. I. & Novikova, M. A. (1955), *Lancet*, **2**, 169-171.
Law, L. W., *Ann. N.Y. Acad. Sci.* (1958), **71**, 976-992.
Lawley, P. D. & Martin, C. W. (1975), *Biochem. J.*, **145**, 85-91.
Lednicer, D., Babcock, J. C., Lyster, S. C. & Duncan, G. W. (1963), *Chem. Ind. London*, 408-409.
Lee, K. H., Ibuka, T., Sims, D., Muraoka, O., Kiyokawa, H. & Hall, I. H. (1981), *J. Med. Chem.*, **24**, 924-927.

Lee, K. H., Mar, E. C. & Hall, I. H. (1978), *J. Med. Chem.*, **21**, 698.
Lee, W. W., Benitez, A., Goodman, L. & Baker, B. R. (1960), *J. Am. Chem. Soc.*, **82**, 2648-2649.
Lee, W. W., Martinez, A. P., Tong, G. L. & Goodman, L. (1963), *Chem. Ind. London*, **52**, 2007-2008.
Lelieveld, P. & van Putten, L. M. (1981), *12th. Int. Congr. Chemotherapy*, Florence, Italy, July 1981, Abstacts of the Meeting, No. 21.
LePage, G. A. (1971), *Nat. Cancer Inst. Monograph*, **34**, 184-187.
LePage, G. A., Bell, J. P. & Wilson, M. J. (1969), *Proc. Soc. Exp. Biol.*, **131**, 1038-1041.
LePage, G. A., Worth, L. S. & Kimball, A. P. (1976), *Cancer Res.*, **36**, 1481-1485.
Le Pecq, J. B. & Paoletti, C. (1967), *J. Mol. Biol.*, **27**, 87-106.
Lerman, L. S. (1961), *J. Mol. Biol.*, **3**, 18-30.
Levenberg, B., Melnick, I. & Buchanan, J. M. (1957), *J. Biol. Chem.*, **225**, 163-176.
Levine, A. S., Sharp, H. L., Mitchell, J., Krivit, W. & Nesbit, M. E. (1969), *Cancer Chemother. Rep.*, **53**, 53-57.
Levis, A. G., Danieli, G. A., & Piccinni, E. (1965), *Nature London*, **207**, 608-610.
Liao, L. L., Krysehan, S. M. & Horwitz, S. B. (1976), *Mol. Pharmacol.*, **12**, 167-176.
Lin, A. J., Cosby, L. A., Shansky, C. W. & Sartorelli, A. C. (1972), *J. Med. Chem.*, **15**, 1247-1251.
Lin, A. J., Shansky, C. W. & Sartorelli, A. C. (1974), *J. Med. Chem.*, **17**, 558.
Lindell, T. J., O'Malley, A. F. & Puglisi, B. (1978), *Biochemistry*, **17**, 1154-1160.
Lockwood, A. H. (1979), *Proc. Nat. Acad. Sci. U.S.A.*, **76**, 1184-1188.
Lown, J. W. (1979), in *Bleomycin: Chemical, Biochemical & Biological Aspects* (Hecht, S., Ed.), p. 184, Springer-Verlag, Berlin.
Lown, J. W., Begleiter, A. Johnson, D. & Morgan, A. R. (1976), *Can. J. Biochem.*, **54**, 110-119.
Lown, J. W., Gunn, B. C., Chang, R. Y., Majumdar, K. C. & Lee, J. S. (1978a), *Can. J. Biochem.*, **56**, 1006-1015.
Lown, J. W. & Joshua, A. V. (1979), *Biochem. Pharmacol.*, **28**, 2017-2026.
Lown, J. W., Joshua, A. V. & McLaughlin, L. W. (1980), *J. Med. Chem.*, **23**, 798-805.
Lown, J. W. & McLaughlin, L. W. (1979), *Biochem. Pharmacol.*, **28**, 2123-2128.
Lown, J. W., McLaughlin, L. W. & Change, Y. M. (1978b), *Bioorg. Chem.*, **7**, 97-110.
Lown, J. W. & Sim, S. K. (1976), *Can. J. Chem.*, **54**, 2563-2572.
Lown, J. W., Sim, S. K., Majumdar, K. C. & Chang, R. Y. (1977), *Biochem. Biophys. Res., Commun.*, **76**, 705-710.
Ludlum, D. B. (1965), *Biochem. Biophys. Acta*, **95**, 674-676.
Ludlum, D. B. (1967), *Biochem. Biophys. Acta*, **142**, 282-284.
Ludlum, D. B. (1974), in *Handbook of Experimental Pharmacology*, Vol. 38, Part 2, pp. 6-17, Springer-Verlag, Berlin.
Ludlum, D. B., Kramer, B. S., Wang, J. & Fenselau, C. (1975), *Biochemistry*, **14**, 5480-5485.
Lukens, L. N. & Herrington, K. A. (1957), *Biochem. Biophys. Acta*, **24**, 432-433.

Lyttle, D. A. & Petering, H. G. (1958), *J. Am. Chem. Soc.*, **80**, 6459–6460.
Maeda, H., Ichimura, H., Satoh, H. & Ohtsuki, K. (1978), *J. Antibiot. Tokyo*, **31**, 468–472.
Marquez, V. E., Lui, P. S., Kelley, J. A., McCormack, J. J. & Driscoll, J. S. (1980), *179th National Meeting, American Chemical Society*, Houston, March 24, 1980, Abstracts MEDI 52.
Marshall, V. P., McGovern, J. P., Richard, F. A., Richard, R. E. & Wiley, P. F. (1978), *J. Antibiot. Tokyo*, **31**, 336–342.
Martin, D. G., Duchamp, D. J. & Chidester, C. G. (1973), *Tetrahedron Lett.*, 2549–2552.
Martin, D. G., Hanka, L. J. & Neil, G. L. (1974), *Cancer Chemother. Rep. Part 1)*, **58**, 935–937.
Mashburn, L. T. & Wriston, J. C., Jr. (1964), *Arch. Biochem. Biophys.*, **105**, 450–452.
Mathe, G., Hayat, M., deVassal, F., Schwartzenberg, M., Schneider, M., Schlumberger, J. R., Iasmin, C. & Rosenfeld, C. (1970), *Rev. Eur. Etud. Clin. Biol.*, **15**, 541–545.
Matsushita, S. & Fanburg, B. L. (1970), *Circ. Res.*, **27**, 415–428.
Mauger, A. B. (1980), in *Topics in Antibiotic Chemistry* (Sammes, P. G., Ed.), Vol. 5, pp. 223–312, Ellis Horwood Limited, Chichester, England.
Mautner, H. G. (1956), *J. Am. Chem. Soc.*, **78**, 5292–5294.
McCoy, T. A., Maxwell, M. & Kruse, P. F., Jr. (1959), *Cancer Res.*, **19**, 591–595.
McGhee, J. D. & Felsenfeld, G. (1979), *Proc. Nat. Acad. Sci. U.S.A.*, **76**, 2133–2137.
McLean, E. K. (1970), *Pharmacol. Rev.*, **32**, 429–483.
Mead, J. A. R. (1975), in *Handbook of Experimental Pharmacology* (Sartorelli, A. C. and Johns, D. G., Eds.), Vol. 38, Part 1, pp. 52–75, Springer-Verlag, Berlin.
Mead, J. A. R., Goldin, A., Kisliuk, R. L. Friedkin, M., Plante, L., Crawford, E. J. & Kwok, G. (1966), *Cancer Res.*, **26**, 2374–2379.
Mead, J. A. R., Venditti, J. M., Schrecker, A. W., Goldin, A. & Kisluik, R. L. (1961), *Nature London*, **189**, 937–939.
Meich, R. P., York, R. & Parks, R. E., Jr. (1969), *Mol. Pharmacol.*, **5**, 30–37.
Meienhofer, J. & Atherton, E. (1973), *Adv. Appl. Microbiol.*, **16**, 203–300.
Meienhofer, J. & Atherton, E. (1977), in *Structure-Activity Relationships Among the Semisynthetic Antibiotics* (Perlman, D., Ed.), pp. 427–529, Academic, New York.
Meienhofer, J., Maeda, H., Glaser, C. B., Czombos, J. & Kuromizu, K. (1972), *Science*, **178**, 875–876.
Meienhofer, J. & Patel, R. P. (1971), *Int. J. Protein Res.*, **3**, 347–350.
Meyers, A. I., Reider, P. J. & Campbell, A. L. (1980), *J. Am. Chem. Soc.*, **102**, 6597–6598.
Michaud, R. L. & Sartorelli, A. C. (1968), *155th Am. Chem. Soc. Nat. Meet.*, San Francisco, Abstracts No. 54.
Mihich, E. (1963), *Pharmacologist*, **5**, 270.
Mikolajczak, K. L. & Smith, C. R. (1978), *J. Org. Chem.*, **43**, 4762–4765.
Miller, D. S., Laszlo, J., McCarty, K. S., Guild, W. R. & Hochstein, P. (1967), *Cancer Res.*, **27**, 632–638.

Mishra, L. C. & Mead, J. A. R. (1972), *Chemotherapy*, **17**, 283-292.
Mizuno, N. S. & Gilboe, D. P. (1970), *Biochem. Biophys. Acta*, **224**, 319-327.
Momparler, R. L. (1972), *Mol. Pharmacol.*, **8**, 362-370.
Momparler, R. L. & Fischer, G. A. (1968), *J., Biol. Chem.* **243**, 4298-4303.
Montgomery, J. A. (1970a), *Prog. Med. Chem.*, **7**, 69-122.
Montgomery, J. A. (1970b), in *Medicinal Chemistry*, 3rd ed., p. 680, Wiley, New York.
Montgomery, J. A. & Hewson, K. (1967), *J. Med. Chem.*, **10**, 665-667.
Montgomery, J. A. & Holum, L. B. (1957), *J. Am. Chem. Soc.*, **79**, 2185-2188.
Montgomery, J. A., James, R., McCaleb, G. S. & Johnston, T. P. (1967), *J. Med. Chem.*, **10**, 668-674.
Montgomery, J. A., Johnston, T. P., Gallagher, A., Stringfellow, C. R. & Schabel, F. M., Jr. (1961), *J. Med. Chem.*, **3**, 265-288.
Montgomery, J. A., Johnston, T. P. & Shealy, Y. F. (1970), in *Medicinal Chemistry*, 3rd ed., (Burger, A., Ed.), Wiley, New York.
Montgomery, J. A., Mayo, J. G. & Hansch, C. (1974), *J. Med. Chem.*, **17**, 477-480.
Moore, E. C. (1969), *Cancer Res.*, **29**, 291-295.
Moore, E. C. & LePage, G. A. (1958), *Cancer Res.*, **18**, 1078-1083.
Moore, H. W. (1977), *Science*, **197**, 527-532.
Moore, S., Kondo, M., Copeland, M., Meienhofer, J. & Johnston, R. K. (1975), *J. Med. Chem.*, **18**, 1098-1101.
Morton, D. (1974), *Report to the American Association for the Advancement of Science*, San Francisco, February meeting.
Mueller, W. E. G., Zahn, R. K. & Seidel, H. J. (1971), *Nature New Biol.*, **232**, 143-145.
Mujamoto, M., Kondo, S., Naganawa, N., Maeda, K., Ohno, M. & Umezawa, H. (1977), *J. Antibiot. Tokyo*, **30**, 340-343.
Muller, W. (1962), *Naturwissenschaften*, **49**, 156-157.
Muraoka, Y., Kobayashi, H., Fujii, A., Kunishima, M., Fujii, T., Nakayama, Y., Takita, T. & Umezawa, H. (1976), *J. Antibiot.*, **29**, 853-840.
Murdock, K. C., Child, R. G., Fabio, P. F., Angier, R. B., Wallace, R. E., Durr, F. E. & Citarella, R. V. (1979), *J. Med. Chem.*, **22**, 1024-1030.
Murdock, K. C., Child, R. G., Lin, Y. I., Warren, J. D., Fabio, P. F., Lee, V. J., Izzo, P. T., Lang, S. A., Jr., Angier, R. B., Citarella, R. V., Wallace, R. E. & Durr, F. E. (1982), *J. Med. Chem.*, **25**, 505-518.
Murfree, S. A., Cunningham, L. S., Hwang, K. M. & Sartorelli, A. C. (1976), *Biochem. Pharmacol.*, **25**, 1227-1231.
Murthy, Y. K. S., Theimann, J. E., Coronelli, C. & Sensi, P. (1966), *Nature*, **211**, 1198-1199.
Nakamura, H., Koyama, G., Jitaka, Y., Ohno, M., Yagisawa, N., Kondo, S., Maeda, K. & Umezawa, H. (1974), *J. Am. Chem. Soc.*, **96**, 4327-4328.
Nakayama, U., Kunishima, M., Omoto, S., Takita, T. & Umezawa, H. (1973), *J. Antibiot. Tokyo*, **26**, 400-401.
Nayak, R., Sirsi, M. & Podder, S. K. (1973), *FEBS Lett.*, **30**, 157-162.
Nishimura, H., Katagiri, K., Sato, K., Mayama, M. & Shimaoka, N. (1956), *J. Antibiot. Tokyo*, **9**, 60-62.

Nishimura, J. S. & Bowers, W. F. (1967), *Biochem. Biophys. Res. Commun.*, **28**, 665–670.
Nobile, A., Charney, W., Perlman, P. L., Herzog, H. L., Payne, C. C., Tully, M. E., Jernik, M. A. & Hershberg, E. B. (1955), *J. Am. Chem. Soc.*, **77**, 5184A.
Noell, C. W. & Robins, R. K. (1958), *J. Org. Chem.*, **23**, 1547–1550.
O'Brien, P. H. (1972), *J. South Carolina Med. Assoc.*, **68**, 466–467.
Ohashi, Y., Abe, H., Kawabe, S. & Ito, Y. (1973), *Agric. Biol. Chem.*, **37**, 2387–2391.
Ohno, M. (1980), in *Anticancer Agents Based on Natural Product Models* (Cassady, J. M. & Douros, J. D., Eds.), pp. 73–131, Academic, New York.
Oki, T. (1977), *J. Antibiot. Tokyo*, **30**, Suppl., S-70-S-84.
Pan, P. C., Pan, S. Y., Tu, Y. H., Wang, S. Y. & Owen, T. Y. (1975), *Hua Hsueh Hsueh Pao*, **30**, 71.
Panthananickal, A., Hansch, C. & Leo, A. (1979), *J. Med. Chem.*, **22**, 1267–1269.
Panthananickal, A., Hansch, C., Leo, A. & Quinn, F. R. (1978), *J. Med. Chem.*, **21**, 16–25.
Panzica, R. P., Robins, R. K. & Townsend, L. B. (1971), *J. Med. Chem.*, **14**, 259–260.
Paoletti, C., Cros, S., Dat-Xuong, N., Lecointe, P. & Moisand, A. (1979), *Chem. Biol. Interact.*, **25**, 45–58.
Paoletti, C., Lecointe, P., Lesca, P., Cros, S., Mansuy, D. & Dat-Xuong, N. (1978), *Biochemie.*, **60**, 1003–1009.
Papanastassiou, Z. B., Bruni, R. J., White, E. & Levins, P. L. (1966), *J. Med. Chem.*, **9**, 725–729.
Parham, W. E. & Wilbur, J. M., Jr. (1961), *J. Org. Chem.*, **26**, 1569–1572.
Parke-Davis & Company (1967), Belg. patent 671,557.
Pasternak, C. A., Fischer, G. A. & Handschumacher, R. E. (1961), *Cancer Res.*, **21**, 110–117.
Patterson, R. J., Jr. (1974), *Clin. Neurosurg.*, **21**, 60–67.
Paudler, W. W., Kerley, G. I. & McKay, J. (1963), *J. Org. Chem.*, **28**, 2194–2197.
Pearson, O. H. (1972), in *Estrogen Target Tissues & Neoplasia* (Dao, T. L., Ed.), pp. 287–305, University of Chicago Press, Chicago.
Pearson, O. H., Manni, A., Trujillo, J. E., Marshall, J. S. & Brodkey, J. (1978), in *Proc. Int. Symp. Endocrine-Related Cancer*, Lyon, France, June 1977 (Mayer, M., Saez, S. and Stoll, B. A., Eds.), pp. 137–149, Imperial Chemical Industries, London.
Penco, S., Angelucci, F., Vigevani, A., Arlandini, E. & Arcamone, F. (1977), *J. Antibiot.* (Tokyo) **30**, 764–766.
Perlman, D., Mauger, A. B. & Weissbach, H. (1966), *Biochem. Biophys. Res. Commun.*, **24**, 513–525.
Phillips, G. R. (1969), *Nature London*, **223**, 374–377.
Pigram, W., Fuller, W. & Hamilton, L. D. (1971), *Nature New Biol.*, **234**, 78–80.
Piskala, A. & Sorm, F. (1964), *Coll. Czech. Chem. Commun.*, **29**, 2060–2076.
Plattner, J. J., Gless, R. D., Cooper, G. K. & Rapoport, H. (1974), *J. Org. Chem.*, **39**, 303–311.
Podwyssotski, V. (1880), *Arch. Exp. Pathol. Pharmakol.*, **13**, 29.

Porter, J. N., Hewitt, R., Hesseltine, C. W., Krupka, G., Lowery, J. A., Wallace, W. S., Bohonos, N. & Williams, J. H. (1952), *Antibiot. Chemother.*, 2, 409-410.
Povirk, L. F., Hogan, M. & Dattagupta, N. (1979), *Biochemistry*, 18, 96-101.
Powell, R. G., Weisleder, D. & Smith, C. R. (1972), *J. Pharm. Sci.*, 61, 1227-1230.
Pressman, B. C. (1963), *J. Biol. Chem.*, 238, 401-409.
Price, C. C. (1974), in *Handbook of Experimental Pharmacology* (Sartorelli, A. C. & Johns, D. J., Eds.), Vol. 38, Part 2, p. 4, Springer-Verlag, Berlin.
Prusoff, W. H. & Goz, B. (1975), in *Handbook of Experimental Pharmacology* (Sartorelli, A. C. & Johns, D. G., Eds.), Vol. 38, Part 2, pp. 272-347, Springer-Verlag, Berlin.
Radding, C. M. (1963), *Proc. XI Int. Congr. Genet.*, The Hague, Netherlands, p. 22.
Rahman, A., Cradock, J. C. & Davignon, J. P. (1978), *J. Pharm. Sci.*, 67, 611-614.
Rao, K. V. (1977), *J. Heterocycli. Chem.*, 14, 653-659.
Rao, K. V., Brooks, S. C., Kugelman, M. & Romano, A. A. (1960), *Antibiot. Ann.*, **1959-1960**, 943-949.
Rao, K. V. & Cullen, W. P. (1968), *Antibiot. Ann.*, **1959-1960**, 950-953.
Rao, K. V. & Renn, D. W. (1963), *Antimicrob. Agents Chemother.*, 77-79.
Reed, D. G. (1975), in *Handbook of Experimental Pharmacology* (Sartorelli A. C & Johns, D. G., Eds.), Vol. 38, Part 2, pp. 747-765, Springer-Verlag, Berlin.
Reich, E., Franklin, R. M. & Schatkin, A. J. (1961), *Science*, 134, 556-557.
Remers, W. (1979), *The Chemistry of Antitumor Antibiotics*, Wiley, New York.
Remy, C. N. (1963), *J. Biol. Chem.*, 238, 1078-1084.
Reyes, P. (1969), *Biochemistry*, 8, 2057-2062.
Reyes, P. & Heidelberger, C. (1965), *Mol. Pharmacol.*, 1, 14-30.
Richter, G. V. (1973), Belg. Patent BE811-110.
Ringold, H. J., Batres, E., Halpern, O. & Necoechea, J. (1959), *J. Am. Chem. Soc.*, 81, 427-432.
Rivers, S. L., Whittington, R. M. & Medrik, T. J. (1966), *Cancer*, 19, 1377-1385.
Roberts, E. C. & Shealy, Y. F. (1971), *J. Med. Chem.*, 14, 125-130.
Roberts, J. J. & Warwick, G. P. (1961), *Biochem. Pharmacol.*, 6, 217-227.
Robins, R. K. (1956), *J. Am. Chem. Soc.*, 78, 784-790.
Robins, R. K., Furcht, F. W., Grauer, A. D. & Jones, J. W. (1956), *J. Am. Chem. Soc.*, 78, 2418-2422.
Roblin, R. O., Jr., Lampen, J. O., English, J. P., Cole, Q. P. & Vaughan, J. R. (1945), *J. Am. Chem. Soc.*, 67, 290-294.
Rose, W. C., Trader, M. W., Laster, W. R., Jr. & Schabel, F. M., Jr. (1978), *Cancer Treat. Rep.*, 62, 779-789.
Rosenberg, B. (1980), *ACS Symp. Ser.*, 140, 143-146.
Rosenberg, B., Renshaw, E., Van Camp, L., Hartwick, J. & Drobnik, J. (1967), *J. Bacteriol.*, 93, 716-721.
Rosenberg, B., Van Camp, L. & Krigas, T. (1965), *Nature London*, 205, 698-699.
Rosowsky, A., Papthanasopoulos, N., Lazarus, H., Foley, G. E. & Modest, E. J. (1974), *J. Med. Chem.*, 17, 672-676.
Ross, W. C. J. (1950), *J. Chem. Soc.*, 2257-2272.

Ross, W. C. J. (1959), *Biochem. Pharmacol.*, **2**, 215–220.
Ross, W. C. J. (1962), *Biological Alkylating Agents*, Butterworths, London.
Ross, W. C. J. (1964), *Biochem. Pharmacol.*, **13**, 969–982.
Ross, W. C. J. & Warwick, G. P. (1956), *J. Chem. Soc.*, 1719–1723, 1724–1732.
Ross, W. C. J. & Wilson, J. G. (1959), *J. Chem. Soc.*, 3616–3622.
Roth, B., Smith, J. M. & Hultquist, M. E. (1950), *J. Am. Chem. Soc.*, **72**, 1914–1918.
Roth, B., Smith, J. M. & Hultquist, M. E. (1951a), *J. Am. Chem. Soc.*, **73**, 2864–2868.
Roth, B., Smith, J. M. & Hultquist, M. E. (1951b), *J. Am. Chem. Soc.*, **73**, 2869–2871.
Rozencweig, M. (1979), in *Cancer Chemotherapy 1979* (Pinedo, H. M., Ed.), pp. 107, Excerpta Medica, Amsterdam.
Ruzicka, L., Goldberg, M. W. & Rosenberg, H. R. (1935), *Helv. Chim. Acta*, **18**, 1487–1498.
Saez, S., Brunat, M., Cheix, F., Colon, J. & Mayer, M. (1978), in *Proc. Int. Symp. Endocrine-Related Cancer*, Lyon, France, June 1977 (Mayer, M., Saez, S. & Stoll, B. A., Eds.), pp. 37–50, Imperial Chemical Industries, London.
Saijo, N., Nishiwaki, Y., Kawase, I., Kobayashi, T., Suzuki, A. & Nutani, H. (1978), *Cancer Treat. Rep.*, **62**, 139–141.
Saito, T., Wakui, A., Yokoyama, M., Takahashi, H., Ishigaki, H., Watanabe, K. & Tada, S. (1974), *J. Imp. Soc. Cancer Ther.*, **9**, 395.
Sanders, H. J. (1974), *Chem. Eng. News*, p. 74, December 23.
Sanders, M. E., Ames, M. M. & Tiede, W. S. (1982), 184th National Meeting of the American Chemical Society, Kansas City, Missouri, September 1982. Abstracts MEDI 55.
Sandoz Ltd. (1968), *Chem. Abstr.*, **68**, 2894x.
Santi, D. V. & McHenry, C. S. (1972), *Proc. Nat. Acad. Sci. U.S.A.*, **69**, 1855–1857.
Santi, D. V. & Sakai, T. T. (1971), *Biochemistry*, **1**, 3598–3607.
Sartiano, G. P., Lynch, W. E. & Bullington, W. D. (1979), *J. Antibiot Tokyo*, **32**, 1038–1045.
Sato, S., Iwaizumi, M., Handa, K. & Tamura, Y. (1977), *Gann*, **68**, 603–608.
Saunders, P. P. & Saunders, G. F. (1971), *Mol. Pharmacol.*, **6**, 335–344.
Sausville, E. A., Piesach, J. & Horwitz, S. B. (1976), *Biochem. Biophys. Res. Commun.*, **73**, 814–822.
Sausville, E. A., Stein, R. W., Presach, J. & Horwitz, S. B. (1978), *Biochemistry*, **17**, 2746.
Sava, G., Giraldi, T., Lassiani, L. & Nisi, c. (1979), *Cancer Treat. Rep.*, **63**, 93–98.
Sawada, H., Tatsumi, K., Sasada, M., Shirakawa, S., Nakamura, T. & Wakisaka, G. (1974), *Cancer Res.*, **34**, 3341–3346.
Scannel, J. P. & Hitchings, G. H. (1966), *Proc. Soc. Expt. Biol.*, **122**, 627–629.
Schaeffer, H. J. & Schwender, C. F. (1974), *J. Med. Chem.*, **17**, 6–8.
Schmidt, L. H. (1960), *Cancer Chemother. Rep.*, **9**, 56–58.
Schmidt, L. H., Fradkin, R., Sullivan, R. & Flowers, (1965), *Cancer Chemother. Rep.*, Suppl. **2**, 1–1528.
Schroeder, W. & Hoeksema, H. (1959), *J. Am. Chem. Soc.*, **81**, 1767–1768.
Schulte, G. (1952), *Z. Krebsforsch.*, **58**, 500–503.

Schwartz, S. H., Bardos, T. J., Burgess, G. H. & Klein, E. (1970), *J. Med. Chem.* 1, 174–179.
Sedov, K. A., Sorokina, I. B., Berlin, Yu. A. & Kosov, M. N. (1969), *Antibiotiki Moscow*, 14, 721–725.
Seeger, D. R., Cosulich, D. B., Smith, J. M. & Hultquist, M. E. (1949), *J. Am.Chem. Soc.*, 71, 1753–1758.
Seeger, D. R., Smith, J. M. & Hultquist, M. E. (1947), *J. Am. Chem. Soc.*, 69, 2567.
Semmelhack, M. F., Chong, B. P., Stauffer, R. D., Rogerson, T. D., Chong, A. & Jones, L. D. (1975), *J. Am. Chem. Soc.*, 97, 2507–2516.
Sengupta, S. K., Tinter, S. K., Lazarus, H., Brown, B. L. & Modest, E. J. (1975), *J. Med. Chem.*, 18, 1175–1181.
Sengupta, S. K., Tinter, S. K., Ramsey, P. G. & Modest, E. J. (1971), *Fed. Proc. Fed. Am. Soc. Exp. Biol.*, 30, 342.
Sengupta, S. K., Trites, D. H., Madhavarao, M. S. & Beltz, W. R. (1979), *J. Med. Chem.*, 22, 797–802.
Shealey, Y. F., Krauth, C. A. & Montgomery, J. A. (1962), *J. Org. Chem.*, 27, 2150–2154.
Shelten, R. S. & Van Campen, M. G. (1947), U.S. Patent 2,430,891.
Shimazaki, M., Kondo, S., Maeda, K., Ohno, M. & Umezawa, H. (1979), *J. Antibiot. Tokyo*, 32, 654–658.
Sigg, H. P., Mauli, R., Flury, E. & Hauser, D. (1965), *Helv. Chim. Acta*, 48, 962–988.
Silvestrini, R., Di Marco, A. & Dasdia, T. (1970), *Cancer Res.*, 30, 966–973.
Singer, B. (1974), *Progr. Nucl. Acid. Res. Mol. Biol.* 15, 219–284.
Sirotnak, F. M., DeGraw, J. I., Moccio, D. M. & Dorick, D. M. (1978), *Cancer Treat. Rep.*, 62, 1047–1052.
Skibba, J. L., Beal, D. D., Ramirez, G. & Bryan, G. T. (1970), *Cancer Res.*, 30, 147–150.
Skoda, J. (1963), in Progress in *Nucleic Acid Research & Molecular Biology*, Vol. 2 (Davidson, J. N. & Cohn, W. E., Eds.), pp. 197–221, Academic, New York.
Sköld, O. (1958), *Biochim. Biophys. Acta*, 29, 651.
Slavik, M., Keiser, H. R., Lovenberg, W. & Sjoerdsma, A. (1971), *Life Sci.*, 10, 1293–1295.
Slavik, K. & Zakrzewski, S. F. (1967), *Mol. Pharmacol.*, 3, 370–377.
Sluyser, M. (1978), in *Proc. Int. Symp. Endocrine-Related Cancer*, Lyon, France, June 1977 (Mayer, M., Saez, S. & Stoll, B. A., Eds.), pp. 285–297, Imperial Chemical Industries, London.
Smith, C. R., Jr., Mikolajczak, K. L. & Powell, R. G. (1980), in *Anticancer Agents Based on Natural Product Models* (Cassady, J. M. & Douros, J. D., Eds.), pp. 391–416, Academic, New York.
Smith, D. A., Roy-Burman, P. & Visser, D. W. (1966), *Biochem. Biophys. Acta*, 119, 221–228.
Smith, T. H., Fijiwara, A. N. & Henry, D. W. (1978), *J. Med. Chem.*, 21, 280–283.
Smith, T. H., Fijiwara, A. N., Henry, D. W. & Lee, W. W. (1976), *J. Am. Chem. Soc.*, 98, 1969–1971.
Snyder, J. A. & McIntosh, J. R. (1976), *Ann. Rev. Biochem.*, 45, 699–720.
Sobell, H. M. & Jain, S. C. (1972), *J. Mol. Biol.*, 68, 21–34.

Sorm, F., Piskala, A., Cihak, A. & Vesely, J. (1964), Experientia, 20, 202–203.
Sowa, J. R. & Price, C. C. (1969), J. Org. Chem., 34, 474–476.
Stahelin, H. (1973), Eur. J. Cancer, 9, 215–221.
Stearns, B., Loses, K. A. & Bernstein, J. (1963), J. Med. Chem., 6, 201.
Steuart, C. D. & Burke, P. J. (1971), Nature London, 233, 109–110.
Stoll, B. A. (1978), in Proc. Int. Symp. Endocrine-Related Cancer, Lyon, France, June 1997 (Mayer, M., Saez, S. & Stoll, B. A., Eds.), pp. 237–248, Imperial Chemical Industry, London.
Struck, R. F., Shealy, Y. F. & Montgomery, J. A. (1971), J. Med. Chem., 14, 693–698.
Sugasawa, T., Toyoda, T., Uchida, N. & Yamaguchi, K. (1976), J. Med. Chem., 19, 675–679.
Suguira, Y. & Kikuchi, T. (1978), J. Antibiot. Tokyo, 31, 1310–1312.
Suguira, Y., Muraoka, Y., Fujii, A., Takita, T. & Umezawa, H. (1979), J. Antibiot. Tokyo, 32, 756–758.
Suzuki, H., Nagai, K., Yamaki, H. & Umezawa, H. (1968), J. Antibiot. Tokyo, 21, 379–386.
Svoboda, G. H. (1961), Lloydia, 24, 173–178.
Svoboda, G. H., Johnson, I. S., Gorman, M. & Neuss, N. (1962), J. Pharm. Sci., 51, 707–720.
Sweeney, M. J., Boder, G. B., Culliman, G. J., Culp, H. W., Daniels, W. D., Dyke, R. W., Gerzon, K., McMahon, R. E., Nelson, R. L., Poore, G. A. & Todd, G. C. (1978), Cancer Res., 38, 2886.
Szybalski, W. (1964), Abh. Dsch Akad. Wiss. Berlin, 4, 1–19.
Szybalski, W. & Iyer, V. N. (1967), in Antibiotics I, Mechanism of Action (Gottlieb, D. & Shaw, P. D., Eds.), pp. 221–245, Springer-Verlag, Berlin.
Takahashi, K., Yoshioka, O., Matsuda, A. & Umezawa, H. (1977), J. Antibiot. Tokyo, 30, 861–869.
Takamizawa, A., Matsumoto, S., Iwata, T., Tochino, Y., Katagiri, K., Yamaguchi, K. & Shiratori, O. (1975), J. Med. Chem., 18, 376–383.
Takita, T., Fujii, A., Fukuoka, T. & Umezawa, H. (1973), J. Antibiot. Tokyo, 26, 252.
Takita, T., Muraoka, Y., Nakatani, T., Fujii, A., Iitaka, Y. & Umezawa, H. (1978), J. Antibiot. Tokyo, 31, 1073–1076.
Takita, T., Muraoka, Y., Nakatani, T., Fujii, A., Umezawa, Y., Naganawa, H. & Umezawa, H. (1978), J. Antibiot. Tokyo, 31, 801–804.
Tanaka, H., Yoshioka, T., Shimauchi, Y., Matsuzawa, Y., Oki, T. & Inui, T. (1980), J. Antibiot. Tokyo, 33, 1323–1330.
Tatsumi, K. & Nishioka, H. (1977), Mutat. Res., 48, 195–204.
Taylor, E. C., Vogl, O. & Cheng, C. C. (1959), J. Am. Chem. Soc., 81, 2442–2448.
Teicher, B. A. & Sartorelli, A. C. (1980), J. Med. Chem., 22, 955–960.
Thurber, T. C. & Townsend, L. B. (1971), 3rd Int. Cong. Heterocycl. Chem., Sendai, Japan, August 1971.
Timmis, G. M. (1958), Ann. N.Y. Acad. Sci., 68, 727.
Timmis, G. M. (1959), U.S. Patent 2,917,432.
Tomasz, M. (1970), Biochem. Biophys. Acta, 213, 288–295.

Tong, G. L., Cory, M., Lee, W. W. & Henry, D. W. (1978), *J. Med. Chem.*, **21**, 732-737.
Tong, G. L., Wie, H. Y., Smith, T. H. & Henry, D. W. (1979), *J. Med. Chem.*, **22**, 912-918.
Townsend, L. B. & Tipson, R. S. (1978), *Nucleic Acid Chemistry*, Wiley, New York.
Trimble, R. B. & Maley, F. (1971), *J. Bacteriol.*, 145-153.
Tritton, T. R., Murphree, S. A. & Sartorelli, A. C. (1978), Biochem. Biophys. Res. Commun., **84**, 802-808.
Tritton, T. R. & Yee, G. (1982), *Science*, **217**, 248-250.
Tscherne, J. S. & Pestka, S. (1975), *Antimicrob. Agents Chemother.*, **8**, 479-487.
Tso, J. Y., Bower, S. G. & Zalkin, H. (1980), *J. Biol. Chem.*, **255**, 6734-6738.
Tsuji, T. & Kosower, E. M. (1971), *J. Am. Chem. Soc.*, **93**, 1991-1999.
Tsurugi, K., Morita, T. & Ogata, K. (1972), *Eur. J. Biochem.*, **29**, 585-592.
Tullius, T. D. & Lippard, S. J. (1981), *J. Am. Chem. Soc.*, **103**, 4620-4622.
Tyagi, A. K. & Cooney, D. A. (1980), *Cancer Res.*, **40**, 4390-4397.
Ueno, Y. (1977), *Pure Appl. Chem.*, **49**, 1737-1745.
Ueno, Y. & Fukushima, K. (1968), *Experientia*, **24**, 1032-1033.
Umezawa, H. (1976), in Bleomycin, Fundamental & Clinical Studies, *Gann Monogr. Cancer Res.*, **19**, 3-36.
Umezawa, H., Muraoka, Y., Fujii, A., Naganawa, H. & Takita, T. (1980), *J. Antibiot. Tokyo*, **33**, 1079-1082.
Umezawa, H., Takahashi, Y., Kinoshita, M., Naganawa, H., Masuda, T., Ishizuka, M., Tatsuta, K. & Takeuchi, T. (1979), *J. Antibiot. Tokyo*, **32**, 1082-1084.
Umezawa, H., Takahashi, Y., Shirai, T. & Fujii, A. (1972), *Ger. Offen.*, **2**, 223, 535.
Umezawa, H., Takeuchi, S., Hori, T., Sawa, T., Ishizuka, T., Ichikawa, T. & Komai, T. (1972), *J. Antibiot. Tokyo*, **25**, 409-420.
Umezawa, H., Takita, T., Fujii, A. & Ito, H. (1973), *Ger. Offen.*, **2**, 144, 299.
Urtasun, R., Band, P., Chapman, J. D., Feldstein, M. L., Mielke, B. & Fryer, C. (1976), *N. Engl. J. Med.*, **294**, 1364-1367.
Urtasun, R., Band, P., Champan, J. D., Rabin, H. R., Wilson, A. F. & Fryer, C. G. (1977), Radiology, **122**, 801-804.
Van-Bac, N., Moisand, C., Gouyette, A., Muzard, G., Dat-Xuong, N., Le Pecq, J. B. & Paoletti, C. (1980), *Cancer Treat. Rep.*, **64**, 879.
Vargha, L., Toldy, L., Feher, O. & Lendvai, S. (1957), *J. Chem. Soc.*, 805-809.
Vasil'Eva, M. N., Martynov, V. S. & Berlin, Yu. A. (1970), *Zh. Org. Khim.*, **6**, 1677-1682.
Vesely, J., Cihak, A. & Sorm, F. (1968a), *Biochem. Pharmacol.*, **17**, 519-524.
Vesely, J., Cihak, A. & Sorm, F. (1968b), *Cancer Res.*, **28**, 1995-2000.
Vesely, J., Cihak, A. & Sorm, F. (1969), *Coll. Czech. Chem. Commun.*, **34**, 901-909.
Vickers, S., Hebborn, P., Moran, J. F. & Triggle, D. J. (1969), *J. Med. Chem.*, **12**, 491-494.
Visser, D. W. (1955), in *Antimetabolites & Cancer* (Rhoades, C. P., Ed.), p. 47, American Association for the Advance of Science, Washington.
Vlasov, G. P., Lashkov, V. N. & Glibin, E. N. (1979), *Zh. Org. Khim.*, **15**, 983-990.
Wakelin, L. P. G. & Waring, M. J. (1976), *Biochem. J.*, **157**, 721.
Waksman, S. A. & Woodruff, H. B. (1940), *Proc. Soc. Exp. Biol. Med.*, **45**, 609.

Walker, R. T. (1979), *Comp. Org. Chem.* **5**, 53–104.
Wall, M. E., Campbell, H. F., Wani, M. C. & Levine, S. G. (1972), *J. Am. Chem. Soc.*, **94**, 3632–3633.
Wall, M. E. & Wani, M. C. (1980), in *Anticancer Agents Based on Natural Product Models* (Cassady, J. M. & Douros, J. D., Eds.), pp. 417–436, Academic, New York.
Wall, M. E., Wani, M. C., Cook, C. E., Palmer, K. H., McPhail, A. T. & Sim, G. A. (1966), *J. Am. Chem. Soc.*, **88**, 3888–3890.
Walpole, A. L. (1958), *Ann. N.Y. Acad. Sci.*, **68**, 750.
Wang, M. C. & Bloch, A. (1972), *Biochem. Pharmacol.*, **21**, 1063–1073.
Wani, M. C., Ronman, P. E., Lindley, J. T. & Wall, M. E. (1980), *J. Med. Chem.*, **23**, 554–560.
Wani, M. C., Taylor, H. L., Wall, M. E., Coggon, P. & MacPhail, A. T. (1971), *J. Am. Chem. Soc.*, **93**, 2325–2327.
Wani, M. C. & Wall, M. E. (1969), *J. Org. Chem.*, **34**, 1364–1367.
Waring, M. (1965), *J. Mol. Biol.*, **13**, 269–282.
Waring, M. (1970), *J. Mol. Biol.*, **54**, 247–279.
Waring, M. (1975), in *Antibiotics III* (Corcoran, J. W. & Hahn, F. E., Eds.), pp. 141–165, Springer-Verlag, Berlin.
Way, J. L., Dahl, J. L. & Parks, R. E., Jr. (1959), *J. Biol. Chem.*, **234**, 1241–1243.
Way, J. L. & Parks, R. E., Jr. (1958), *J. Biol. Chem.*, **234**, 467–480.
Wechter, W. J., Johnson, M. A., Hall, C. M., Warner, D. T., Berger, A. E., Wenzel, A. M., Gish, D. T. & Neil, G. L. (1975), *J. Med. Chem.*, **18**, 339–344.
Wei, C. & McLaughlin, C. S. (1974), *Biochem. Biophys. Res. Commun.*, **57**, 838.
Weinkam, R. J. & Lin, H. S. (1979), *J. Med. Chem.*, **22**, 1193–1198.
Weinreb, S. M. & Auerbach, J. (1975), *J. Am. Chem. Soc.*, **97**, 2503–2506.
Weinstock, L. T., Grabowski, B. F. & Cheng, C. C. (1970), *J. Med. Chem.*, **13**, 995–997.
Wheeler, G. P. (1974), in *Handbook of Experimental Pharmacology* (Sartorelli, A. C. & Johns, D. G., Eds.), Vol. 38, Part 2, p. 7, Springer-Verlag, Berlin.
White, F. R. (1959), *Cancer Chemother. Rep.*, **4**, 52–55.
White, H. L. & White, J. R. (1968), *Mol. Pharmacol.*, **4**, 549–565.
White, J. R. & White, H. L. (1964), *Science*, **145**, 1312–1313.
Wiley, P. F., Johnson, J. L. & Houser, D. J. (1977a), *J. Antibiot. Tokyo*, **30**, 628–629.
Wiley, P. F., Kelly, R. B., Caron, E. L., Wiley, V. H., Johnson, J. H., MacKellar, F. A. & Mizsak, S. A. (1977b), *J. Am. Chem. Soc.*, **9**, 542–549.
Witiak, D. T. (1978), *J. Med. Chem.*, **21**, 1194–1197.
Wolpert-DeFilippes, M. K., Adamson, R. H., Cysyk, R. L. & Johns, D. G. (1975), *Biochem. Pharmacol.*, **24**, 751–754.
Wystrach, V. P., Kaiser, D. W. & Schaefer, F. C. (1955), *J. Am. Chem. Soc.*, **77**, 5915–5918.
Yesair, D. W. & Kensler, C. J. (1975), in *Handbook of Experimental Pharmacology* (Sartorelli, A. C. & Johns, D. G., Eds.), Vol. 38, Part 2, pp. 820–828, Springer-Verlag, Berlin.
Yoshimoto, A., Matsuzawa, Y., Oki, T., Naganawa, H., Takeuchi, T. & Umezawa, H. (1980), *J. Antibiot. Tokyo*, **33**, 1150–1157.

Young, C. W. & Hodas, S. (1965), *Biochem. Pharmacol.*, **14**, 205-214.
Yuhas, J. M. (1973), *J. Nat. Cancer Inst.*, **51**, 69-78.
Yuntsen, H., Yonehara, H. & Vi, H. (1954), *J. Antibiot. Tokyo,* , 113-115.
Zee-Cheng, K. Y. & Cheng, C. C. (1973), *J. Heterocycl. Chem.*, **10**, 85-88.
Zee-Cheng, K. Y. & Cheng, C. C. (1978), *J. Med. Chem.*, **21**, 291-294.
Zee-Cheng, K. Y., Paull, K. D. & Cheng, C. C. (1974), *J. Med. Chem.*, **17**, 347-351.
Zorbach, W. W. & Tipson, R. S. (1968), *Synthetic Procedures in Nucleic Acid Chemistry*, Wiley-Interscience, New York.
Zubrod, C. G. (1974), in *Handbook of Experimental Pharmacology, Part I* (Sartorelli, A. C. & Johns, D. G., Eds.), pp. 1-11, Springer-Verlag, Berlin.
Zunino, F., Gambetta, R., Di Marco, A. & Zaccara, A. (1972), *Biochem. Biophys. Acta*, **277**, 489-498.

Index

Acivicin, 131, 132, 160
  mode of action, 160
Aclacinomycin, 178
Acrolein, 91, 92
Actinomycin D, 167
Actinomycins, 167, 172
  analogs, 174, 175, 177
  names, 172
  structures, 173
  synthesis, 177
Adenine, 96
Adenine arabinoside, 137, 138, 144
Adenosine deaminase, 153
  inhibitors, 153
Adriamycin, see Doxorubicin
AD-32, 184
Alanosine, 132, 161
Alazopeptin, 131
Alkylating agent, 7
  design, 98
  mode of action, 93, 94
  structure-activity relationships, 98
  synthesis, 124
Alkylation, 86
  bifunctional, 95
  bioreductive, 89
  chemistry of, 86
  of enzymes, 97
Amethopterin, 5, 131, 154
  analogs, 155, 156, 157
  synthesis, 165
3'-Amino-3'-deoxyadenosine, 144
Aminopterin, 154
5-Aminouridine, 134, 151
Androgens, 224
Anguidine, 207

Aniline mustard, 116
  analogs, 116, 121
Ansamitocins, 219
Anthracyclines, 178
  antitumor activities, 183, 186
  biotransformation, 187
  classification, 179
  mode of action, 180
  structures, 178
Anthramycins, 108
Anthraquinones, 188, 189
Antibiotics, 7
Antimetabolite, 7, 85, 127
  biochemistry, 127
  design, 140
  discovery, 140
Antimitotic, 7
Aryldiazonium ions, 89
Aryltriazines, 89, 123
L-Asparaginase, 163
Aureolic acids, 197
Autochthonous tumors, 66
8-Azaadenine, 140
5-Azacytidine, 138, 146, 148
  mode of action, 146
  synthesis, 164
5-Aza-2'-deoxycytidine, 147
8-Azaguanine, 140, 141
Azaserine, 131, 132, 159
Azathioprine, 163
5-Azauracil, 147
6-Azauracil, 138, 147
5-Azauridine, 138
6-Azauridine, 134, 147
  triacetyl derivative, 147
6-Azauridylate, 147

263

Azetomycin, 174
Aziridinium ion, 86, 98
Azotomycin, 131

Bacillus Calmette-Guerin, 235
BCNU, 88, 102
 chloroethylation of DNA, 96
 decomposition pathways, 88
 synthesis, 125
Benzoquinones
 aziridinyl, 105
 mustards, 118
Biochemical screens, 53
 enzymes, 53
 macromolecular synthesis, 57
Bioreductive alkylation, 89, 91
Bis(hydroxymethyl)-1H-pyrolizine dicarbamates, 110
Bis(hydroxymethyl)pyrrole dicarbamates, 110
Bleomycin, 4, 190
 analogs, 193, 194
 chemistry, 191, 193
 metal complexes, 191, 192
 mode of action, 191
 structure, 190
Bleomycinic acid, 196
Bisantrene, 189
 synthesis, 190
Bisintercalating agents, 170
Bone marrow:
 toxicity of drugs, 4
 transplantation, 18
Bouvardin, 210
Breast cancer, hormonal dependence, 224
Bruceantin, 221
Busulfan, 95

Camptothecin, 199
 analogs, 200, 201
 mode of action, 201
 synthesis, 200
Carbazilquinone, 105
Carbinolamines, 108
 vinylogous, 109
Carcinogen-induced tumors, 67
Carcinomas, 4
Carminomycin, 178
Carrier groups for alkylating agents, 111
Catharanthus alkaloids, 210, 214
 analogs, 217
 antitumor activity, 213

 structures, 213
CCNU, 97, 102
Cell:
 clonogenic, 32
 cycle, 5, 7, 93
 differentiation, 3
 division process, 5, 6
 invasiveness, 3
 kinetic measurement, 7
 loss, 26
 proliferation, 3, 5
 stem, 13
Cell-kill concept, 29
Cell population:
 expanding, 9
 hierarchy, 13
 organization, 13
 renewal, 10, 11
 self-renewal, 13
 static, 9
Cellular selection, 35
Cephalotoxine, 207
Chemotherapy, 3
 adjuvant, 3
 combination, 3
 predictive, 34
 synergistic, 7
Chlorambucil, 98
Chloroquine mustard, 114
Chlorotrianisine, 226
 synthesis, 229
Chlorozotocin, 125
Choriocarcinoma, 4
Chromomycins, 197
Cisplatin, 4, 203
Cliomycins, 195
Clonal evolution, 35
Clone, 11, 12
 human tumor screens, 51
 progression, 18
Clonogenic cell, 30, 31
Colchicine, 210, 211, 214
Coralyne, 171
Cordycepin, 144
Coformycin, 153
Crosslinking of DNA, 95
Cyclocytidine, 153
Cyclophosphamide, 4, 91, 117
 bioactivation, 91
 synthesis, 124
Cytosine, 96

INDEX

Cytosine arabinoside, 137, 138, 152
  antitumor activity, 152
  deamination, 152
  derivatives, 152
  synthesis, 164
Cytidine deaminase, 138, 152
  inhibitors, 152, 154
Cytidine kinase, 152
Cytotoxicity screens, 49

Dacarbazine, 89
  hydrolysis, 89
  synthesis, 126
Daunorubicin, 167, 178, 182
  analogs, activity, 183
3-Deazauridine, 147
Decoynine, 144
5-deoxyinosine, cleavage to dialdehyde, 144
Deoxyribonucleic acid (DNA), 94
  alkylation of, 94, 95
  crosslinking, 95
2'-Deoxyuridylate, 135
Di(aminomethyl)cyclohexylplatinum (II), 202
Diamminecyclobutanedicarboxylatoplatinum, 202
Diazomycin, 131
6-Diazo-5-oxo-norleucine (DON), 131, 132, 159
  analogs, 160
Dibromomannitol, 112
Diethylstilbestrol, 226
  synthesis, 229
Dihydrofolate reductase, 8, 36
  inhibitors, 136, 156, 157, 158
Doxorubicin, 178, 182
  analogs, activity, 183
  toxicity, 182
Drugs:
  distribution in tumor, 5
  resistance by tumor, 36
  selectivity, 5
  sensitivity, inherent, 34

Echinomycin, 170
EHNA, 153
Ellipticene, 171
  analogs, 171, 172
Enzyme screens, 23
  adenosine deaminase, 56
  aminopeptidases, 54
  dihydrofolate reductase, 56
  DNA gyrase, 55
  esterase, 54
  galactosidase, 55
  ornithine decarboxylase, 55
  phosphodiesterase, 54
  phospholipase, 56
  reverse transcriptase, 55
  tyrosine hydroxylase, 57
Enzymes, alkylation of, 97
Epoxides, 104
Estradiol, 224
  propionate ester, 224
Estrogen, 224, 225
  receptors, 224
Ethidium, 167
Ethinyl estradiol, 226
  synthesis, 229
Ethylenimines, 105
  antitumor activity, 107
Ethyl nitrosourea, 95
Etiposide (VP 16-213), 214
  synthesis, 215

5-Fluorocytidine, 149
  derivatives, 149
5-Fluorouracil, 5, 85, 135, 138
  anabolism, 148
  analogs, 149, 150, 151
  2'-deoxyribonucleotide, 149
  design, 148, 148
  synthesis, 163
Fluoxymestrone, 225
  synthesis, 229
Folic acid, 155
Ftorafur, 149

GANU, 113
Gene amplification, 36
Glucocorticoids, 226
Glutamine antagonists, 131
Glycoprotein, membrane, 35
Gompertz equation, 21
Gompertzian growth kinetics, 28
Growth fraction, 26
Guanine, 94, 95, 203

Hansch equations, 122
  aniline mustards, 122
  aryldialkyltriazines, 123
  nitrosoureas, 123
Harringtonine, 207, 209
  synthesis, 208

Helenalin, 92, 226
Hexamethylmelamine, 109
Homoharringtonine, 207, 209
Hormones, 223
  effects on tumors, 223
Host response modification screens, 68
Human tumor clone screens, 51
Hycanthone, 170
Hydroxyaniline mustard, 115
Hydroxyurea, 161
Hyperplasia, 10
Hypoxanthine-guanine phosphoribosyltransferase, 138, 141

Immune-deprived mice, 30
Immunostimulants, 233, 235
Immunotherapy, 235
In vitro screening strategies, 57
Indicine-N-oxide, 221
Intercalating agents, 85, 167
Intercalation, 165
  effects on DNA, 167
Interferon, 235
Iphosphamide, 121
Isocoformycin, 153

Kinetics:
  cell, 7, 23
  Gompertzian growth, 28
  tumor growth, 19

Labeling index, 24
Leucovorin, 136
Leukemia:
  acute lymphocytic, 18
  acute melanocytic, 18
  childhood, 4
  chronic myelocytic, 14
  screening systems:
    AKR spontaneous, 66
    L-1210, 62, 157
    P-388, 61, 119
Levamisole, 235
Lymphoma:
  Burketts, 4
  cells from AKR mice, 30

Macromolecular synthesis screens, 57
  DNA methylation, 57
  nucleic acid polymerases, 57
  protein synthesis, 57

Malignant transformation screens, 52
Mammary carcinoma screen, 66
D-Mannitol mustard, 112
D-Mannitol myleran, 112
Marcellomycin, 185
Maytansine, 212
  structure, 218
  synthesis, 218
Maytansinoids, 210, 219
Mazethramycin, 108
Mechlorethamine, 95, 98
  synthesis, 124
Medroxyprogesterone acetate, 226
  synthesis, 232
Melanoma, B-16 screen, 62
Melphalan, 111
  isomers, 111
  peptide derivatives, 111
  synthesis, 124
5-Mercapto-2'-deoxyuridine, 151
6-Mercaptopurine, 127, 138, 139, 141
  synthesis, 162
Metaplasia, 10
Metastasis, 3
Methanesulfonates:
  antitumor activity, 104
  mode of action, 106
  toxicity, 106
5,10-Methenyltetrahydrofolate, 131
Methotrexate, *see* Amethopterin
Methyl CCNU, 124
α-Methylenelactones, 92, 110
5,10-Methylenetrahydrofolate, 134, 148
Methylglyoxal bis(guanylhydrazone), 223
Methylnitrosourea, 87
5-Methyltetrahydrohomofolate, 155
6-(Methylthio)purine, 130, 141
Metronidazole, 237
Microbial screens, 44
  microbial inhibition, 44
  prophage induction, 48
  viral inhibition, 47
Microtubules, 212
Misonidazole, 237
Mithramycin, 197
Mitomyin C, 90
  activation, 90
  analogs, 119
  antitumor activity, 120
  reduction potential, 119
Mitosis, 5, 6

INDEX

percent labeled, 25
Mitotane, 227
  synthesis, 233
Mitotic index, 23
Mitotic inhibitors, 210
Mitoxantrone, see Dihydroxy-anthracenedione
Mode of action:
  alkylating agents, 94
  antitumor agents, 85
  intercalating agents, 167
Mustards:
  nitrogen, 86, 115
  phosphoramide, 91, 117
  sulfur, 87
Mutagenesis, 94, 101
Mutation, 35
Myleran, 104

Nafoxidine, 225
Neocarcinostatin, 203
Neothramycins, 108
Nitidine, 171
0-Nitrobenzyl chloride, 120
Nitrogen mustards, 86
  antitiumor activity, 100
  aryl, 99
  chemical reactivity, 99
Nitromin, 121
Nitrosoureas, 87, 101
  antitumor activity, 103
  quantitative structure-activity relationships, 123
Nogalamycin, 187
  derivatives, 187
Nor-nitrogen mustard, 98
Nucleic acid synthesis, 6
Nucleotidyl transferase, 98

Olivomycins, 197
  derivatives, 197
Organ specific tumors, 64
  human, 65
  rodent, 64
Orotidylate decarboxylase, 147
Oxazinomycin, 147

Pharmacokinetics, 5
Phase-specific screens, 51
Phenestrin, 114
Phleomycins, 193
Phorbol, 221

Phosphoramide mustard, 91
Phthalanilides, 223
Phytohaemagglutinin, 31
Piperazinedione, 99
Pipobroman, 223
Platinum complexes, 203
  discovery, 204
  toxicity, 204
Podophylline, 212
  analogs, 212
  synthesis, 212
Podophyllotoxins, 210, 211
Polymerase, DNA, 137, 167, 180, 203
Polypeptide antibiotics, 202
Prednisone, 233
Primary screening, 70
Procarbazine, 92
  synthesis, 126
Progesterone, 226
  17α-hydroxy, caproate, 226
Prolactin, 224, 227
Prophage induction screens, 48
Protein synthesis inhibitors, 205
Psicofuranine, 144
Purine antimetabolites, 146
Purine nucleotides, 127
Puromycin, 144
Pyrazolo[3,4-d]pyrimidines, 141
Pyrimethamine, 158
Pyrimidine antimetabolites, antitumor activity, 154
Pyrimidine nucleotides, 132, 133
Pyrrolo(1,4)benzodiazepines, 108
  mode of action, 109
  toxicities, 109

Quassinoids, 221
Quinonimine, 119
Quinoxalin antibiotics, 170

Radioprotective compounds, 236
Radiosensitizing agents, 236
Resistance, drug, 35
Ribonucleoside diphosphate reductase inhibitors, 136, 161
Rubidazone, 178
Rudolfomycin, 185

Sangivamycin, 141
Screening strategies, 69

Screens:
  autochthonous tumors, 65
  biochemical, 53
  carcinogen-induced tumors, 67
  host response modification, 68
  in vitro, 43
  in vivo, 59
  microbial, 44
  organ specific, 64
  primary, 70
  secondary, 73
  side effects, 69
  strategies, 69
  syngeneic tumors, 61
  tissue culture, 49
Sibiromycin, 108
Spleen-cell-conditioned medium, 30
Stathmokinetic technique, 23
Stem cell, 11, 12
  biology, 19, 29
  kinetics, 19
  origin of human tumors, 15
  pluripotent, 14
Steroidal nitrogen mustards, 114
Streptonigrin, 201
  analogs, 201
  biosynthesis, 202
  mode of action, 202
  synthesis, 202
Streptozotocin, 113
  analogs, 113
  synthesis, 125
Sulfonium ions, 87
Sulfur mustard, 87
  analogs, 101, 117
Sulfur stripping, 97
Syngeneic tumor screens, 61

Tallysomycins, 195
Tamoxifen, 224, 225
  synthesis, 229
Taxol, 220
Teniposide (VM-26), 214
  synthesis, 215
Terpenes, 220
Testolactone, 225
Testosterone, 225
  17α-methyl, 225
  2α-methyldihydro, 225
  19-nor,17α-methyl, 225
  propionic ester, 225

Tetrahydrohomofolate, 157
Tetrahydrouridine, 153
Tetramin, 105
Therapeutic ratio, 33
Therapy, 4
  schedule dependent, 9
  systemic, 4
  tumor specific, 4
6-Thioguanine, 127, 138
Thioinosinate, 127
Thiosemicarbazones, 161
  bis, 223
Thio-TEPA, 126
Thymidine kinase, 151
Thymidylate, 134
Thymidylate synthetase, 138, 152
  inhibitors, 135, 136, 148, 158
Thymine, 96, 136
Tilorone, 235
  synthesis, 236
Tissue culture, 6
  screens, 49
    cytotoxicity, 49
    human tumor clones, 51
    malignant transformation, 52
    phase-specific, 51
Tomaymycin, 108
Toyocamycin, 141
Transplanted tumor screens, 60
  Ehrlich tumor, 60
  leukemia L-1210, 62
  leukemia P-388, 61
  Lewis lung carcinoma, 63
  melanoma B-16, 62
  sarcoma 180, 60
Trenimon, 105
Tricothecanes, 205
  classification, 206
  mode of action, 206
Triethylenemelamine, 105
  synthesis, 126
Trifluorothymidine, 136, 158
Trimethoprim, 158
Tripdiolide, 220
Trophosphamide, 121
Tryptophan mustard, 111
Tubercidin, 141
Tubulin, 210
  inhibitors, 211
Tumor:
  growth, 19

INDEX

    exponential, 21
    Gompertzian, 22
    volume doubling time, 22
  size, measurement, 19
  stem cell origin of, 17
  testicular, 4
  Wilms, 4
Tumor-specific proteins:
  carcino-embryonic antigen, 20
  α-foetaprotein, 20
  human chorionic gonadotrophin, 20
  paraproteins, 20

Uracil mustard, 113
Urethane mustard, 118

Variamycins, 197
Vinblastine, 4, 212, 215, 216
  analogs, 217
  toxicity, 216
Vincristine, 8, 210, 212, 216
  analogs, 218
  antitumor activity, 214
  toxicity, 216
Vindesine, 216
Viral inhibition screens, 47

Walker tumor, 113
  reduction potential of cells, 118
Withanolide E, 221

Xanthine oxidase, 139

Zorbamycin, 195
Zorbonamycins, 196